Transport Processes Primer

Constantine Pozrikidis

Transport Processes Primer

Springer

Constantine Pozrikidis
College of Engineering
University of Massachusetts Amherst
Amherst, MA, USA

ISBN 978-1-4939-9911-8 ISBN 978-1-4939-9909-5 (eBook)
https://doi.org/10.1007/978-1-4939-9909-5

This Springer imprint is published by the registered company Springer Science+Business Media, LLC, part of Springer Nature.
The registered company address is: 233 Spring Street, New York, NY 10013, U.S.A.

Preface

Transport phenomena or *transport processes* is a concept proposed by chemical engineers in an effort to unify fluid mechanics, heat transfer, and mass transfer in a stationary material or moving fluid. The core procedure prescribed in texts and elsewhere involves defining a finite or infinitesimal control volume, and then performing an integral or differential momentum, heat, mass, or some other type of balance of a transportable entity to derive a governing equation.

The balance typically states that the rate of accumulation of a certain extensive property of interest, such as mass, momentum, or total energy, is determined by the rates of convective and diffusive transport across the boundaries of the control volume, as well as by appropriate rates of interior and surface loss or production. The control volume itself may be stationary or evolve in an arbitrary fashion.

One subtlety of the aforementioned approach is that mass, momentum, and energy balances are consequences of the principle of mass conservation, Newton's second law of motion, and the first law of thermodynamics. In their classical form, these natural laws apply to well-defined bodies or pieces of material (closed systems), as opposed to control volumes that allow matter to cross their boundaries (open systems).

The principle of mass conservation is pertinent to the mass of a single-species material parcel or species X parcel,

$$\iiint \rho \, \mathrm{d}V, \qquad \iiint \rho_X \, \mathrm{d}V,$$

where ρ is the density; Newton's second law of motion is pertinent to the momentum of a single-species material parcel or species X parcel,

$$\iiint \rho \, \mathbf{u} \, \mathrm{d}V, \qquad \iiint \rho_X \, \mathbf{u}_X \, \mathrm{d}V,$$

where $\mathbf{u}$ is the fluid velocity; the first law of thermodynamics is pertinent to the total energy of a single-species material parcel or species X parcel,

$$\iiint \rho \left(\tfrac{1}{2} \, |\mathbf{u}|^2 + v + u\right) \mathrm{d}V, \qquad \iiint \rho_X \left(\tfrac{1}{2} \, |\mathbf{u}_X|^2 + v_X + u_X\right) \mathrm{d}V,$$

where v is the specific potential energy and u is the specific internal energy. Consequently, introducing fluid dynamics based on momentum transport over a control volume could be hard to justify in foresight though straightforward to explain in hindsight.

A way out is to rewrite the natural laws so that they apply to open systems that allow solid or fluid material to enter or exit a control volume through inlets and outlets. However, the dual restatement may appear like a band-aid that undermines the omnipotence of the classical approach and unnecessarily complicates the logistics.

A further complication is that a system may be closed with regard to one property but open with regard to another. For example, a system may be closed with respect to mass but open with respect to energy. A system that is closed with regard to every possible transportable entity or transmittable field is completely isolated. The notion of open, closed, and isolated systems has been discussed in the natural sciences and under the auspices of information theory, sociology, biology, anthropology, linguistics, history, political science, and philosophy.

To ensure accuracy and scientific rigor, the governing equations of transport phenomena are best derived from the classical natural laws applied to material parcels. The transport approach may then be validated and employed as a practical method of formulating equations and obtaining solutions in science and engineering applications. The recommended procedure involves the following steps:

1. Write an expression for a property of interest attributed to a material parcel, such as mass, momentum, total specific energy, or species mass.
2. Use the Reynolds transport equation to express the rate of change of the parcel property in terms of accumulation over the parcel volume and associated flux integrated over the parcel surface.
3. Introduce a physical law for the rate of change of the parcel property. For example, the rate of change of mass is zero and the rate of change of momentum is given by Newton's second law of motion.
4. Regard the parcel as a control volume, or else consider the parcel occupying a control volume of interest at a particular instant to obtain an integral transport balance.
5. Apply the divergence theorem to convert all boundary integrals into volume integrals, and discard the integral signs to derive governing differential balances in the form of differential equations written in conservative or nonconservative form.

My main goal in this book is to review the basic concepts and notions of transport processes and illustrate the origin of the governing equations by deriving and summarizing the equations of mass, momentum, energy, enthalpy, entropy, and other related transport for homogeneous fluids and mixtures of fluids in the context of mechanical, chemical, biological, biomedical, and other mainstream science and engineering.

Noteworthy features of the discourse with regard to mass transport includes the interpretation of diffusion in terms of species parcel kinematics, the discussion of Fick's and fractional diffusion laws, the introduction of partial stresses and associated equations of motion for individual species in a mixture, and the study of species and mixture energetics.

A summary of transport equations in differential and integral forms are presented and unified in Appendix A. All necessary relations from thermodynamics employed in the text are derived in Appendix B for a self-contained discourse.

Matlab[1] programs performing numerical simulations of random walks that illustrate the nature of ordinary and fractional diffusion are listed in the text.

Amherst, MA, USA

Constantine Pozrikidis

[1]Matlab® is a proprietary computing environment for numerical computation and data visualization. Matlab and Simulink are registered trademarks of The MathWorks, Inc. For product information, please contact: The MathWorks, Inc., 3 Apple Hill Drive, Natick, MA 01760-2098, USA, Tel: 508-647-7000, Fax: 508-647-7001, E-mail: info@mathworks.com, Web: www.mathworks.com.

Contents

1 Homogeneous Fluids 1
1.1 The Concept of Fluid Velocity 1
1.1.1 Continuum Approximation 1
1.1.2 Velocity Field 2
1.1.3 Differentiability 2
1.1.4 Material Point Particles 2
1.1.5 Material Parcel 3
1.1.6 Exercise 3
1.2 Density and Other Intensive Fields 3
1.2.1 Density as a Field Function 3
1.2.2 Extensive and Intensive Properties 4
1.2.3 Temperature Field 4
1.2.4 Exercise 4
1.3 Fundamental Decomposition of Kinematics 4
1.3.1 Velocity-Gradient Tensor 5
1.3.2 Divergence of the Velocity 5
1.3.3 Fundamental Decomposition of Kinematics 6
1.3.4 Vorticity Vector 6
1.3.5 The Vorticity Is the Curl of the Velocity 6
1.3.6 Four Types of Local Motion 7
1.3.7 Translation 7
1.3.8 Rotation 7
1.3.9 Deformation 8
1.3.10 Contraction and Expansion 8
1.3.11 Exercises 8
1.4 Lagrangian Mapping of Point Particles 9
1.4.1 RGB 9
1.4.2 Point-Particle Position 9
1.4.3 Point-Particle Velocity 9
1.4.4 Labeling and Physical Spaces 10
1.4.5 Jacobian Matrix 10
1.4.6 Material Vectors 10
1.4.7 Volumetric Coefficient 11
1.4.8 Exercise 11
1.5 The Material Derivative 11
1.5.1 Rate of Change Following a Point Particle 11
1.5.2 Relation to Eulerian Derivatives 12

1.5.3 Directional Derivative ... 12
1.5.4 Point-Particle Velocity ... 13
1.5.5 Point-Particle Acceleration ... 13
1.5.6 Properties of the Material Derivative ... 13
1.5.7 Fundamental Law of Kinematics ... 13
1.5.8 Volumetric Evolution Equation ... 14
1.5.9 Expansion or Contraction of a Material Parcel ... 14
1.5.10 Exercises ... 14
1.6 Material Parcels and Natural Laws ... 14
1.6.1 Parcel Properties ... 15
1.6.2 Integration in Labeling Space ... 15
1.6.3 Evolution of Parcel Properties ... 15
1.6.4 Natural Laws ... 16
1.6.5 Surface Flux and Volume Production ... 16
1.6.6 Conductive Flux and Surface Traction ... 16
1.6.7 Narrative Form ... 17
1.6.8 Transport Balances ... 17
1.6.9 Generalized Laws ... 17
1.6.10 Exercises ... 17
1.7 The Reynolds Transport Equation ... 17
1.7.1 Formal Derivation ... 18
1.7.2 Powers of the Metric Coefficient ... 18
1.7.3 Reynolds Transport Equation ... 18
1.7.4 Convection by Normal Motion ... 19
1.7.5 Orientation of the Unit Normal Vector ... 19
1.7.6 Evolution of a Parcel's Volume ... 19
1.7.7 Essential Step ... 20
1.7.8 Evolution Equations ... 20
1.7.9 Exercise ... 20
1.8 Integral Balance Over a Control Volume ... 20
1.8.1 Control Volume Occupied by a Parcel ... 21
1.8.2 Evolving Control Volume ... 22
1.8.3 Accumulation ... 22
1.8.4 Convective Transport ... 22
1.8.5 Control-Volume Balance ... 23
1.8.6 Balance Over an Open System ... 23
1.8.7 Exercise ... 23
1.9 Mass Conservation and Material Continuity ... 23
1.9.1 Mass Conservation for a Closed System ... 23
1.9.2 Instance of a General Law ... 24
1.9.3 Continuity Equation ... 24
1.9.4 Material Continuity ... 25
1.9.5 Eulerian Form ... 25
1.9.6 Incompressible Fluids ... 25
1.9.7 Evolution of a Specific Parcel Property ... 26
1.9.8 Traveler's Time Derivative ... 26
1.9.9 Exercises ... 27

1.10 Rankin–Hugoniot Density Condition ... 27
1.10.1 Ensuring Regularity ... 27
1.10.2 One-Dimensional Flow ... 28
1.10.3 Exercise ... 29
1.11 Integral Mass Balance ... 29
1.11.1 Mass Accumulation ... 29
1.11.2 Mass Accumulation Due to Convection ... 29
1.11.3 Stationary or Recirculating Fluid ... 30
1.11.4 Engineering Analysis ... 30
1.11.5 Mass Conservation for an Open System ... 30
1.11.6 Exercise ... 30
1.12 Tracers and Packets ... 30
1.12.1 Labeling Tracers ... 30
1.12.2 Reynolds Transport Equation ... 31
1.12.3 Rate of Change of a Packet Property ... 32
1.12.4 Flux and Production ... 33
1.12.5 Packet Property Conservation ... 33
1.12.6 Accumulation ... 34
1.12.7 Control Volume Occupied by a Packet ... 34
1.12.8 Mixtures ... 35
1.12.9 Exercise ... 35

2 Momentum and Forces ... 37
2.1 Traction and the Stress Tensor ... 37
2.1.1 Force in Terms of the Traction ... 38
2.1.2 Traction in Terms of the Stress Tensor ... 38
2.1.3 Hydrostatics ... 39
2.1.4 Pressure in Hydrodynamics ... 39
2.1.5 Deviatoric Stress Tensor ... 39
2.1.6 Incompressible Newtonian Fluids ... 39
2.1.7 Exercise ... 40
2.2 Parcel Momentum ... 40
2.2.1 Momentum in Terms of the Point-Particle Acceleration ... 40
2.2.2 Alternative Derivations ... 41
2.2.3 Exercise ... 41
2.3 Cauchy's Equation of Motion ... 41
2.3.1 Instance of a General Form ... 41
2.3.2 Point-Particle Acceleration ... 42
2.3.3 Euler, Bernoulli, and Navier–Stokes Equations ... 42
2.3.4 Eulerian Form ... 43
2.3.5 Conservative Form ... 43
2.3.6 Stress–Momentum Tensor ... 43
2.3.7 Exercise ... 44
2.4 Rankin–Hugoniot Momentum Condition ... 44
2.4.1 One-Dimensional Flow ... 45
2.4.2 Exercise ... 46

2.5 Noninertial Frames 46
2.5.1 Position, Velocity, and Acceleration 46
2.5.2 Equation of Motion 47
2.5.3 Exercise 47
2.6 Integral Momentum Balance 48
2.6.1 Physical Interpretation 48
2.6.2 Newton's Second Law of Motion for an Open System 48
2.6.3 Engineering Analysis 49
2.6.4 Exercise 49
2.7 Angular Momentum 49
2.7.1 Center of Momentum 49
2.7.2 Rate of Change of Parcel Angular Momentum 49
2.7.3 Evolution in Terms of Eulerian Derivatives 50
2.7.4 Evolution in Terms of Point-Particle Acceleration 50
2.7.5 Exercises 51
2.8 Newton's Law for the Angular Momentum 51
2.8.1 Instance of a General Law 51
2.8.2 Stress Tensor Spin 52
2.8.3 Cauchy Angular Momentum Equation 53
2.8.4 Symmetry of the Stress Tensor 53
2.8.5 Eulerian Form 53
2.8.6 Conservative Form 53
2.8.7 Angular Momentum Transport 54
2.8.8 Balance on an Open System 54
2.8.9 Exercises 55
2.9 Mechanical Energy Balance 55
2.9.1 Velocity Projected on Acceleration 55
2.9.2 Cauchy Equation 55
2.9.3 Velocity Projected on Acceleration 55
2.9.4 Double-Dot Product 56
2.9.5 Mechanical Energy Loss or Gain 56
2.9.6 Lagrangian Form 56
2.9.7 Conservative Form 57
2.9.8 Deviatoric Stress Tensor 57
2.9.9 Integral Mechanical Energy Balance 57
2.9.10 Energy Conversion 58
2.9.11 Exercises 58

3 Energy Balances in a Homogeneous Fluid 59
3.1 Mechanical Energy of a Fluid Parcel 59
3.1.1 Potential Energy 59
3.1.2 Rate of Change of the Potential Energy 60
3.1.3 Potential Energy Due to Gravity 60
3.1.4 Potential Energy Due to an Arbitrary Force Field 61
3.1.5 Kinetic Energy 61
3.1.6 Mechanical Energy 62
3.1.7 Mechanical Energy Loss or Gain 62

3.1.8 Energy Conversion ... 62
3.1.9 Exercises ... 63
3.2 First Law of Thermodynamics ... 63
3.2.1 Work Performed and Work Received ... 63
3.2.2 First Law of Thermodynamics for a Fluid Parcel ... 63
3.2.3 Heat Added and Work Received ... 64
3.2.4 Conductive Heat Flux ... 64
3.2.5 Interior Heat Generation ... 64
3.2.6 Surface Heat Generation ... 65
3.2.7 Specific Internal Energy ... 65
3.2.8 Energy Balance ... 65
3.2.9 First Law of Thermodynamics for a Closed System ... 66
3.2.10 Specific Form ... 66
3.2.11 Instance of a General Form ... 66
3.2.12 Exercise ... 67
3.3 Total-Energy Transport ... 67
3.3.1 Parcel Balance Equation ... 67
3.3.2 Eulerian Form ... 68
3.3.3 Evolution Equations ... 68
3.3.4 Lagrangian Form ... 68
3.3.5 Interpretation ... 69
3.3.6 Deviatoric Stress Tensor ... 69
3.3.7 Integral Energy Balance ... 69
3.3.8 Energy Balance for an Open System ... 70
3.3.9 Exercises ... 70
3.4 Internal-Energy Transport ... 70
3.4.1 Evolution Equations ... 70
3.4.2 Deviatoric Stress Tensor ... 71
3.4.3 Integral Internal-Energy Balance ... 71
3.4.4 Thermodynamics ... 72
3.4.5 Exercises ... 72
3.5 Kinetic–Internal Energy Transport ... 72
3.5.1 Evolution Equation ... 73
3.5.2 Deviatoric Stress Tensor ... 73
3.5.3 Integral Kinetic–Internal Energy Balance ... 73
3.5.4 Exercise ... 74
3.6 Enthalpy Transport ... 74
3.6.1 Parcel Enthalpy ... 74
3.6.2 Evolution of the specific enthalpy ... 74
3.6.3 Simplification ... 75
3.6.4 Final Act ... 75
3.6.5 Eulerian Conservative Form ... 76
3.6.6 Integral Enthalpy Balance ... 76
3.6.7 Exercise ... 76
3.7 Temperature Evolution ... 76
3.7.1 In Terms of c_v ... 77
3.7.2 In Terms of c_p ... 78
3.7.3 Reconciliation ... 79

3.7.4 Integral Thermal Energy Balance ... 79
3.7.5 Exercise ... 80
3.8 Entropy Transport ... 80
3.8.1 Gibbs Fundamental Equation ... 80
3.8.2 Gibbs Fundamental Equation in Terms of the Enthalpy ... 81
3.8.3 Entropy Evolution Equation ... 81
3.8.4 Evolution of Parcel Entropy ... 81
3.8.5 Why Entropy? ... 82
3.8.6 Integral Entropy Balance ... 82
3.8.7 Exercise ... 82
3.9 Second Law of Thermodynamics ... 82
3.9.1 Clausius–Duhem Inequality ... 83
3.9.2 Clausius–Duhem Inequality for a Symmetric Stress Tensor ... 83
3.9.3 Exercises ... 84
3.10 Fourier's Law of Conduction ... 84
3.10.1 Divergence of the Flux ... 84
3.10.2 Isotropic Materials ... 85
3.10.3 Clausius–Duhem Inequality ... 85
3.10.4 Onsager Principle ... 85
3.10.5 Exercise ... 85
3.11 Helmholtz Free Energy ... 85
3.11.1 Evolution Equations ... 86
3.11.2 Exercise ... 86
3.12 Gibbs Free Energy ... 86
3.12.1 Evolution Equations ... 87
3.12.2 Exercise ... 87

4 Solutions and Mixtures ... 89
4.1 Molecular, Mass, and Other Fractions ... 89
4.1.1 Molecular (Molar) Fractions ... 89
4.1.2 Probabilities ... 90
4.1.3 Continuum Approximation ... 90
4.1.4 Mole and Molar Mass ... 90
4.1.5 Mixture Molar Mass ... 91
4.1.6 Molar-Mass Concentrations ... 91
4.1.7 Mass Fractions ... 91
4.1.8 Mass Densities ... 92
4.1.9 Molar Concentrations ... 92
4.1.10 Partial Volumes ... 93
4.1.11 Why So Many? ... 94
4.1.12 Exercise ... 94
4.2 Gradients in a Binary Mixture ... 94
4.2.1 Molecular Fractions, Molar-Mass Fractions, and Mass Concentrations ... 94
4.2.2 Relation Between $\nabla\phi_R$ and ∇f_R ... 96
4.2.3 Relation Between $\nabla\omega_R$ and ∇f_R ... 97

4.2.4 Relation Between ∇f_R and ∇c_R 97
4.2.5 Exercise 97
4.3 Gradients in a Multicomponent Mixture 97
4.3.1 First Relation Between $\nabla \phi_R$ and ∇f_X 97
4.3.2 Second Relation Between $\nabla \phi_R$ and ∇f_X 98
4.3.3 Relation Between ∇f_R and $\nabla \phi_X$ 99
4.3.4 Relation Between $\nabla \omega_R$ and ∇f_X 99
4.3.5 Exercise 100
4.4 Species Velocity and Momentum 100
4.4.1 Mass (Barycentric) Velocity 100
4.4.2 Significance of the Mixture Mass Velocity 100
4.4.3 Species Mass Flux 101
4.4.4 Momentum Velocity 101
4.4.5 Mixture Governing Equations 101
4.4.6 Significance of the Species Velocities 102
4.4.7 Mixture Lag Velocities 102
4.4.8 Multicomponent Mixtures 103
4.4.9 Diffusion Velocities 103
4.4.10 Exercises 103
4.5 Solutions vs. Multicomponent Materials 103
4.5.1 Coffee 103
4.5.2 Blood 104
4.5.3 Diffusion in Molecular Mixtures 104
4.5.4 Kinetic Energy 104
4.5.5 Exercises 105
4.6 Fick–Stefan Framework 105
4.6.1 Point Particles 106
4.6.2 The Boundaries of Red and Blue Parcels Are Sharp and Well Defined 106
4.6.3 Red and Blue Parcel Separation 106
4.6.4 Diffusion as the Result of Red and Blue Parcel Separation 107
4.6.5 Diffusion in a Box 107
4.6.6 Parcel Expansion 108
4.6.7 Parcel Advection 108
4.6.8 Exercises 108
4.7 Species Kinematics 108
4.7.1 Red and Blue Lagrangian Frameworks 109
4.7.2 Red and Blue Material Derivatives 109
4.7.3 Mixture Material Derivative 110
4.7.4 Acceleration 110
4.7.5 Reynolds Transport Equation 111
4.7.6 Parcel Superposition 112
4.7.7 Exercise 112
4.8 Fundamental Decomposition of Kinematics 112
4.8.1 Velocity-Gradient Tensor 112
4.8.2 Rate of Expansion 113
4.8.3 Vorticity 113

4.8.4 Antisymmetric Part of the Velocity Gradient Tensor 114
4.8.5 Symmetric Part of the Velocity Gradient Tensor 114
4.8.6 Rate-of-Deformation Tensor 115
4.8.7 Exercise 115
4.9 Essence of Diffusion 115
4.9.1 Mixture of Isotopes in a Box 115
4.9.2 Random Crossings 116
4.9.3 Macroscopic State of Rest 116
4.9.4 Diffusion Alone Is Responsible for Different Species Velocities 116
4.9.5 Diffusion in a Gas 117
4.9.6 Diffusion in a Liquid or Solid 117
4.9.7 Further Types of Diffusion 117
4.9.8 Exercise 117
4.10 Random Walkers 117
4.10.1 Walking Particles 117
4.10.2 Particle Population 118
4.10.3 Population Balance 118
4.10.4 Convection–Diffusion Equation 119
4.10.5 Green's Function 119
4.10.6 Numerical Simulation 120
4.10.7 Exercises 123
4.11 Random Walks 123
4.11.1 Particle Position After *n* Steps 124
4.11.2 *m* Steps to the Right 124
4.11.3 dpd for *m* 125
4.11.4 Expected Value and Variance for *m* 126
4.11.5 De Moivre–Laplace Theorem 126
4.11.6 dpd for *k* 127
4.11.7 Unbiased Walk 128
4.11.8 Walk with Pauses 128
4.11.9 Exercise 129
4.12 Random Walkers with Continuous Steps 129
4.12.1 Balance Equation 130
4.12.2 Local Expansion 130
4.12.3 Notched Positions 131
4.12.4 Mapped pdf 131
4.12.5 Intercepting Particles 131
4.12.6 Exercise 132

5 Species Balances 133
5.1 Species Mass Balances 133
5.1.1 Red Material Derivative 134
5.1.2 Reynolds Transport Equation 134
5.1.3 Eulerian Mass Balance 135
5.1.4 Control Volumes 135
5.1.5 Parcel Superposition 135
5.1.6 Y Mass in an X Parcel 136

5.1.7 Exercise 136
5.2 Species Density Evolution Equations 136
5.2.1 Multicomponent Solution 137
5.2.2 From Species to Mixture 137
5.2.3 Evolution of Mass Fractions 137
5.2.4 Diffusion in a Box 138
5.2.5 Exercise 138
5.3 Rankin–Hugoniot Species Density Condition 139
5.3.1 One-Dimensional Flow 139
5.3.2 Exercise 141
5.4 Species Molar Balances 141
5.4.1 Number of Red and Blue Molecules 141
5.4.2 Exercise 142
5.5 Species Concentration Evolution Equations 142
5.5.1 Mixture Molar Concentration 143
5.5.2 Uniform Mixture Concentration 143
5.5.3 Evolution of Parcel Number of Molecules 143
5.5.4 Predator–Prey Equations 144
5.5.5 Exercise 144
5.6 Species Parcel Momentum 144
5.6.1 Species Particle Acceleration 144
5.6.2 Conservative Form 145
5.6.3 Reconciliation 146
5.6.4 Exercise 146
5.7 Species Parcel Angular Momentum 146
5.7.1 Evolution in Terms of Acceleration 147
5.7.2 Evolution in Terms of Eulerian Derivatives 147
5.7.3 Reconciliation 148
5.7.4 Exercise 149
5.8 Diffusive Mass Flux 149
5.8.1 Species Parcel Mass Conservation 149
5.8.2 Binary Mixture 149
5.8.3 Blue Parcels 150
5.8.4 Binary Mixture 150
5.8.5 The Sum of the Species Diffusive Mass Fluxes Is Zero 150
5.8.6 Control Volumes 150
5.8.7 Mass Balances Over a Mixture Parcel 151
5.8.8 Stationary Mixture 152
5.8.9 Stationary Species 152
5.8.10 Validation of Diffusion Flux Laws 152
5.8.11 Multicomponent Mixture 153
5.8.12 Exercise 153
5.9 Diffusion Velocities 153
5.9.1 Species Velocities in Terms of Diffusion Fluxes 154
5.9.2 Stationary Species 154
5.9.3 Kinetic Energy 154
5.9.4 Point-Particle Motion 155

5.9.5 Chaotic Advection 155
5.9.6 Exercise 155
5.10 Fundamental Decomposition of Kinematics 155
5.10.1 Velocity-Gradient Tensor 156
5.10.2 Rate of Expansion 156
5.10.3 Incompressible Mixtures 157
5.10.4 Exercise 157
5.11 Density Evolution Equations Redux 157
5.11.1 Eulerian Form 157
5.11.2 Lagrangian Form 158
5.11.3 Evolution of Mass Fractions 158
5.11.4 Alternative Derivation 159
5.11.5 Stationary Mixture 159
5.11.6 Exercise 159
5.12 Red and Blue Concentrations 159
5.12.1 Evolution of Species Molar Concentrations 160
5.12.2 Evolution of Mixture Concentration 161
5.12.3 Evolution of Molar Fractions 161
5.12.4 Diffusive Molar Flux Validation 162
5.12.5 Exercise 162

6 Diffusion Laws 163
6.1 Fick's Law 163
6.1.1 Fick's Law in Terms of Mass Fractions 163
6.1.2 Fick's Law in Terms of Molar Fractions 164
6.1.3 Fick's Law in Terms of Molar-Mass Concentrations 164
6.1.4 Diffusive Flux Orientation 165
6.1.5 Diffusion Velocities 165
6.1.6 Difference in Species Velocities 166
6.1.7 Fick's Law for the Molar Flux 166
6.1.8 Multicomponent Mixture 166
6.1.9 Exercise 167
6.2 Fick's Diffusion Kinematics 167
6.2.1 Velocity-Gradient Tensor 167
6.2.2 Rate of Expansion 168
6.2.3 Vorticity 168
6.2.4 Exercise 168
6.3 Species Fraction Evolution Equations 168
6.3.1 Mass Fraction Evolution Equation 168
6.3.2 Stationary Mixture 169
6.3.3 Molar Fraction Evolution Equation 169
6.3.4 Exercise 170
6.4 Coyote 170
6.4.1 Species and Mixture Velocities 170
6.4.2 Uniform c 171
6.4.3 Evolution Equations 171
6.4.4 Fick's Law 172
6.4.5 Summary 172

6.4.6 Motion of Point Particles ... 173
6.4.7 Depletion of Whiskey Layer ... 174
6.4.8 Exercise ... 174
6.5 Red and Blue Gas ... 174
6.5.1 Fick's Law ... 174
6.5.2 Chemical Activity ... 175
6.5.3 Self-diffusion ... 176
6.5.4 Exercise ... 176
6.6 Generalized Local Diffusive Flux Laws ... 176
6.6.1 Comprehensive Laws ... 177
6.6.2 Exercise ... 177
6.7 Fractional Diffusion ... 177
6.7.1 Fractional Laplacian ... 177
6.7.2 Fractional Diffusion in One Dimension ... 178
6.7.3 Principal-Value Integral ... 178
6.7.4 Second Moment of the Gaussian Distribution ... 179
6.7.5 Fractional Laplacian in One Dimension ... 180
6.7.6 Population Evolution ... 180
6.7.7 Green's Function ... 181
6.7.8 Fractional Diffusion in Two Dimensions ... 182
6.7.9 Fractional Diffusion in Three Dimensions ... 182
6.7.10 Exercises ... 183
6.8 Random Jumpers ... 183
6.8.1 Expected Displacement and Variance ... 183
6.8.2 Population Balances ... 184
6.8.3 Power-Law Probability Distribution ... 184
6.8.4 Population Balance for the Power-Law Distribution ... 185
6.8.5 Fractional Diffusion Equation ... 185
6.8.6 Numerical Simulation ... 186
6.8.7 Code ... 187
6.8.8 Exercises ... 192

7 Species and Mixture Hydrodynamics ... 193
7.1 Species Stress Tensor ... 193
7.1.1 Interpretation ... 194
7.1.2 Species Pressures ... 194
7.1.3 Deviatoric Stress Tensor ... 195
7.1.4 Exercise ... 195
7.2 Equation of Species Parcel Motion ... 195
7.2.1 Equation of Species Parcel Motion ... 195
7.2.2 Constraints ... 196
7.2.3 Species Momentum Transport ... 196
7.2.4 Integral Red Momentum Balance ... 197
7.2.5 Suspensions, Emulsions, and Granular Materials ... 197
7.2.6 Exercise ... 197
7.3 Species Interaction Force ... 198
7.3.1 Action and Reaction ... 198
7.3.2 Interaction Laws ... 198

7.3.3 Scaling ... 199
7.3.4 Darcy and Brinkmann Laws ... 199
7.3.5 Binary Mixture ... 199
7.3.6 Fick's Law ... 199
7.3.7 Generalization ... 199
7.3.8 Multicomponent Mixture ... 200
7.3.9 Exercise ... 201
7.4 Exchange of Momentum Due to Species Conversion ... 201
7.4.1 Kinetics ... 201
7.4.2 Molecules Carrying Momentum ... 201
7.4.3 Multicomponent Mixtures ... 202
7.4.4 Other Interspecies Momentum Transfer Laws ... 203
7.4.5 Exercise ... 203
7.5 Cauchy Species Equation of Motion ... 203
7.5.1 Lagrangian Form ... 203
7.5.2 Blue Species ... 204
7.5.3 Momentum Exchange ... 204
7.5.4 Stationary Species ... 205
7.5.5 From Species to Mixture ... 205
7.5.6 Exercise ... 206
7.6 Rankin–Hugoniot Momentum Condition ... 206
7.6.1 Exercise ... 206
7.7 Species Angular Momentum Balance ... 207
7.7.1 Stress Tensor Spin ... 207
7.7.2 Cauchy Angular Momentum Equation ... 208
7.7.3 Implicit Effect of the Torque Field ... 209
7.7.4 Symmetry of the Species Stress Tensor ... 209
7.7.5 Exercise ... 209
7.8 Parcel Superposition ... 209
7.8.1 Species and Mixture Parcels ... 209
7.8.2 Mixture Momentum of Parcels ... 210
7.8.3 From Species to Mixture ... 210
7.8.4 Y Momentum of X Parcel ... 211
7.8.5 Exercise ... 211
7.9 From Species to Mixture Stresses ... 211
7.9.1 Species Modified Stress Tensors ... 212
7.9.2 Mixture Stress and Equation of Motion ... 212
7.9.3 Multicomponent Mixture ... 213
7.9.4 Binary Mixture ... 213
7.9.5 Stress–Momentum Tensor ... 213
7.9.6 Symmetry of the Stress Tensor ... 214
7.9.7 Pressure ... 214
7.9.8 Mechanical Energy Loss or Gain ... 214
7.9.9 Constitutive Equations ... 215
7.9.10 Exercises ... 215
7.10 Stationary Mixtures ... 216
7.10.1 Species Equations of Motion ... 216
7.10.2 Fick's Law ... 216

7.10.3 Mixture Stress Tensor ... 218
7.10.4 One-Dimensional Self-diffusion ... 218
7.10.5 Exercise ... 219

8 Mixture Energy Balances ... 221
8.1 Kinetic Energy Balance ... 221
8.1.1 Effective Species Force ... 222
8.1.2 Mixture Kinetic Energy ... 223
8.1.3 Exercise ... 223
8.2 Kinetic–Internal Energy Balance ... 223
8.2.1 Species Kinetic–Internal Energy in Terms of Diffusion Fluxes ... 223
8.2.2 Mixture-Specific Kinetic–Internal Energy ... 224
8.2.3 Interspecies Energy Transfer ... 225
8.2.4 From Species to Mixture ... 225
8.2.5 Summary ... 226
8.2.6 Exercise ... 226
8.3 Total Energy Flux ... 227
8.3.1 Definitions ... 227
8.3.2 Inviscid Species Tensors ... 228
8.3.3 Exercise ... 228
8.4 Internal-Energy Balance ... 229
8.4.1 Thermal Part of the Mixture Internal Energy ... 229
8.4.2 Mixture Internal Energy ... 230
8.4.3 Species Internal Energy Balance ... 230
8.4.4 Mixture Thermal Internal Energy Balance ... 231
8.4.5 Temperature Evolution ... 231
8.4.6 Exercise ... 231
8.5 Enthalpy Balance ... 232
8.5.1 First Leap of Faith ... 232
8.5.2 Second Leap of Faith ... 233
8.5.3 Temperature Evolution ... 234
8.5.4 Exercise ... 234
8.6 Partial Species Enthalpies ... 234
8.6.1 Multicomponent Mixtures ... 236
8.6.2 First Leap of Faith ... 237
8.6.3 Exercise ... 239
8.7 Partial Molar Enthalpies ... 239
8.7.1 Exercise ... 240
8.8 Entropy Balance ... 241
8.8.1 Entropy Transport ... 242
8.8.2 Enthalpy, Helmholtz Free Energy, and Gibbs Energy ... 243
8.8.3 Advanced Formulations ... 243
8.8.4 Exercises ... 243

A Summary of Transport Equations ... 245
A.1 Eulerian and Lagrangian Evolution Equations ... 245
A.2 Density ... 245

A.3 Momentum ... 246
A.4 Angular Momentum ... 246
A.5 Kinetic Energy ... 246
A.6 Total Energy ... 246
A.7 Internal Energy ... 247
A.8 Kinetic–Internal Energy ... 247
A.9 Enthalpy ... 247
A.10 Entropy ... 247
A.11 Unification ... 248
A.12 Integral Balances ... 248
A.13 Summary ... 249
A.14 Exercise ... 249

B Elements of Thermodynamics ... 257
B.1 Equations of State ... 257
B.1.1 Ideal Gas ... 257
B.1.2 PvT Surface ... 258
B.1.3 pv Diagram ... 258
B.1.4 Euler's Cyclic Relation ... 258
B.1.5 Moduli of Pressure and Volume ... 260
B.1.6 Redlich–Kwong Equation of State ... 261
B.1.7 Exercises ... 261
B.2 State Functions ... 261
B.2.1 Functions $\psi(v, T)$ and $\psi(p, T)$ ... 262
B.2.2 Internal Energy ... 263
B.2.3 Heat Capacity Under Constant Volume ... 263
B.2.4 Ideal Gas ... 263
B.2.5 Exercise ... 264
B.3 Entropy ... 264
B.3.1 Ideal Gas ... 264
B.3.2 Power Laws ... 265
B.3.3 Entropy of an Ideal Gas ... 265
B.3.4 Exercise ... 266
B.4 First Law of Thermodynamics ... 266
B.4.1 Expansion of a Gas ... 266
B.4.2 Reversibility ... 267
B.4.3 Compression ... 267
B.4.4 Entropy and Reversibility ... 268
B.4.5 Exercise ... 268
B.5 Carnot Cycle ... 268
B.5.1 Exercise ... 269
B.6 Enthalpy ... 269
B.6.1 Heat Capacity Under Constant Pressure ... 270
B.6.2 Relations Between c_v and c_p ... 270
B.6.3 Ideal Gas ... 271
B.6.4 Dependence of Enthalpy on Pressure ... 271
B.6.5 Exercise ... 272

B.7 Free Energies 272
B.7.1 Gibbs Free Energy 273
B.7.2 Exercise 273
B.8 Maxwell Relations 274
B.8.1 Exercise 274

Index 275

Notation

Scalars are set in italic; vectors and matrices are set in bold face. A dot over a variable indicates the rate of change.

$\mathbf{a}$	Acceleration
$\mathbf{b}$	Distributed body force
$\mathbf{E}$	Rate-of-deformation tensor
$\mathbf{f}$	Traction
$\mathbf{g}$	Acceleration of gravity
h	Specific enthalpy
$\mathbf{I}$	Identity matrix
$\mathbf{L}$	Velocity-gradient tensor
$\mathbf{n}$	Normal unit vector
p	Pressure
$\dot{q}$	Local volumetric rate of interior heat generation
$\dot{Q}$	Rate of interior heat production
S	Entropy
s	Specific entropy
T	Temperature
u	Specific internal energy
$\mathbf{u}$	Fluid velocity
$\mathbf{v}$	Velocity
$\mathbf{x}$	Position
$\mathbf{X}$	Position
α	Rate of expansion
$\boldsymbol{\alpha}$	Lagrangian label
η	Specific enthalpy
ι	Physical constant associated with a distributed body force
μ	Viscosity
$\boldsymbol{\Xi}$	Vorticity tensor
ρ	Density
$\boldsymbol{\sigma}$	Stress tensor
$\boldsymbol{\omega}$	Vorticity vector
$\boldsymbol{\Omega}$	Angular velocity
$\mathcal{J}$	Metric of Lagrangian mapping
$\mathrm{D}/\mathrm{D}t$	Material derivative
$\cdot$	Inner vector product (projection)
$\times$	Outer vector product
$\otimes$	Tensor product of two vectors
$\boldsymbol{\nabla}$	Gradient (nabla) operator

Homogeneous Fluids

1

By definition, all molecules of a homogeneous fluid are identical in the absence of phase coexistence, phase separation, or any other unusual behavior. However, molecular number density variations responsible for fluid density inhomogeneities are allowed. Thus, a homogeneous fluid is not necessarily a uniform-density fluid.

In the first chapter, we study the motion of a homogeneous fluid and introduce the notions of fluid velocity, density, and other similar physical fields identified as intensive properties in the context of the continuum approximation. The concept of a material point particle and associated material derivative allow us to develop evolution, balance, and transport equations for suitable fields based on physical laws and principles of thermodynamics.

The derivations of transport equations presented in this chapter are pivoted on the Reynolds transport equation, also known as the Reynolds transport theorem, which can be traced to the Leibniz integral rule for integration volumes that evolve in time. In the present context, the integration volumes are material fluid parcels moving and deforming under the influence of a specified flow.

1.1 The Concept of Fluid Velocity

Consider the motion of a small material fluid parcel consisting of a single molecular species. The velocity of translation of the parcel in a certain direction is defined as the arithmetic average, also called the mean or expected value, of the instantaneous component of the velocity of all molecules residing inside the parcel in that direction.

1.1.1 Continuum Approximation

Now we consider a small parcel with spherical shape of radius ϵ centered at a certain point, $\mathbf{x}$, at a certain observation time, t. The parcel velocity at time t depends on the parcel radius, ϵ. Taking the limit as ϵ tends to zero, we find that the average of the molecular velocities tends to a well-defined limit, until ϵ becomes comparable to intermolecular separations, as shown in Fig. 1.1. At that stage, we observe pronounced oscillations that are manifestations of random molecular motions inherent in any material.

C. Pozrikidis, *Transport Processes Primer*,
https://doi.org/10.1007/978-1-4939-9909-5_1

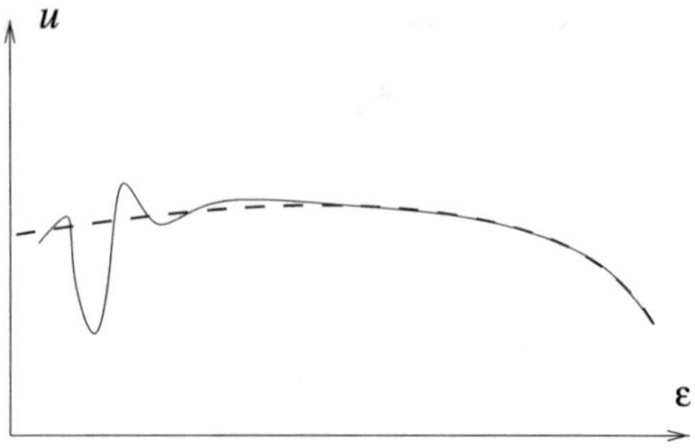

Fig. 1.1 The fluid velocity at a point in a fluid is defined as the outer limit of the parcel velocity with respect to the parcel radius, ϵ, drawn with the broken line. The oscillations are due to random molecular fluctuations

1.1.2 Velocity Field

The velocity of the fluid, $\mathbf{u}$, at the parcel center, $\mathbf{x}$, at time, t, is defined as the apparent or outer limit of the velocity of the parcel as the parcel radius ϵ tends to zero, immediately before the discrete nature of the fluid becomes apparent. Accordingly, we write

$$\mathbf{u}(\mathbf{x}, t) \tag{1.1.1}$$

to indicate that the velocity is a function of position, $\mathbf{x}$, and time, t. This definition is the cornerstone of the *continuum approximation* in fluid mechanics, solid mechanics, and more general continuum mechanics.

Similar approximations can be made for the kinetic energy and other properties attributed to molecules, as discussed in Sect. 1.2.

1.1.3 Differentiability

Under normal conditions, the velocity, $\mathbf{u}(\mathbf{x}, t)$, is an infinitely differentiable function of position, $\mathbf{x}$, and time, t. This means that the derivatives of each component of $\mathbf{u}$ with respect to Cartesian or other curvilinear coordinates are well defined.

For example, the second partial derivative of the x velocity component with respect to y, and the time derivative,

$$\left(\frac{\partial^2 u_x}{\partial y^2}\right)_{z,x,t}, \qquad \left(\frac{\partial u_x}{\partial t}\right)_{x,y,z}, \tag{1.1.2}$$

are both well defined in the bulk of a fluid as well as in the limit as we approach the boundaries.

Spatial or temporal discontinuities of the velocity may arise under extreme conditions in high-speed flows or else appear due to mathematical idealization. For example, a velocity discontinuity may be forced by the impulsive motion of a plate in a viscous fluid or by the parallel merging of two streams with different velocities. Both discontinuities are mathematical idealizations that cannot be realized in practice.

1.1.4 Material Point Particles

The aforementioned limit of infinitesimal parcel size leads us to the notion of a material point particle with infinitesimal volume and mass. A fluid in three-dimensional space can be regarded as a continuous medium consisting of a triply infinite collection of material point particles, even though the number of molecules comprising the material is finite.

By definition, a point particle located at $\mathbf{X}(t)$ moves with the local and instantaneous fluid velocity,

$$\mathbf{u}(\mathbf{X}(t), t). \tag{1.1.3}$$

Note that time dependence enters explicitly and implicitly through the generally evolving point-particle position. The acceleration of a point particle is relevant to the dynamics of the flow with regard to Newton's second law of motion, as discussed in Chap. 2.

For simplicity, a material point particle will be referenced simply as a point particle. Nonmaterial point particles named tracers are defined and discussed in Sect. 1.12.

1.1.5 Material Parcel

A material parcel consists of a permanent and contiguous infinite collection of point particles. In the absence of singularities, a point particle at the surface of a material parcel remains at the surface of the parcel at all times.

1.1.6 Exercise

1.1.1 Discuss whether the continuum approximation is appropriate for dry sand.

1.2 Density and Other Intensive Fields

We have introduced the fluid velocity, $\mathbf{u}$, in terms of the average velocity of the molecules constituting a small fluid parcel by taking the outer limit as the size of the parcel becomes infinitesimal. This point of view led us to the Eulerian description, $\mathbf{u}(\mathbf{x}, t)$, where $\mathbf{u}$ is regarded a field function of position, $\mathbf{x}$, and time, t.

1.2.1 Density as a Field Function

Repeating the limiting procedure, we compute the ratio between the mass and volume of a small parcel, take the outer limit as the size of the parcel becomes infinitesimal, and thus obtain the fluid density, ρ, as a function of position, $\mathbf{x}$, and time, t,

$$\rho(\mathbf{x}, t). \tag{1.2.1}$$

Spatial and time derivatives of the density are well defined under the auspices of the continuum approximation in an ordinary density field. Discontinuities arise as mathematical idealizations.

The density of a material point particle located at $\mathbf{X}(t)$ at time t arises from the density field as

$$\rho(\mathbf{X}(t), t). \tag{1.2.2}$$

Time dependence enters explicitly and implicitly through the generally evolving point-particle position.

1.2.2 Extensive and Intensive Properties

The limiting procedure can be repeated for any extensive material property that can be attributed to a material parcel, such as the internal energy, enthalpy, or entropy. By definition, an extensive property is proportional to the parcel's volume or mass.

Under the auspices of the continuum approximation, as the size of a parcel becomes infinitesimal, the ratio of an extensive property to the parcel volume or mass tends to an outer limit that is an intensive property regarded a function of position, $\mathbf{x}$, and time, t. An intensive property can be subsequently scaled by a dimensionless, physical, or universal constant, as required.

An intensive property is truly a point-particle property that characterizes the kinematic, dynamic, thermodynamic, or some other state of an infinitesimal fluid parcel.

1.2.3 Temperature Field

If the extensive property is the kinetic energy of the fluctuation velocity of the molecular constituents of a parcel of a gas, the corresponding intensive property is the temperature, T, regarded as a function of position and time,

$$T(\mathbf{x}, t). \tag{1.2.3}$$

Space and time derivatives of the temperature are well defined. Conversely, the temperature at a point in a gas is an indicator of the kinetic energy of the fluctuating velocity of the molecular constituents of a small parcel of gas.

1.2.4 Exercise

1.2.1 Discuss a pair of extensive–intensive properties of your choice.

1.3 Fundamental Decomposition of Kinematics

To deduce the nature of a fluid motion, we consider spatial variations of the velocity, $\mathbf{u}$, in the neighborhood of a chosen fixed point, $\mathbf{x}_0$. Expanding the jth velocity component of the velocity in a Taylor series with respect to $\mathbf{x}$ about $\mathbf{x}_0$, and keeping only the linear terms, we find that

$$u_j(\mathbf{x}, t) \simeq u_j(\mathbf{x}_0, t) + \widehat{x}_i \left(\frac{\partial u_j}{\partial x_i}\right)_{\mathbf{x}_0, t}, \tag{1.3.1}$$

where j is a free index, summation is implied the repeated index, i, and the vector $\widehat{\mathbf{x}} = \mathbf{x} - \mathbf{x}_0$ connects the chosen fixed point, $\mathbf{x}_0$, to the variable field point, $\mathbf{x}$.

Note that $\widehat{x}_i$ was placed intentionally before the partial derivative on the right-hand side of (1.3.1).

1.3.1 Velocity-Gradient Tensor

The partial derivatives of the velocity on the right-hand side of (1.3.1) are the components of the velocity-gradient tensor, sometimes also called the rate of relative displacement tensor, denoted and defined as

$$\mathbf{L} \equiv \nabla \otimes \mathbf{u}, \tag{1.3.2}$$

where ∇ is the gradient operator and $\otimes$ denotes the tensor product. The Cartesian components of the velocity gradient tensor are given by

$$L_{ij} = \frac{\partial u_j}{\partial x_i}. \tag{1.3.3}$$

For convenience, the velocity gradient tensor is routinely denoted without the tensor product symbol as

$$\mathbf{L} \equiv \nabla \mathbf{u} \equiv \nabla \otimes \mathbf{u}. \tag{1.3.4}$$

In writing $\nabla \mathbf{u}$, the tensor product symbol between the gradient operator, ∇, and the velocity, $\mathbf{u}$, is implied. Expressions for the components of $\mathbf{L}$ in non-Cartesian, rectilinear or curvilinear coordinates involving corresponding velocity components and derivatives can be derived.

In vector notation, Eq. (1.3.1) takes the form

$$\mathbf{u}(\mathbf{x}, t) \simeq \mathbf{u}(\mathbf{x}_0, t) + \widehat{\mathbf{x}} \cdot \mathbf{L}(\mathbf{x}_0, t). \tag{1.3.5}$$

Note the left-to-right vector–matrix multiplication in the second term on the right-hand side.

1.3.2 Divergence of the Velocity

The trace of the velocity gradient tensor, $\mathbf{L}$, is the sum of the diagonal elements. We find that the trace is equal to the divergence of the velocity,

$$\alpha \equiv \mathrm{trace}(\mathbf{L}) = \frac{\partial u_i}{\partial x_i} = \nabla \cdot \mathbf{u}, \tag{1.3.6}$$

where summation is implied over the repeated index, i. Expressions for the divergence in non-Cartesian, rectilinear or curvilinear coordinates involving the corresponding velocity components can be derived.

1.3.3 Fundamental Decomposition of Kinematics

It is useful to decompose the velocity gradient, $\mathbf{L}$, into an antisymmetric component, $\boldsymbol{\Xi}$, a symmetric component with vanishing trace, $\mathbf{E}$, and an isotropic component with generally nonzero trace,

$$\mathbf{L} = \boldsymbol{\Xi} + \mathbf{E} + \tfrac{1}{3}\,\alpha\,\mathbf{I}, \tag{1.3.7}$$

where $\mathbf{I}$ is the identity matrix,

$$\boldsymbol{\Xi} \equiv \tfrac{1}{2}\,(\mathbf{L} - \mathbf{L}^{\mathrm{T}}), \qquad \Xi_{ij} = \tfrac{1}{2}\Big(\frac{\partial u_j}{\partial x_i} - \frac{\partial u_i}{\partial x_j}\Big) \tag{1.3.8}$$

is the *vorticity tensor*,

$$\mathbf{E} \equiv \tfrac{1}{2}\,(\mathbf{L} + \mathbf{L}^{\mathrm{T}}) - \tfrac{1}{3}\,\alpha\,\mathbf{I}, \qquad E_{ij} = \tfrac{1}{2}\Big(\frac{\partial u_j}{\partial x_i} + \frac{\partial u_i}{\partial x_j}\Big) - \tfrac{1}{3}\,\alpha\,\delta_{ij} \tag{1.3.9}$$

is the *rate-of-deformation tensor*, the superscript T denotes the matrix transpose, and δ_{ij} is Kronecker's delta defined such that $\delta_{ij} = 1$ when $i = j$ or 0 otherwise.

1.3.4 Vorticity Vector

Because the vorticity tensor is antisymmetric, it has only three independent components that can be compiled into a vorticity vector, $\boldsymbol{\omega}$, such that

$$\Xi_{ij} = \tfrac{1}{2}\,\epsilon_{ijk}\,\omega_k, \tag{1.3.10}$$

where ϵ_{ijk} is the Levi–Civita symbol representing the Cartesian components of the alternating tensor, and summation is implied over the repeated index k.

The Levi–Civita symbol, ϵ_{ijk}, is defined such that $\epsilon_{ijk} = 1$ in cyclic permutation of the three indices, -1 in anti-cyclic permutation, and 0 otherwise. For example, $\epsilon_{xyz} = 1$, $\epsilon_{yxz} = -1$, and $\epsilon_{xxy} = 0$.

Explicitly, the vorticity tensor and vorticity vector are related by

$$\boldsymbol{\Xi} = \tfrac{1}{2}\begin{bmatrix} 0 & \omega_z & -\omega_y \\ -\omega_z & 0 & \omega_x \\ \omega_y & -\omega_x & 0 \end{bmatrix}. \tag{1.3.11}$$

Note that the diagonal elements are all zero. Only the top 2×2 diagonal block involving ω_z survives in a two-dimensional flow in the xy plane where ω_x and ω_y are both identically zero.

1.3.5 The Vorticity Is the Curl of the Velocity

Conversely, the vorticity vector derives from the vorticity tensor as

$$\omega_k = \epsilon_{kij}\,\Xi_{ij} = \tfrac{1}{2}\,\epsilon_{kij}\Big(\frac{\partial u_j}{\partial x_i} - \frac{\partial u_i}{\partial x_j}\Big) = \epsilon_{kij}\,\frac{\partial u_j}{\partial x_i}, \tag{1.3.12}$$

where summation is implied over the repeated indices, i and j. The last expression shows that the vorticity is the curl of the velocity,

$$\boldsymbol{\omega} = \nabla \times \mathbf{u}. \tag{1.3.13}$$

Explicitly, the vorticity vector is given by

$$\boldsymbol{\omega} = \Big(\frac{\partial u_z}{\partial y} - \frac{\partial u_y}{\partial z}\Big)\,\mathbf{e}_x + \Big(\frac{\partial u_x}{\partial z} - \frac{\partial u_z}{\partial x}\Big)\,\mathbf{e}_y + \Big(\frac{\partial u_y}{\partial x} - \frac{\partial u_x}{\partial y}\Big)\,\mathbf{e}_z, \tag{1.3.14}$$

where $\mathbf{e}_x$, $\mathbf{e}_y$, and $\mathbf{e}_z$ are unit vectors along the x, y, and z axes. The components of the vorticity vector in non-Cartesian coordinates are given by more involved expressions.

1.3.6 Four Types of Local Motion

Substituting (1.3.10) into (1.3.7) and the resulting expression into (1.3.1), we derive an expression for the spatial distribution of the velocity in the neighborhood of a point, $\mathbf{x}_0$, in terms of the vorticity vector, $\boldsymbol{\omega}$, the rate-of-deformation tensor, $\mathbf{E}$, and the divergence of the velocity field, α,

$$u_j(\mathbf{x}, t) \simeq u_j(\mathbf{x}_0, t) + \tfrac{1}{2}\,\epsilon_{jki}\,\omega_k(\mathbf{x}_0, t)\,\widehat{x}_i + \widehat{x}_i\,E_{ij}(\mathbf{x}_0, t) + \tfrac{1}{3}\,\alpha\,\widehat{x}_j, \tag{1.3.15}$$

where $\widehat{\mathbf{x}} = \mathbf{x} - \mathbf{x}_0$ and summation is implied over the repeated indices, i and k. Four types of local motion are conveyed by the right-hand side of (1.3.15).

1.3.7 Translation

The first term on the right-hand side of (1.3.15) expresses rigid-body translation. Under the action of this term, a point particle at the field position $\mathbf{x}$ moves with the velocity of a point particle at position $\mathbf{x}_0$. If the rest of the terms on the right-hand side of (1.3.15) were absent, the fluid in the neighborhood of $\mathbf{x}_0$ would translate as a rigid body.

1.3.8 Rotation

The second term on the right-hand side of (1.3.15) is the indicial representation of the jth component of the outer vector product

$$\boldsymbol{\Omega}(\mathbf{x}_0, t) \times \widehat{\mathbf{x}}, \tag{1.3.16}$$

where

$$\boldsymbol{\Omega} = \tfrac{1}{2}\,\boldsymbol{\omega}. \tag{1.3.17}$$

This expression shows that a point particle at $\mathbf{x}$ rotates about the point $\mathbf{x}_0$ with angular velocity that is equal to half the vorticity evaluated at $\mathbf{x}_0$. Conversely, the vorticity vector is parallel to the point-particle angular velocity vector. The magnitude of the vorticity vector is equal to twice that of the point-particle angular velocity vector.

1.3.9 Deformation

Because the rate-of-deformation tensor **E** is real and symmetric, it has three real eigenvalues and three mutually orthogonal eigenvectors. Because the trace of **E** is zero by construction, the sum of the eigenvalues is zero.

Under the action of the velocity field expressed by the third term on the right-hand side of (1.3.15), three infinitesimal fluid parcels resembling slender needles aligned with the eigenvectors elongate or compress in their respective directions while preserving their mutually orthogonal orientations. Consequently, an ellipsoidal fluid parcel whose axes are aligned with the eigenvectors deforms, increasing or decreasing its aspect ratios, while preserving its initial orientation.

These observations suggest that the third term on the right-hand side of (1.3.15) expresses parcel deformation that preserves orientation. Moreover, because the trace of **E** is zero, the motion also preserves the parcel volume.

1.3.10 Contraction and Expansion

The last term on the right-hand side of (1.3.15) represents isotropic expansion or contraction. Under the action of this term, a small spherical parcel of radius b and volume

$$\delta V = \frac{4\pi}{3}\, b^3(t) \tag{1.3.18}$$

centered at the point $\mathbf{x}_0$ expands or contracts isotropically, undergoing neither translation, nor rotation, nor deformation.

To compute the rate of expansion, we note that, after a small time interval $\mathrm{d}t$ has elapsed, the radius of the spherical parcel has become

$$(1 + \tfrac{1}{3}\,\alpha\, \mathrm{d}t)\, b, \tag{1.3.19}$$

and the parcel volume has changed by the differential amount

$$\mathrm{d}\,\delta V = \frac{4\pi}{3}\left(1 + \tfrac{1}{3}\,\alpha\, \mathrm{d}t\right)^3 b^3(t) - \frac{4\pi}{3}\, b^3(t). \tag{1.3.20}$$

Expanding the cubic power of the binomial, linearizing the resulting expression with respect to $\mathrm{d}t$, and rearranging, we find that

$$\frac{1}{\delta V}\,\frac{\mathrm{d}\,\delta V}{\mathrm{d}t} = \alpha. \tag{1.3.21}$$

When $\alpha > 0$, the fluid expands; when $\alpha < 0$, the fluid contracts. This expression justifies calling the divergence of the velocity, $\alpha \equiv \nabla \cdot \mathbf{u}$, the rate of expansion or rate of dilatation.

1.3.11 Exercises

1.3.1 Derive (1.3.21) from (1.3.20).

1.3.2 Compute the velocity-gradient tensor in a simple shear flow with velocity components $u_x = \xi\, y$, $u_y = 0$, and $u_z = 0$, where ξ is a constant with units of inverse time called the shear rate.

1.4 Lagrangian Mapping of Point Particles

Under the auspices of the continuum approximation discussed in Sect. 1.1, a parcel of fluid in three-dimensional space can be regarded as a continuous material consisting of a triply infinite collection of material point particles identified by three scalar labels, α_1, α_2, and α_3. The three labels can be interpreted as point particle coordinates in labeling space and accommodated into a three-dimensional Lagrangian vector label,

$$\boldsymbol{\alpha} = (\alpha_1, \alpha_2, \alpha_3). \tag{1.4.1}$$

In two dimensions, only two scalar labels are employed. In one dimension, only one scalar label is employed.

1.4.1 RGB

Pictorially, the fluid is a three-dimensional array of billiard balls whose color, encoded by $\boldsymbol{\alpha}$, changes smoothly in space. The three components of $\boldsymbol{\alpha}$ can be the red–green–blue (RGB) encoding. As the billiard balls move in a domain of flow, their individual colors remain constant in time.

A possible choice for the components of $\boldsymbol{\alpha}$ is the point particle Cartesian coordinates at a specified time, t_0,

$$\boldsymbol{\alpha} = \mathbf{X}(t_0). \tag{1.4.2}$$

Non-Cartesian coordinates at a specified time, t_0, may also serve as point-particle labels.

1.4.2 Point-Particle Position

The instantaneous position of a point particle, $\mathbf{X}$, can be regarded as a function of the label, $\boldsymbol{\alpha}$, and time, t,

$$\mathbf{x} = \mathbf{X}(\boldsymbol{\alpha}, t). \tag{1.4.3}$$

Partial derivatives of $\mathbf{X}$ with respect to t and partial derivatives with respect to each of the three components of $\boldsymbol{\alpha}$ are well defined. The derivative with respect to t under constant $\boldsymbol{\alpha}$ should be distinguished from the Eulerian derivative taken at a fixed position, $\mathbf{x}$.

1.4.3 Point-Particle Velocity

By definition, a point particle located at $\mathbf{x}$ moves with the local and instantaneous fluid velocity,

$$\mathbf{U}(\boldsymbol{\alpha}, t) = \mathbf{u}(\mathbf{X}(\boldsymbol{\alpha}, t), t). \tag{1.4.4}$$

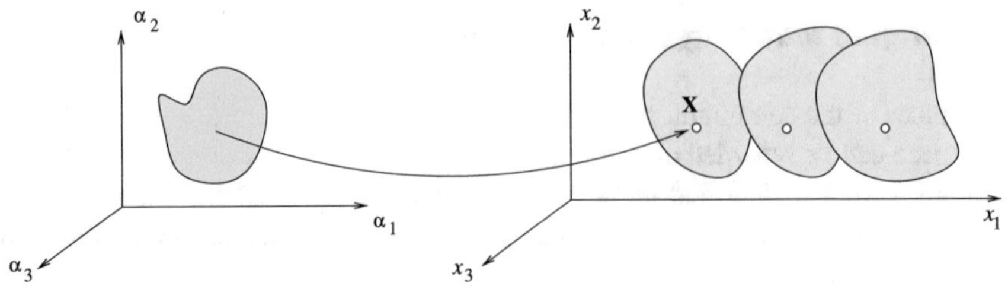

Fig. 1.2 Lagrangian mapping of the parametric space, $\boldsymbol{\alpha}$, to the physical space, $\mathbf{x}$. The position of a point particle in physical space is denoted by $\mathbf{X}$. The image of an evolving physical parcel remains fixed in parameter space. A moving parcel typically has different physical shapes at different times

The right-hand side is the fluid velocity evaluated at the position of the point particle at time t. Note that the right-hand side depends on t explicitly as well as implicitly by way of the point-particle position, $\mathbf{X}$.

1.4.4 Labeling and Physical Spaces

The functional dependence shown in (1.4.3) can be formalized in terms of a generally time-dependent mapping of the labeling space of $\boldsymbol{\alpha}$ to the physical space,

$$\boldsymbol{\alpha} \to \mathbf{X}, \tag{1.4.5}$$

as shown in Fig. 1.2. The mapping can be expressed in a standard symbolic form that displays explicitly the time dependence,

$$\mathbf{X}(t) = \boldsymbol{\mathcal{C}}_t(\boldsymbol{\alpha}). \tag{1.4.6}$$

The subscript, t, emphasizes that the mapping function $\boldsymbol{\mathcal{C}}_t$ may change in time.

1.4.5 Jacobian Matrix

The Jacobian matrix of a Lagrangian mapping function is the matrix of partial derivatives of $\boldsymbol{\mathcal{C}}_t(\boldsymbol{\alpha})$ with respect to the Lagrangian label with elements

$$\mathcal{J}_{ij} = \frac{\partial\, [\boldsymbol{\mathcal{C}}_t]_j}{\partial \alpha_i}, \tag{1.4.7}$$

where $[\boldsymbol{\mathcal{C}}_t]_j$ is the jth component of $\boldsymbol{\mathcal{C}}_t$. Note the order of indices on the right-hand side. In solid mechanics, the transpose of $\mathcal{J}$ is called the deformation gradient.

1.4.6 Material Vectors

A differential vector in labeling space, $\mathrm{d}\boldsymbol{\alpha}$, is mapped to a differential material vector in physical space, $\mathrm{d}\mathbf{X}$, by the equation

$$\mathrm{d}\mathbf{X} = \mathrm{d}\boldsymbol{\alpha} \cdot \mathcal{J} = \mathcal{J}^{\mathrm{T}} \cdot \mathrm{d}\boldsymbol{\alpha}, \tag{1.4.8}$$

where the superscript T denotes the matrix transpose. In index notation, Eq. (1.4.8) takes the from

$$dX_i = d\alpha_j \, \mathcal{J}_{ji} = \mathcal{J}^{\mathrm{T}}_{ij} \, d\alpha_j, \tag{1.4.9}$$

where summation is implied over the repeated index, j.

1.4.7 Volumetric Coefficient

A differential volume in labeling space, $dV(\boldsymbol{\alpha})$, is mapped to a differential material volume in physical space, $dV(\mathbf{X})$, according to (1.4.8). The two volumes are related by the equation

$$dV(\mathbf{X}) = \mathcal{J}(\boldsymbol{\alpha}) \, dV(\boldsymbol{\alpha}), \tag{1.4.10}$$

where $\mathcal{J}$ is the Lagrangian volumetric coefficient given by

$$\mathcal{J} \equiv (\frac{\partial \boldsymbol{C}_t}{\partial \alpha_1} \times \frac{\partial \boldsymbol{C}_t}{\partial \alpha_2}) \cdot \frac{\partial \boldsymbol{C}_t}{\partial \alpha_3} = \det(\boldsymbol{\mathcal{J}}) = \det(\boldsymbol{\mathcal{J}}^{\mathrm{T}}), \tag{1.4.11}$$

and det denotes the determinant. The precise form of $\mathcal{J}(\boldsymbol{\alpha}, t)$ depends on the Lagrangian mapping function, $\boldsymbol{C}_t(\boldsymbol{\alpha})$.

If a fluid is compressible, $\mathcal{J}(\boldsymbol{\alpha}, t)$ may change in time. If a fluid is incompressible, $\mathcal{J}(\boldsymbol{\alpha})$ remains constant in time.

1.4.8 Exercise

1.4.1 Discuss the Lagrangian mapping of point particles in one dimension over the x axis.

1.5 The Material Derivative

Consider a scalar, vectorial, or tensorial field, $\phi(\mathbf{x}, t)$, that can be attributed to a point particle. For example, ϕ can be the fluid density, ρ, the temperature, T, the position of a point particle, $\mathbf{X}$, the velocity of a point particle, $\mathbf{U}$, or the fluid density multiplied by the velocity, $\rho \, \mathbf{u}$.

1.5.1 Rate of Change Following a Point Particle

The material derivative expresses the rate of change of the property expressed by ϕ following a material point particle. By definition, a material point particle moves with the local and current fluid velocity.

The point-particle property, ϕ, can be regarded a function of Lagrangian label, $\boldsymbol{\alpha}$, and time, t. The material derivative is the partial derivative with respect to time under constant Lagrangian label $\boldsymbol{\alpha}$,

$$\frac{D\phi}{Dt} \equiv \left(\frac{\partial \phi}{\partial t}\right)_{\boldsymbol{\alpha}}. \tag{1.5.1}$$

Partial derivatives with respect to the constituents of $\boldsymbol{\alpha}$ at a certain time may also be defined. By contrast, the Eulerian time derivative, $\partial\phi/\partial t$, is taken under constant position, $\mathbf{x}$, and the Eulerian spatial derivative, $\partial\phi/\partial x_i$, is taken under constant time, t.

1.5.2 Relation to Eulerian Derivatives

Using the differentiation chain rule, we find that the material derivative can be expressed in terms of Eulerian derivatives with respect to time, t, and point particle position, $\mathbf{X}(t)$, as

$$\frac{\mathrm{D}\phi}{\mathrm{D}t} = \frac{\partial\phi}{\partial t} + \left(\frac{\partial X_i}{\partial t}\right)_{\boldsymbol{\alpha}} \frac{\partial\phi}{\partial x_i}, \tag{1.5.2}$$

where summation is implied over the repeated index, i. Since a point particle moves with the local and current fluid velocity, we set

$$\left(\frac{\partial X_i}{\partial t}\right)_{\boldsymbol{\alpha}} = u_i \tag{1.5.3}$$

in the second term on the right-hand side, where $\mathbf{u}$ is the fluid velocity evaluated at position $\mathbf{X}$ at time t, and obtain

$$\frac{\mathrm{D}\phi}{\mathrm{D}t} = \frac{\partial\phi}{\partial t} + \mathbf{u}\cdot\nabla\phi, \tag{1.5.4}$$

where $\nabla\phi$ is the gradient of ϕ.

Equation (1.5.4) relates the material derivative on the left-hand side to Eulerian derivatives with respect to $\mathbf{x}$ and t on the right-hand side. The first term on the right-hand side is zero in a stationary field, ϕ, while the second term is zero in a spatially homogeneous field where $\nabla\phi = \mathbf{0}$.

1.5.3 Directional Derivative

The second term on the right-hand side of (1.5.4) is parallel to the directional derivative along the local and instantaneous velocity,

$$\mathbf{u}\cdot\nabla\phi = |\mathbf{u}|\,\frac{\partial\phi}{\partial\ell}, \tag{1.5.5}$$

where $|\mathbf{u}|$ is the magnitude of the velocity and ℓ is arc length in space measured in the direction of $\mathbf{u}$. We may introduce the unit vector in the direction of the velocity, $\mathbf{e}_u \equiv \mathbf{u}/|\mathbf{u}|$, and define the directional derivative

$$\mathbf{e}_u\cdot\nabla\phi = \frac{\partial\phi}{\partial\ell}. \tag{1.5.6}$$

When ϕ does not change in the direction of the velocity at a point, this derivative is zero. In a steady flow, the velocity points in the direction of a streamline.

1.5.4 Point-Particle Velocity

By definition, the material derivative of the point-particle position, $\mathbf{X}$, is the point-particle velocity,

$$\mathbf{U} \equiv \frac{\mathrm{D}\mathbf{X}}{\mathrm{D}t}, \tag{1.5.7}$$

where $\mathbf{U} = \mathbf{u}(\mathbf{X}(t), t)$. This expression affirms that the point-particle velocity is the fluid velocity evaluated at the position of the point particle at a certain time. Conversely, the point-particle velocity can be used to reconstruct a velocity field at a given time.

1.5.5 Point-Particle Acceleration

The material derivative of the point-particle velocity is the point-particle acceleration, given by

$$\mathbf{a} \equiv \frac{\mathrm{D}\mathbf{U}}{\mathrm{D}t} = \frac{\partial \mathbf{u}}{\partial t} + \mathbf{u} \cdot \mathbf{L}, \tag{1.5.8}$$

where $\mathbf{L} \equiv \nabla \otimes \mathbf{u}$ is the velocity-gradient tensor. In index notation, the jth component of the acceleration is

$$a_j = \frac{\partial u_j}{\partial t} + u_i \frac{\partial u_j}{\partial x_i}, \tag{1.5.9}$$

where summation is implied over the repeated index, i. Expressions for the components of the acceleration in non-Cartesian coordinates can be derived based on (1.5.8).

1.5.6 Properties of the Material Derivative

The material derivative obeys familiar rules of function sum and product differentiation,

$$\frac{\mathrm{D}(f+g)}{\mathrm{D}t} = \frac{\mathrm{D}f}{\mathrm{D}t} + \frac{\mathrm{D}g}{\mathrm{D}t}, \qquad \frac{\mathrm{D}(fg)}{\mathrm{D}t} = f\,\frac{\mathrm{D}g}{\mathrm{D}t} + g\,\frac{\mathrm{D}f}{\mathrm{D}t}. \tag{1.5.10}$$

To prove these properties, we merely substitute the definition of the material derivative in terms of Eulerian derivatives. Derivatives of the material derivative with respect to time or spatial coordinates can be taken, as required.

1.5.7 Fundamental Law of Kinematics

The Lagrangian volumetric coefficient, $\mathcal{J}$, defined in (1.4.10) and computed by the expression shown in (1.4.11) as $\mathcal{J} = \det(\mathcal{J})$, evolves according to the fundamental law of kinematics expressed by

$$\frac{1}{\mathcal{J}} \frac{\mathrm{D}\mathcal{J}}{\mathrm{D}t} = \nabla \cdot \mathbf{u}, \tag{1.5.11}$$

where $\nabla \cdot \mathbf{u}$ is the divergence of the velocity expressing the rate of expansion of an infinitesimal fluid parcel. To derive (1.5.11), we take the material derivative of (1.4.11), expand the derivative of the determinant, and recall that $\mathrm{D}X_i/\mathrm{D}t = u_i(\mathbf{X}, t)$.

Using the properties of the material derivative, we write

$$\frac{\mathrm{D} \ln \mathcal{J}}{\mathrm{D}t} = \nabla \cdot \mathbf{u}. \tag{1.5.12}$$

We see that the logarithm appears naturally in equations involving extensive properties, such as a parcel's volume.

1.5.8 Volumetric Evolution Equation

Resolving the material derivative into Eulerian derivatives on the left-hand side of (1.5.11), and rearranging, we obtain the evolution equation

$$\frac{\partial \mathcal{J}}{\partial t} + \nabla \cdot (\mathcal{J} \mathbf{u}) = 2\, \mathcal{J}\, \nabla \cdot \mathbf{u}\,. \tag{1.5.13}$$

An evolution equation provides us with an expression for the rate of change in time expressed by the Eulerian time derivative, $\partial/\partial t$.

1.5.9 Expansion or Contraction of a Material Parcel

The volume of a small material parcel, δV, consisting of the same material point particles, evolves according to the equation

$$\frac{1}{\delta V} \frac{\mathrm{D}\delta V}{\mathrm{D}t} = \nabla \cdot \mathbf{u}. \tag{1.5.14}$$

Using the properties of the material derivative, we write

$$\frac{\mathrm{D} \ln \delta V}{\mathrm{D}t} = \nabla \cdot \mathbf{u}. \tag{1.5.15}$$

1.5.10 Exercises

1.5.1 Express the material derivative of the ratio $\mathrm{D}(f/g)/\mathrm{D}t$ in terms of the material derivatives of two functions, f and g.

1.5.2 Derive Eq. (1.5.13).

1.6 Material Parcels and Natural Laws

A material parcel is a contiguous piece of a fluid or solid consisting of the same infinite collection of material point particles. Conversely, a material point particle is an infinitesimal material parcel with permanent identity that arises by ignoring the molecular nature of a fluid or solid under the auspices of the continuum approximation.

1.6.1 Parcel Properties

Consider an intensive scalar, vectorial, or tensorial field, $\phi(\mathbf{x}, t)$, that can be attributed or assigned to a point particle; an example is the fluid density, ρ. The corresponding property of a material fluid parcel is given by

$$\iiint_{\text{parcel}} \phi \, \mathrm{d}V. \tag{1.6.1}$$

If ϕ is the fluid density, ρ, the parcel property is the parcel mass; if ϕ is the fluid density multiplied by the velocity, $\rho\, \mathbf{u}$, the parcel property is the parcel momentum.

1.6.2 Integration in Labeling Space

Recalling the definition of the Lagrangian volumetric coefficient, $\mathcal{J}$, defined with respect to a Lagrangian label, $\boldsymbol{\alpha}$, we write $\mathrm{d}V = \mathcal{J}\, \mathrm{d}V(\boldsymbol{\alpha})$ and obtain

$$\iiint_{\text{parcel}} \phi \, \mathrm{d}V = \iiint_{\text{parcel}} \phi \, \mathcal{J} \, \mathrm{d}V(\boldsymbol{\alpha}). \tag{1.6.2}$$

The domain of integration on the right-hand side is the *fixed* region occupied by the parcel in the labeling space.

1.6.3 Evolution of Parcel Properties

The rate of change of a parcel property as the parcel moves and evolves under the action of a flow is given by an ordinary time derivative,

$$\frac{\mathrm{d}}{\mathrm{d}t} \iiint_{\text{parcel}} \phi \, \mathrm{d}V. \tag{1.6.3}$$

Replacing the ordinary time derivative, $\mathrm{d}/\mathrm{d}t$, with the Eulerian time derivative, $\partial/\partial t$, is not appropriate. We recall that the material derivative, $\mathrm{D}/\mathrm{D}t$, is the rate of change following a material point particle, and write

$$\frac{\mathrm{d}}{\mathrm{d}t} \iiint_{\text{parcel}} \phi \, \mathrm{d}V = \iiint_{\text{parcel}} \frac{\mathrm{D}(\phi\, \delta V)}{\mathrm{D}t}, \tag{1.6.4}$$

where δV is a differential material volume consisting of a permanent collection of point particles. Note that the ordinary derivative of the parcel has been transferred inside the integral as a material derivative operating on a product that includes the infinitesimal material volume, δV.

Since a differential material volume, δV, evolves under the action of the fluid velocity, it should not be confused with a generic stationary differential volume, $\mathrm{d}V$, when a time derivative of some kind is encountered.

1.6.4 Natural Laws

In the context of mass, momentum, and energy transport unified under the auspices of physical transport discussed in this text, the rate of change of a parcel's property is governed by a physical law or principle of thermodynamics, including the following: (a) mass conservation for a closed system, (b) Newton's second law of motion applied to a coherent piece of a material, and (c) the first law of thermodynamics.

For example, when ϕ is the fluid density, ρ, the integral in (1.6.1) is the parcel mass. Mass conservation requires that the corresponding time derivative shown in (1.6.3) is zero, as discussed in Sect. 1.10.

1.6.5 Surface Flux and Volume Production

A natural law typically prescribes that the rate of change of a material parcel's physical property is given by the sum of three terms mediating three distinct physical processes:

(a) A surface process associated with short-range molecular interactions or molecular transport, as in the case of heat conduction, force transmission, or species diffusion.
(b) A volume process associated with interior production or loss due, for example, to an external body force or internal heat generation.
(c) A surface process associated with long-range surface production or loss, as in the case of surface radiation.

Adding these three contributions, we obtain

$$\frac{\mathrm{d}}{\mathrm{d}t}\iiint_{\text{parcel}} \phi \, \mathrm{d}V = -\iint_{\text{parcel}} \mathbf{n}\cdot\boldsymbol{\xi} \, \mathrm{d}S + \iiint_{\text{parcel}} \dot{\pi} \, \mathrm{d}V + \iint_{\text{parcel}} \dot{\nu} \, \mathrm{d}S, \tag{1.6.5}$$

where $\mathbf{n}$ is the outward unit normal vector, subject to the following definitions:

$\boldsymbol{\xi}$ is a flux vector associated with process (a)
$\dot{\pi}$ is a rate of volumetric production associated with process (b)
$\dot{\nu}$ is a rate of surface production associated with process (c)

The rates $\dot{\pi}$ and $\dot{\nu}$ are positive in production and negative in loss. The interpretation of the flux, $\boldsymbol{\xi}$, depends on the particular context under consideration.

1.6.6 Conductive Flux and Surface Traction

In the case of heat transport, $\boldsymbol{\xi}$ is the conductive heat flux and $\mathbf{n}\cdot\boldsymbol{\xi}$ is outward rate of conductive heat flux across the parcel surface. The negative sign in front of the first integral in (1.6.5) ensures that when heat flows outward, $\mathbf{n}\cdot\boldsymbol{\xi} > 0$, the parcel's thermal content decreases in time. Conversely, when heat flows inward, $\mathbf{n}\cdot\boldsymbol{\xi} < 0$, the parcel's thermal content increases in time.

In the case of momentum transport, $\boldsymbol{\xi} = -\boldsymbol{\sigma}$, where $\boldsymbol{\sigma}$ is the stress tensor, as discussed in Sect. 2.3. The negative sign is dictated by Newton's second law of motion. The product $\mathbf{n}\cdot\boldsymbol{\sigma}$ is the traction exerted on the parcel surface, as discussed in Sect. 2.1. Note that $\mathbf{n}\cdot\boldsymbol{\sigma}$ is equal to $\boldsymbol{\sigma}\cdot\mathbf{n}$ only if the stress tensor is symmetric.

1.6.7 Narrative Form

In narrative form, Eq. (1.6.5) reads:

$$
\begin{aligned}
&(\text{ rate of change of a material parcel property })\\
&\quad = (\text{ rate of short-range } surface \text{ action or molecular transport })\\
&\quad + (\text{ rate of interior } volume \text{ production or loss })\\
&\quad + (\text{ rate of long-range } surface \text{ production or loss }).
\end{aligned} \tag{1.6.6}
$$

Two surface terms and one volume term appear on the right-hand side.

1.6.8 Transport Balances

The Reynolds transport equation derived in Sect. 1.7 provides us with an expression for the left-hand side of (1.6.5) in terms of Eulerian time and space derivatives involving the normal component of the velocity over a parcel surface. In Sect. 1.8, we will combine equation (1.6.5) with the Reynolds transport equation to develop balance equations over stationary, moving, or evolving control volumes.

1.6.9 Generalized Laws

An intensive point-particle property, ϕ, may represent a conceptual or nonphysical entity, such as volumetric information density, data usage, household income, karma, and financial or spiritual wealth. Parcels can be interpreted as coherent stationary or migrating virtual communities exhibiting internal dynamics and interacting with their environment. Phenomenological and other laws can be introduced to specify the rate of change of corresponding parcel properties, and thereby achieve closure.

1.6.10 Exercises

1.6.1 List and discuss the laws of thermodynamics.

1.6.2 A school of fish are swimming in the ocean. Discuss the interpretation of Eq. (1.6.5) regarding the fish population.

1.7 The Reynolds Transport Equation

Expanding the material derivative of the product inside the integral on the right-hand side of (1.6.4), and using the evolution equation (1.5.14) for the rate of change of a differential material volume, $\mathrm{D}\,\delta V/\mathrm{D}t = (\nabla \cdot \mathbf{u})\,\delta V$, we obtain

$$
\frac{\mathrm{d}}{\mathrm{d}t} \iiint_{\text{parcel}} \phi \,\mathrm{d}V = \iiint_{\text{parcel}} \Big(\frac{\mathrm{D}\phi}{\mathrm{D}t} + \phi\, \nabla \cdot \mathbf{u} \Big)\, \mathrm{d}V, \tag{1.7.1}
$$

where $\nabla \cdot \mathbf{u}$ is the divergence of the velocity expressing the rate of expansion of the fluid.

1.7.1 Formal Derivation

Equation (1.7.1) can be derived formally by taking the time derivative of both sides of (1.6.2), obtaining

$$\frac{\mathrm{d}}{\mathrm{d}t}\iiint_{\text{parcel}} \phi\, \mathrm{d}V = \frac{\mathrm{d}}{\mathrm{d}t}\iiint_{\text{parcel}} \phi\, \mathcal{J}\, \mathrm{d}V(\boldsymbol{\alpha}). \tag{1.7.2}$$

Since the domain of integration in the $\boldsymbol{\alpha}$ labeling space is fixed, we may transpose time differentiation and space integration on the right-hand side by introducing the material derivative, so that

$$\frac{\mathrm{d}}{\mathrm{d}t}\iiint_{\text{parcel}} \phi\, \mathrm{d}V = \iiint_{\text{parcel}} \frac{\mathrm{D}(\phi\mathcal{J})}{\mathrm{D}t}\, \mathrm{d}V(\boldsymbol{\alpha}). \tag{1.7.3}$$

Expanding the material derivative of the product and using the evolution equation (1.5.11), $\mathrm{D}\mathcal{J}/\mathrm{D}t = \mathcal{J}\, \nabla \cdot \mathbf{u}$, we obtain (1.7.1).

1.7.2 Powers of the Metric Coefficient

For $\phi = 1/\mathcal{J}$, Eq. (1.7.3) yields

$$\frac{\mathrm{d}}{\mathrm{d}t}\iiint_{\text{parcel}} \frac{1}{\mathcal{J}}\, \mathrm{d}V = \frac{\mathrm{d}}{\mathrm{d}t}\iiint_{\text{parcel}} \mathrm{d}V(\boldsymbol{\alpha}) = 0, \tag{1.7.4}$$

which confirms that the parcel volume in the $\boldsymbol{\alpha}$ space, represented by the second integral, remains constant in time.

For $\phi = \mathcal{J}^{\beta}$, where β is an arbitrary exponent, we obtain

$$\begin{aligned}\frac{\mathrm{d}}{\mathrm{d}t}\iiint_{\text{parcel}} \mathcal{J}^{\beta}\, \mathrm{d}V &= \iiint_{\text{parcel}} \frac{\mathrm{D}\mathcal{J}^{\beta+1}}{\mathrm{D}t}\, \mathrm{d}V(\boldsymbol{\alpha}) = (\beta+1)\iiint_{\text{parcel}} \mathcal{J}^{\beta}\frac{\mathrm{D}\mathcal{J}}{\mathrm{D}t}\, \mathrm{d}V(\boldsymbol{\alpha}) \\ &= (\beta+1)\iiint_{\text{parcel}} \mathcal{J}^{\beta-1}\frac{\mathrm{D}\mathcal{J}}{\mathrm{D}t}\, \mathrm{d}V = \frac{\beta+1}{\beta}\iiint_{\text{parcel}} \frac{\mathrm{D}\mathcal{J}^{\beta}}{\mathrm{D}t}\, \mathrm{d}V,\end{aligned} \tag{1.7.5}$$

where the integral is computed over the instantaneous parcel volume. Setting $\beta = -1$, we recover (1.7.4).

1.7.3 Reynolds Transport Equation

We may resolve the material derivative inside the integrand on the right-hand side of (1.7.1) into Eulerian derivatives with respect to time, t, and position, $\mathbf{x}$, and then combine two terms by using an identity to write

$$\mathbf{u} \cdot \nabla\phi + \phi\, \nabla \cdot \mathbf{u} = \nabla \cdot (\phi\, \mathbf{u}), \tag{1.7.6}$$

obtaining

$$\frac{\mathrm{d}}{\mathrm{d}t}\iiint_{\text{parcel}} \phi \,\mathrm{d}V = \iiint_{\text{parcel}} \Big(\frac{\partial \phi}{\partial t} + \nabla \cdot (\phi\,\mathbf{u})\Big)\,\mathrm{d}V. \tag{1.7.7}$$

Equation (1.7.7) essentially expresses the Leibniz integration rule for computing the derivative of volume integrals with time-dependent integration domains. In our case, the integration domain is the volume of a material parcel moving under the influence of a prescribed velocity field, **u**.

Finally, we apply the Gauss divergence theorem for the second term inside the integral on the right-hand side of (1.7.7), and thereby derive the Reynolds transport equation for a fluid parcel,

$$\frac{\mathrm{d}}{\mathrm{d}t}\iiint_{\text{parcel}} \phi \,\mathrm{d}V = \iiint_{\text{parcel}} \frac{\partial \phi}{\partial t}\,\mathrm{d}V + \iint_{\text{parcel}} \phi\,\mathbf{n}\cdot\mathbf{u}\,\mathrm{d}S, \tag{1.7.8}$$

where **n** is the normal unit vector pointing *outward* from the parcel and $\mathbf{n}\cdot\mathbf{u}$ is the normal velocity component. An implied assumption is that the product $\phi\,\mathbf{u}$ is continuous over the parcel volume so that its divergence does not exhibit a singular behavior. Equation (1.7.8) is a direct consequence of kinematics and should not be mistaken for a natural law.

1.7.4 Convection by Normal Motion

If the velocity is tangential over the parcel surface, $\mathbf{n}\cdot\mathbf{u} = 0$, the integrand of the second integral on the right-hand side of (1.7.8) is identically zero. Under these conditions, the fluid inside the parcel exhibits local or global circulatory motion. Only normal motion associated with *convective transport* contributes to this integral, as discussed in Sect. 1.8.

1.7.5 Orientation of the Unit Normal Vector

By convention, the normal unit vector, **n**, points *outward* from a parcel. The orientation can be reversed, provided that the plus sign is replaced by a minus sign before the last integral in (1.7.8). This reversal will be implemented in Sect. 1.8 where Eq. (1.7.8) will be transformed into a balance over a control volume with the unit normal vector pointing into the control volume.

1.7.6 Evolution of a Parcel's Volume

When $\phi = 1$, the integral on the left-hand side of (1.7.8) is the parcel volume, V_{parcel}. Since ϕ is constant, the first integral on the right-hand side of (1.7.8) involving $\partial\phi/\partial t$ is zero, yielding

$$\frac{\mathrm{d}V_{\text{parcel}}}{\mathrm{d}t} = \iint_{\text{parcel}} \mathbf{n}\cdot\mathbf{u}\,\mathrm{d}S. \tag{1.7.9}$$

The volume of a material parcel does not change when the velocity is tangential over the entire parcel surface, that is, in the entire absence of normal motion.

1.7.7 Essential Step

The evaluation of the left-hand side of (1.7.8) by way of a natural law is a crucial and often understated step in developing evolution, balance, and transport equations. Combining the physical law expressed by (1.6.5) with the Reynolds transport equation (1.7.8), and rearranging, we obtain

$$\iiint_{\text{parcel}} \frac{\partial \phi}{\partial t}\,\mathrm{d}V + \iint_{\text{parcel}} \phi\,\mathbf{n}\cdot\mathbf{u}\,\mathrm{d}S = -\iint_{\text{parcel}} \mathbf{n}\cdot\boldsymbol{\xi}\,\mathrm{d}S + \iiint_{\text{parcel}} \dot{\pi}\,\mathrm{d}V + \iint_{\text{parcel}} \dot{\nu}\,\mathrm{d}S. \quad (1.7.10)$$

The second term on the left-hand side and the first term on the right-hand side can be combined into a single term that incorporates a composite flux, typically identified as a convection–diffusion flux or momentum–stress flux,

$$\iiint_{\text{parcel}} \frac{\partial \phi}{\partial t}\,\mathrm{d}V + \iint_{\text{parcel}} \mathbf{n}\cdot(\phi\,\mathbf{u}+\boldsymbol{\xi})\,\mathrm{d}S = \iiint_{\text{parcel}} \dot{\pi}\,\mathrm{d}V + \iint_{\text{parcel}} \dot{\nu}\,\mathrm{d}S, \quad (1.7.11)$$

where convection pertains to $\mathbf{u}$ and diffusion or stress pertains to $\boldsymbol{\xi}$. The term *convection* refers to the transportive action of the velocity field. The term *advection* is sometimes used to indicate passive convection when the transported entity does not affect the velocity distribution.

1.7.8 Evolution Equations

When the last term on the right-hand side of (1.7.10) is absent, $\dot{\nu} = 0$, we may use the divergence theorem to convert the two surface integrals into volume integrals. Discarding the volume integral signs, we obtain an evolution equation stated in the Eulerian conservative form

$$\frac{\partial \phi}{\partial t} + \nabla\cdot(\phi\,\mathbf{u}) = -\nabla\cdot\boldsymbol{\xi} + \dot{\pi}. \quad (1.7.12)$$

The qualifier *conservative* refers to the divergence of the convective flux, $\phi\,\mathbf{u}$ on the left-hand side. When integrating this term over a volume and applying Gauss' divergence theorem, we obtain a surface integral involving a corresponding flux.

All transport evolution equations derived in this book conform with the unifying master form shown in (1.7.12) for appropriate choices for ϕ, $\boldsymbol{\xi}$, and $\dot{\pi}$, as discussed in Appendix A.

1.7.9 Exercise

1.7.1 State and discuss the Reynolds transport equation in one dimension along the x axis.

1.8 Integral Balance Over a Control Volume

Having introduced a physical law to evaluate the rate of change of a material parcel's physical property, we may regard the Reynolds transport equation (1.7.8), repeated below for convenience,

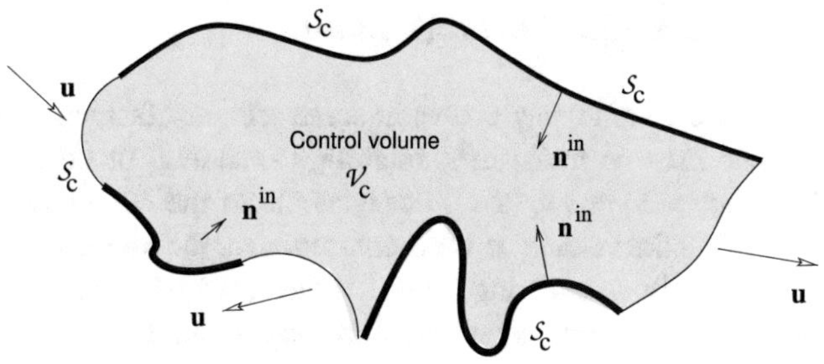

Fig. 1.3 Illustration of a control volume, $\mathcal{V}_c$, in a flow bounded by solid or fluid, stationary or moving surfaces, $\mathcal{S}_c$, where $\mathbf{n}^{\rm in}$ is the inward normal unit vector. The fluid velocity, $\mathbf{u}$, may be zero or nonzero over different parts of $\mathcal{S}_c$

$$\frac{\rm d}{{\rm d}t}\iiint_{\rm parcel}\phi\,{\rm d}V=\iiint_{\rm parcel}\frac{\partial\phi}{\partial t}\,{\rm d}V+\iint_{\rm parcel}\phi\,\mathbf{n}\cdot\mathbf{u}\,{\rm d}S, \tag{1.8.1}$$

as a generic *transport equation* expressing the balance of a property associated with ϕ over a control volume, $\mathcal{V}_c$, identified with the *instantaneous* parcel volume. In fact, we will see that the control volume does not need to be stationary, but may translate, rotate, expand, or deform in an arbitrary fashion in the domain of a flow.

1.8.1 Control Volume Occupied by a Parcel

To display the balance, we merely rearrange equation (1.8.1) into a generic transport equation,

$$\iiint_{\mathcal{V}_c}\frac{\partial\phi}{\partial t}\,{\rm d}V=\iint_{\mathcal{S}_c}\phi\,\mathbf{n}^{\rm in}\cdot\mathbf{u}\,{\rm d}S+\frac{\rm d}{{\rm d}t}\iiint_{\rm parcel}\phi\,{\rm d}V, \tag{1.8.2}$$

where $\mathcal{S}_c$ is the surface of the material parcel, now regarded as the boundary of the control volume, $\mathcal{V}_c$, and $\mathbf{n}^{\rm in}=-\mathbf{n}$ is the normal unit vector pointing *into* the control volume, as shown in Fig. 1.3.

The last integral on the right-hand side of (1.8.2) must be evaluated by invoking a natural law, as discussed in Sect. 1.7. This realization is the cornerstone of transport processes in the physical and engineering sciences. Adopting expression (1.6.5), we obtain

$$\iiint_{\mathcal{V}_c}\frac{\partial\phi}{\partial t}\,{\rm d}V=\iint_{\mathcal{S}_c}\phi\,\mathbf{n}^{\rm in}\cdot\mathbf{u}\,{\rm d}S+\iint_{\mathcal{S}_c}\mathbf{n}^{\rm in}\cdot\boldsymbol{\xi}\,{\rm d}S+\iiint_{\mathcal{V}_c}\dot{\pi}\,{\rm d}V+\iint_{\mathcal{S}_c}\dot{\nu}\,{\rm d}S. \tag{1.8.3}$$

The first and second terms on the right-hand side can be combined into a composite flux incorporating a convective and a diffusive or frictional contribution,

$$\iiint_{\mathcal{V}_c}\frac{\partial\phi}{\partial t}\,{\rm d}V=\iint_{\mathcal{S}_c}\mathbf{n}^{\rm in}\cdot(\phi\,\mathbf{u}+\boldsymbol{\xi})\,{\rm d}S+\iiint_{\mathcal{V}_c}\dot{\pi}\,{\rm d}V+\iint_{\mathcal{S}_c}\dot{\nu}\,{\rm d}S. \tag{1.8.4}$$

The first surface integral on the right-hand side expresses the combined effect of convection–diffusion; in the present context, diffusion stands for a generalized short-range molecular action.

1.8.2 Evolving Control Volume

A control volume may evolve in some arbitrary fashion. This means that the boundary of a control volume may be translating, rotating, expanding, or deforming with a position or time dependent boundary velocity, $\mathbf{u}_c(\mathbf{x}, t)$. For example, if the boundary of a control volume translates as a rigid body, then the velocity $\mathbf{u}_c(t)$ is constant over the boundary of a control volume.

To signify an evolving control volume, we write $\mathcal{V}_c(t)$ and $\mathcal{S}_c(t)$. Since adding time dependence to integration limits is permitted as long time derivatives are confined inside integrals, the balance equation (1.8.4) also applies for an evolving control volume, that is,

$$\iiint_{\mathcal{V}_c(t)} \frac{\partial \phi}{\partial t} \, \mathrm{d}V = \iint_{\mathcal{S}_c(t)} \mathbf{n}^{\mathrm{in}} \cdot (\phi \, \mathbf{u} + \boldsymbol{\xi}) \, \mathrm{d}S + \iiint_{\mathcal{V}_c(t)} \dot{\pi} \, \mathrm{d}V + \iint_{\mathcal{S}_c(t)} \dot{\nu} \, \mathrm{d}S. \tag{1.8.5}$$

Note that the control-volume boundary velocity does not appear explicitly in this equation.

1.8.3 Accumulation

The volume integral on the left-hand side of (1.8.3) or (1.8.5),

$$\iiint_{\mathcal{V}_c(t)} \frac{\partial \phi}{\partial t} \, \mathrm{d}V, \tag{1.8.6}$$

is the rate of accumulation of the physical or conceptual entity represented by ϕ inside a stationary control volume or an evolving control volume whose shape has been frozen to its instantaneous shape at a certain time, t. The rate of accumulation should not be confused with the rate of change of the property associated with ϕ following a material parcel that occupies the control volume at a certain time.

The partial time derivative in (1.8.6) may be transferred outside the integral only under the silent stipulation that the integration domain is fixed or frozen in time, even though the control volume occupied by a material parcel at a certain instant may be evolving in time.

1.8.4 Convective Transport

The surface integral after the equal sign on the right-hand side of (1.8.3) over the boundary of the control volume,

$$\iint_{\mathcal{S}_c(t)} \phi \, \mathbf{n}^{\mathrm{in}} \cdot \mathbf{u} \, \mathrm{d}S, \tag{1.8.7}$$

expresses the rate of convective transport of the property associated with ϕ *into* the control volume.

For example, the integral

$$\iint_{\mathcal{S}_c(t)} \rho \, \mathbf{n}^{\mathrm{in}} \cdot \mathbf{u} \, \mathrm{d}S \tag{1.8.8}$$

expresses the rate of convective transport of mass into the control volume, where ρ is the fluid density. It is interesting to contrast this interpretation with that of the surface integral $\iint \mathbf{n} \cdot \mathbf{u} \, \mathrm{d}S$ on the right-hand side of (1.7.9), which expresses the rate of change of volume of a parcel of fluid whose instantaneous boundary defines a control volume.

Table 1.1 Typical balance of a transported quantity over a stationary, moving, or evolving control volume

(rate of accumulation in a stationary or frozen control volume)
= (rate of inward convection with velocity **u**)
+ (rate of short-range surface action or molecular transport)
+ (rate of interior production or loss)
+ (rate of long-range surface production or loss)

Four terms appear on the right-hand side. A choice on the appropriate term inside each set of parentheses on the right-hand side is made according to context

1.8.5 Control-Volume Balance

A typical balance of a transported quantity takes the form shown in Table 1.1. The implementation and interpretation of the balance will be illustrated in this chapter for mass transport, and in Chaps. 2 and 3 for momentum, angular momentum, mechanical energy, and other types of transport.

1.8.6 Balance Over an Open System

The balance displayed in Table 1.1 is often adopted as a point of departure for writing transport equations, subject to specific choices for the terms inside each set of parentheses on the right-hand side. This approach is essentially pivoted on implied natural laws for open systems that allow matter to enter across boundaries.

1.8.7 Exercise

1.8.1 Explain how mass may accumulate inside a stationary control volume even though the fluid may be incompressible.

1.9 Mass Conservation and Material Continuity

The mass of a material fluid parcel, m, arises by integrating the fluid density, ρ, over the parcel volume,

$$m = \iiint_{\text{parcel}} \rho(\mathbf{x}, t) \, \mathrm{d}V, \tag{1.9.1}$$

where the volume of integration is generally evolving as the parcel is advected and deforms in a specified flow, even in a steady flow. Only in hydrostatics the volume of integration remains constant in time.

1.9.1 Mass Conservation for a Closed System

By definition, a system that is closed with respect to mass does not allow matter to enter or exit through the boundaries. Mass may neither disappear nor be produced inside a closed nonrelativistic system, except under the auspices of the paranormal.

We may regard a material fluid parcel as a closed system with regard to mass, and require that the mass of the parcel remains constant as the parcel moves around in a flow,

$$\frac{\mathrm{d}m}{\mathrm{d}t} = \frac{\mathrm{d}}{\mathrm{d}t} \iiint_{\text{parcel}} \rho \, \mathrm{d}V = 0. \tag{1.9.2}$$

This conservation equation expresses an instance of a natural law. The principle of mass conservation applies to homogeneous as well as heterogeneous and spatially homogeneous molecular mixtures consisting of different chemical species.

1.9.2 Instance of a General Law

A general natural law pertinent to a property of a material parcel was stated in Eq. (1.6.5), repeated below for convenience,

$$\frac{\mathrm{d}}{\mathrm{d}t} \iiint_{\text{parcel}} \phi \, \mathrm{d}V = - \iint_{\text{parcel}} \mathbf{n} \cdot \boldsymbol{\xi} \, \mathrm{d}S + \iiint_{\text{parcel}} \dot{\pi} \, \mathrm{d}V + \iint_{\text{parcel}} \dot{\nu} \, \mathrm{d}S. \tag{1.9.3}$$

Comparing this equation with the mass conservation law expressed by (1.9.2), we set $\phi = \rho$ and

$$\boldsymbol{\xi} = \mathbf{0}, \qquad \dot{\pi} = 0, \qquad \dot{\nu} = 0. \tag{1.9.4}$$

Because the concept of mass diffusion does not apply to a single-species fluid, even in the presence of density variations, the surface mass flux, $\boldsymbol{\xi}$, is zero. Mass neither appears spontaneously inside a material, $\dot{\pi} = 0$, nor radiates across the boundaries, $\dot{\nu} = 0$.

In Chap. 5, we will confirm that *chemical species mass* can be generated by chemical conversion in a binary or multicomponent mixture.

1.9.3 Continuity Equation

Now transferring the time derivative in (1.9.2) inside the integral as a material derivative, we obtain

$$\frac{\mathrm{d}m}{\mathrm{d}t} = \iiint_{\text{parcel}} \frac{\mathrm{D}(\rho \, \delta V)}{\mathrm{D}t} = 0, \tag{1.9.5}$$

where δV is an infinitesimal material volume. Since the material parcel under consideration is arbitrary, we require that

$$\frac{\mathrm{D}(\rho \, \delta V)}{\mathrm{D}t} = 0, \tag{1.9.6}$$

irrespective of the physical properties of the fluid.

Expanding the material derivative of the product inside the integral in (1.9.5) using the usual rules of product differentiation, and setting $\mathrm{D}\delta V/\mathrm{D}t = (\boldsymbol{\nabla} \cdot \mathbf{u}) \, \delta V$ according to the fundamental law of kinematics, we obtain

$$\frac{\mathrm{d}m}{\mathrm{d}t} = \iiint_{\text{parcel}} \Big(\frac{\mathrm{D}\rho}{\mathrm{D}t} + \rho \, \boldsymbol{\nabla} \cdot \mathbf{u} \Big) \, \mathrm{d}V = 0, \tag{1.9.7}$$

where $\nabla \cdot \mathbf{u}$ is the rate of expansion. Setting the integrand to zero, we obtain the continuity equation

$$\frac{\mathrm{D}\rho}{\mathrm{D}t} + \rho\, \nabla \cdot \mathbf{u} = 0. \tag{1.9.8}$$

Rearranging, we obtain

$$\rho\, \frac{\mathrm{D}}{\mathrm{D}t}\Big(\frac{1}{\rho}\Big) = \nabla \cdot \mathbf{u}. \tag{1.9.9}$$

In terms of the specific volume defined as the inverse of the density, $v \equiv 1/\rho$,

$$\frac{1}{v}\, \frac{\mathrm{D}v}{\mathrm{D}t} = \nabla \cdot \mathbf{u}. \tag{1.9.10}$$

All versions of the continuity equation derived in this section apply for compressible as well as incompressible fluids.

1.9.4 Material Continuity

Mass conservation imposes a kinematic constraint by demanding that the structure of the velocity field be such that (a) fluid parcels do not tend to occupy the same volume in space, and (b) contiguous parcels do not separate or tear apart to generate voids or cracks.

1.9.5 Eulerian Form

Resolving the material derivative of the density into Eulerian derivatives,

$$\frac{\mathrm{D}\,\rho}{\mathrm{D}t} = \frac{\partial\,\rho}{\partial t} + \mathbf{u} \cdot \nabla\rho, \tag{1.9.11}$$

and substituting this expression into (1.9.8), we obtain the conservative form of the continuity equation,

$$\frac{\partial\,\rho}{\partial t} + \nabla \cdot (\rho\, \mathbf{u}) = 0, \tag{1.9.12}$$

which can be regarded as an evolution equation for the density: if the local distributions of ρ and $\mathbf{u}$ are available, the rate of change, $\partial\rho/\partial t$, can be evaluated at a chosen point.

Alternatively, Eq. (1.9.12) can be derived by applying the Reynolds transport equation (1.7.7) for property $\phi = \rho$.

1.9.6 Incompressible Fluids

Because the volume of a material point particle of an incompressible fluid remains constant in time, the density also remains constant in time,

$$\frac{\mathrm{D}\rho}{\mathrm{D}t} = 0, \tag{1.9.13}$$

that is, the rate of change of the density following a point particle is zero. Referring to the continuity equation, we find that the velocity field must be solenoidal,

$$\nabla \cdot \mathbf{u} = 0. \tag{1.9.14}$$

The term *solenoidal* owes its origin to electromagnetics. The velocity field $\mathbf{u} = \nabla \times \mathbf{A}$ is solenoidal for any differentiable vector field, $\mathbf{A}$.

An incompressible fluid is not necessarily a uniform-density fluid. The inaccurate statement *an incompressible fluid is a fluid with uniform density* is made frequently in textbooks and other sources.

1.9.7 Evolution of a Specific Parcel Property

Consider a material fluid parcel property, Ψ, defined in terms of an intensive property, ψ, as

$$\Psi = \iiint_{\text{parcel}} \psi \rho \, \mathrm{d}V, \tag{1.9.15}$$

where ρ is the fluid density. For example, ψ can be the velocity, the temperature, or half the square of the magnitude of the velocity.

The rate of change of this parcel property is given by an ordinary time derivative transferred into the integral as a material derivative,

$$\frac{\mathrm{d}\Psi}{\mathrm{d}t} = \iiint_{\text{parcel}} \frac{\mathrm{D}(\psi \rho \, \delta V)}{\mathrm{D}t} = \iiint_{\text{parcel}} \Big(\frac{\mathrm{D}\psi}{\mathrm{D}t} \rho \, \mathrm{d}V + \psi \frac{\mathrm{D}(\rho \, \delta V)}{\mathrm{D}t} \Big), \tag{1.9.16}$$

where δV is an infinitesimal material volume. The second term inside the second integral is zero because of (1.9.6), yielding

$$\frac{\mathrm{d}}{\mathrm{d}t} \iiint_{\text{parcel}} \psi \rho \, \mathrm{d}V = \iiint_{\text{parcel}} \frac{\mathrm{D}\psi}{\mathrm{D}t} \rho \, \mathrm{d}V. \tag{1.9.17}$$

This expression is useful for computing the rate of change of parcel properties involving *specific material properties*, where the term *specific* refers to the multiplicative factor of the fluid density, as shown in (1.9.15). Example are the specific kinetic energy and the specific internal energy.

1.9.8 Traveler's Time Derivative

The rate of change of the density recorded by an observer who moves with velocity $\mathbf{v}$ is given by an ordinary time derivative,

$$\frac{\mathrm{d}\rho}{\mathrm{d}t} = \frac{\partial \rho}{\partial t} + \mathbf{v} \cdot \nabla \rho, \tag{1.9.18}$$

identified as a traveler's time derivative. If $\mathbf{v} = \mathbf{0}$, then this time derivative is the Eulerian time derivative taken at a fixed position in space. If $\mathbf{v}$ is the fluid velocity, $\mathbf{u}$, then this time derivative is the material derivative.

In fact, the traveler's time derivative is the tracer time derivative discussed in Sect. 1.12 with reference to the motion of nonmaterial point particles.

Combining expression (1.9.18) with the continuity equation (1.9.12), we obtain the rate of change of the density recorded by an observer,

$$\frac{d\rho}{dt} = \mathbf{v} \cdot \nabla \rho - \nabla \cdot (\rho \mathbf{u}). \tag{1.9.19}$$

This expression will be used in Sect. 1.10 to derive the Rankin–Hugoniot density condition.

1.9.9 Exercises

1.9.1 Discuss a physical example of an incompressible fluid with varying density.

1.9.2 Discuss the physical interpretation of (1.9.17) for $\phi = \mathbf{u}$.

1.9.3 Apply Eq. (1.9.17) for $\psi = \rho^n$ and simplify the right-hand side using the continuity equation, where n is an arbitrary exponent.

1.10 Rankin–Hugoniot Density Condition

Using Eq. (1.9.19), we find that the rate of change of the density jump across a density discontinuity front moving with velocity $\mathbf{v}$ is

$$\frac{d\Delta\rho}{dt} = \mathbf{v} \cdot \nabla \Delta\rho - \nabla \cdot \big(\Delta(\rho \mathbf{u})\big), \tag{1.10.1}$$

where Δ denotes the discontinuity or jump. All terms in this equation are finite, except for two terms on the right-hand side involving spatial derivatives normal to the evolving front, inherent in the gradient and divergence.

1.10.1 Ensuring Regularity

To ensure regularity, we concentrate on these annoying terms and require that, to leading order,

$$0 = v_n \, (\mathbf{n} \cdot \nabla \Delta\rho) - \mathbf{n} \cdot \nabla\big(\Delta(\rho \, \mathbf{n} \cdot \mathbf{u})\big), \tag{1.10.2}$$

where $\mathbf{n}$ is the unit vector normal to the front and $v_n \equiv \mathbf{v} \cdot \mathbf{n}$ is the normal component of the front velocity. Rearranging, we obtain the Rankin–Hugoniot density condition

$$v_n = \frac{\Delta(\rho \, u_n)}{\Delta\rho}, \tag{1.10.3}$$

where $u_n \equiv \mathbf{u} \cdot \mathbf{n}$ is the normal component of the fluid velocity.

The Rankin–Hugoniot density condition essentially relates the density discontinuity to the normal component of the momentum discontinuity in terms of the velocity of a moving front. If the velocity is continuous across the front, then $v_n = u_n$.

1.10.2 One-Dimensional Flow

In the idealized case of one-dimensional flow along the x axis, we obtain the expression

$$v_x = \frac{\Delta(\rho\, u_x)}{\Delta\rho}, \tag{1.10.4}$$

where u_x is the fluid velocity. If the velocity is continuous across the front, then $v_x = u_x$.

It is instructive to derive Eq. (1.10.4) by performing a mass balance over a one-dimensional control volume identified with a fixed interval, $[a, b]$, that includes the discontinuity, $x_d(t)$, that is, $a < x_d(t) < b$ within a time interval of interest. The mass contained inside this fixed interval is

$$m = \int_a^b \rho\, \mathrm{d}x. \tag{1.10.5}$$

A mass balance requires that

$$\frac{\mathrm{d}m}{\mathrm{d}t} = (\rho u_x)_{x=a} - (\rho u_x)_{x=b}. \tag{1.10.6}$$

Writing

$$m = \int_a^{x_d(t)} \rho\, \mathrm{d}x + \int_{x_d(t)}^b \rho\, \mathrm{d}x,\ v \tag{1.10.7}$$

and differentiating using the Leibnitz integral rule, we find that

$$\frac{\mathrm{d}m}{\mathrm{d}t} = -\Delta\rho\, \frac{\mathrm{d}x_d}{\mathrm{d}t} + \int_a^b \frac{\partial\rho}{\partial t}\, \mathrm{d}x, \tag{1.10.8}$$

where

$$\Delta\rho = \rho_{x_d^+} - \rho_{x_d^-}, \tag{1.10.9}$$

the superscript + indicates evaluation at the upper edge of the discontinuity, and the superscript − indicates evaluation at the lower edge of the discontinuity.

Now using the continuity equation to manipulate the term inside the integral in (1.10.8), we obtain

$$\frac{\mathrm{d}m}{\mathrm{d}t} = -\Delta\rho\, \frac{\mathrm{d}x_d}{\mathrm{d}t} - \int_a^{x_d^-(t)} \frac{\partial(\rho\, u_x)}{\partial x}\, \mathrm{d}x - \int_{x_d^+(t)}^b \frac{\partial(\rho\, u_x)}{\partial x}\, \mathrm{d}x, \tag{1.10.10}$$

where $\mathrm{d}x_d/\mathrm{d}t = v_x$. Performing the integration, we find that

$$\frac{\mathrm{d}m}{\mathrm{d}t} = -\Delta\rho\, v_x + (\rho\, u_x)_{x=a} - (\rho\, u_x)_{x=b} + \Delta(\rho u_x), \tag{1.10.11}$$

where

$$\Delta(\rho u_x) = (\rho u_x)_{x_d^+} - (\rho u_x)_{x_d^-}. \tag{1.10.12}$$

Note that this definition is consistent with that given in (1.10.9). Substituting (1.10.11) into (1.10.6), simplifying, and rearranging, we recover equation (1.10.4).

1.10.3 Exercise

1.10.1 Discuss equation (1.10.3) for a stationary front, $\mathbf{v} = \mathbf{0}$.

1.11 Integral Mass Balance

The conservative form of the continuity equation shown in (1.9.12) is repeated below for convenience,

$$\frac{\partial \rho}{\partial t} + \nabla \cdot (\rho\, \mathbf{u}) = 0, \tag{1.11.1}$$

where ρ is the fluid density and $\mathbf{u}$ is the fluid velocity.

Integrating this equation over a control volume, $\mathcal{V}_c$, that is bounded by one surface or a collection of surfaces, $\mathcal{S}_c$, as shown in Fig. 1.3, and applying the divergence theorem to convert the volume integral of the second term on the left-hand side into a surface integral, we obtain an integral mass balance,

$$\iiint_{\mathcal{V}_c} \frac{\partial \rho}{\partial t}\, \mathrm{d}V = \iint_{\mathcal{S}_c} \rho\, \mathbf{n}^{\mathrm{in}} \cdot \mathbf{u}\, \mathrm{d}S, \tag{1.11.2}$$

where $\mathbf{n}^{\mathrm{in}}$ is the normal unit vector pointing *into* the control volume. Alternatively, the integral mass balance (1.11.2) could have been derived by applying the general transport equation (1.8.2) for the fluid density, $\phi = \rho$, and ensuring mass conservation by setting the last term on the right-hand side to zero.

1.11.1 Mass Accumulation

The left-hand side of the mass transport equation (1.11.2) is the rate of mass accumulation inside the control volume. Mass may accumulate inside a stationary or frozen control volume if the fluid is incompressible but the density varies with time, position, or both, or if the fluid is compressible so that it can undergo compression or dilution.

1.11.2 Mass Accumulation Due to Convection

The mass transport equation (1.11.2) states that the rate of accumulation of mass inside a control volume is balanced by the rate of convective mass transport into the control volume across the boundaries due to the normal component of the fluid velocity, $\mathbf{n}^{\mathrm{in}} \cdot \mathbf{u}$. Because tangential motion associated with circulation does not carry fluid, it is unable to implement mass convection.

1.11.3 Stationary or Recirculating Fluid

In the case of a stationary fluid, $\mathbf{u} = \mathbf{0}$, or an entirely recirculating fluid, $\mathbf{n}^{\text{in}} \cdot \mathbf{u} = 0$, we obtain a simple statement of mass conservation,

$$\iiint_{\mathcal{V}_c} \frac{\partial \rho}{\partial t} \, dV = 0. \tag{1.11.3}$$

Mass may not accumulate inside a stationary or recirculating, compressible or incompressible body of fluid.

1.11.4 Engineering Analysis

In practice, the integral mass balance allows us to develop approximate relations between global properties of a steady or unsteady flow involving the velocity and the density, subject to sensible simplifications at the inlets and outlets. The simplifications are guided by experience and physical intuition.

1.11.5 Mass Conservation for an Open System

By definition, a system that is open with respect to mass allows matter to enter through the boundaries. The balance equation (1.11.2) could have been written at the outset as a fundamental statement of mass conservation for an open system identified with a control volume. The mathematical derivations could then be carried out backward, leading us to the continuity equation. The merits of the forward approach become apparent while attempting to also derive the parcel mass-conservation equation (1.9.2).

1.11.6 Exercise

1.11.1 Write the counterpart of the mass balance equation (1.11.2) in one or two dimensions.

1.12 Tracers and Packets

By definition, a stationary or evolving packet consists of point particles that move with an arbitrary velocity, $\mathbf{v}$, instead of the fluid velocity, $\mathbf{u}$. A *material* packet is a *material* parcel whose point particles are material entities moving with the fluid velocity, $\mathbf{u}$. To distinguish generalized point particles from material point particles, we refer to the former as tracers. A packet consists of an infinite collection of tracers.

1.12.1 Labeling Tracers

The tracers of a packet can be labeled using a vectorial label, $\boldsymbol{\beta}$, that is analogous to the Lagrangian label, $\boldsymbol{\alpha}$, used for material point particles. Tracer derivatives of a suitable field attributed to the tracers, ϕ, may then be defined. An example is the time derivative expressing the rate of change of ϕ following a tracer,

$$\dot{\phi} \equiv \left(\frac{\partial \phi}{\partial t}\right)_{\beta}, \tag{1.12.1}$$

named the tracer derivative. For example,

$$\dot{\mathbf{X}} = \mathbf{V} = \mathbf{v}(\mathbf{X}, t), \qquad \dot{\mathbf{V}} = \mathbf{A} = \mathbf{a}(\mathbf{X}, t), \tag{1.12.2}$$

where $\mathbf{X}$ is the tracer position, $\mathbf{V}$ is the tracer velocity, $\mathbf{A}$ is the tracer acceleration, and $\mathbf{a}(\mathbf{x}, t)$ is the tracer acceleration field.

Working as in Sect. 1.5, we find that

$$\dot{\phi} = \frac{\partial \phi}{\partial t} + \mathbf{v} \cdot \nabla \phi. \tag{1.12.3}$$

Invoking the definition of the material derivative, $\mathrm{D}/\mathrm{D}t$, we find that

$$\dot{\phi} = \frac{\mathrm{D}\phi}{\mathrm{D}t} + (\mathbf{v} - \mathbf{u}) \cdot \nabla \phi. \tag{1.12.4}$$

When the tracers are material entities, $\mathbf{v} = \mathbf{u}$, or when the difference $\mathbf{v} - \mathbf{u}$ is normal to the gradient $\nabla \phi$, the second term on the right-hand side is zero and the tracer derivative reduces to the material derivative.

The continuity equation (1.9.8) can be restated as

$$\frac{\mathrm{D}\rho}{\mathrm{D}t} + \rho \, \nabla \cdot \mathbf{u} = \dot{\rho} - \mathbf{v} \cdot \nabla \rho + \nabla \cdot (\rho \, \mathbf{u}) = 0, \tag{1.12.5}$$

where ρ is the density. When $\mathbf{v}$ is defined such that $\mathbf{v} \cdot \nabla \rho = \nabla \cdot (\rho \mathbf{u})$, the density remains constant along tracer paths, $\dot{\rho} = 0$.

1.12.2 Reynolds Transport Equation

A Reynolds transport equation similar to that shown in (1.7.8) can be written for a packet. We find that

$$\frac{\mathrm{d}}{\mathrm{d}t} \iiint_{\text{packet}} \phi \, \mathrm{d}V = \iiint_{\text{packet}} \frac{\partial \phi}{\partial t} \, \mathrm{d}V + \iint_{\text{packet}} \phi \, \mathbf{n} \cdot \mathbf{v} \, \mathrm{d}S, \tag{1.12.6}$$

where $\mathbf{n}$ is the normal unit vector pointing *outward* from the packet. A packet with a non-evolving shape exhibits a purely tangential surface velocity field, $\mathbf{n} \cdot \mathbf{v} = 0$.

Setting $\phi = 1$, we find that the rate of change of a packet's volume is given by

$$\frac{\mathrm{d}V_{\text{packet}}}{\mathrm{d}t} = \iint_{\text{packet}} \mathbf{n} \cdot \mathbf{v} \, \mathrm{d}S, \tag{1.12.7}$$

which is the integral statement of the fundamental law of kinematics. When $\mathbf{n} \cdot \mathbf{v} = 0$, the shape of the packet remains unchanged and the packet volume remains constant in time.

Setting $\phi = \rho\,\mathbf{u}$, we find that the rate of change of a packet's momentum is given by

$$\frac{d}{dt}\iiint_{\text{packet}} \rho\,\mathbf{u}\,dV = \iiint_{\text{packet}} \frac{\partial(\rho\,\mathbf{u})}{\partial t}\,dV + \iint_{\text{packet}} \rho\,\mathbf{n}\cdot(\mathbf{v}\otimes\mathbf{u})\,dS, \tag{1.12.8}$$

where $\mathbf{n}\cdot(\mathbf{v}\otimes\mathbf{u}) = (\mathbf{n}\cdot\mathbf{v})\,\mathbf{u}$. The second integral on the right-hand side does not appear when $\mathbf{n}\cdot\mathbf{v} = 0$. Note that momentum is defined with respect to the fluid velocity, $\mathbf{u}$, as opposed to the packet point-particle velocity, $\mathbf{v}$.

In applications of the Reynolds transport equations, the point-particle property ϕ typically depends on physical material, kinematic, or thermodynamic fields, such as the fluid density, ρ, velocity, $\mathbf{u}$, or a specific energy.

1.12.3 Rate of Change of a Packet Property

The left-hand side of the Reynolds transport equation (1.12.6), repeated below for convenience,

$$\frac{d}{dt}\iiint_{\text{packet}} \phi\,dV, \tag{1.12.9}$$

expresses the rate of change of a packet's property. To evaluate this term, we note that a packet and a material parcel can be made to coincide at a particular instant in time, but not necessarily at subsequent times. Combining the Reynolds transport equations (1.12.6) and (1.7.8), we find that

$$\frac{d}{dt}\iiint_{\text{packet}} \phi\,dV = \frac{d}{dt}\iiint_{\text{parcel}} \phi\,dV + \iint_{\text{packet}} \phi\,\mathbf{n}\cdot(\mathbf{v}-\mathbf{u})\,dS \tag{1.12.10}$$

for a packet that coincides with a material parcel at a certain time. The first term on the right-hand side pertaining to a material parcel is dictated by a natural law, including mass conservation, Newton's second law of motion, or the first principle of thermodynamics.

The last term on the right-hand side of (1.12.10) does not appear when $\mathbf{n}\cdot\mathbf{v} = \mathbf{n}\cdot\mathbf{u}$ over the packet surface, that is, when the normal velocity of the packet's point particles is the same as that of the material point particles.

For $\phi = 1$, we use (1.7.9) to evaluate the first term on the right-hand side of (1.12.10) and simplify to recover (1.12.7). For $\phi = \rho$, we set

$$\frac{d}{dt}\iiint_{\text{parcel}} \rho\,dV = 0 \tag{1.12.11}$$

to ensure mass conservation, and obtain an evolution equation for the mass contained inside a packet,

$$\frac{d}{dt}\iiint_{\text{packet}} \rho\,dV = \iint_{\text{packet}} \rho\,\mathbf{n}\cdot(\mathbf{v}-\mathbf{u})\,dS. \tag{1.12.12}$$

For $\phi = \rho\mathbf{u}$, we evaluate the first term on the right-hand side of (1.12.10) using Newton's second law of motion, as discussed in Chap. 2.

Rearranging equation (1.12.10), we obtain

$$\frac{\mathrm{d}}{\mathrm{d}t}\iiint_{\text{parcel}} \phi \,\mathrm{d}V = \frac{\mathrm{d}}{\mathrm{d}t}\iiint_{\text{packet}} \phi \,\mathrm{d}V - \iint_{\text{packet}} \phi\, \mathbf{n}\cdot(\mathbf{v}-\mathbf{u})\,\mathrm{d}S \tag{1.12.13}$$

for a packet that coincides with a material parcel at a certain time.

1.12.4 Flux and Production

Using expression (1.12.13) for the rate of change of a material parcel's property to evaluate the first term on the right-hand side of (1.12.10), we obtain the generic equation

$$\frac{\mathrm{d}}{\mathrm{d}t}\iiint_{\text{packet}} \phi \,\mathrm{d}V = -\iint_{\text{packet}} \mathbf{n}\cdot\boldsymbol{\xi}\,\mathrm{d}S + \iiint_{\text{packet}} \dot{\pi}\,\mathrm{d}V + \iint_{\text{packet}} \dot{\nu}\,\mathrm{d}S + \iint_{\text{packet}} \phi\,\mathbf{n}\cdot(\mathbf{v}-\mathbf{u})\,\mathrm{d}S, \tag{1.12.14}$$

involving surface fluxes and volume integrals on the right-hand side. The first and fourth terms on the right-hand side can be combined into a single surface integral,

$$-\iint_{\text{packet}} \mathbf{n}\cdot\big(\boldsymbol{\xi} + \phi\,(\mathbf{u}-\mathbf{v})\big)\,\mathrm{d}S. \tag{1.12.15}$$

The term enclosed by the outer parentheses inside the integral constitutes a combined diffusive–convective flux.

1.12.5 Packet Property Conservation

The packet surface velocity, $\mathbf{v}$, may be adjusted so that

$$\phi\,\mathbf{n}\cdot\mathbf{v} = \mathbf{n}\cdot(\boldsymbol{\xi} + \phi\,\mathbf{u}), \tag{1.12.16}$$

where $\boldsymbol{\xi}$ can be a vector or tensor field. As a consequence, the combined integrand in (1.12.15) is identically zero and Eq. (1.12.14) simplifies to

$$\frac{\mathrm{d}}{\mathrm{d}t}\iiint_{\text{packet}} \phi \,\mathrm{d}V = \iiint_{\text{packet}} \dot{\pi}\,\mathrm{d}V + \iint_{\text{packet}} \dot{\nu}\,\mathrm{d}S. \tag{1.12.17}$$

In the absence of volume and surface production, $\dot{\pi} = 0$ and $\dot{\nu} = 0$, the integral on the left-hand side remains constant in time.

In the case of mass, $\phi = \rho$, $\boldsymbol{\xi} = \mathbf{0}$, $\dot{\pi} = 0$, and $\dot{\nu} = 0$. Condition (1.12.16) requires that $\mathbf{n}\cdot\mathbf{v} = \mathbf{n}\cdot\mathbf{u}$.

In the case of momentum, $\phi = \rho\,\mathbf{u}$ and $\boldsymbol{\xi} = -\boldsymbol{\sigma}$, where $\boldsymbol{\sigma}$ is the stress tensor, as discussed in Chap. 2. Condition (1.12.16) requires that

$$\rho\,\mathbf{n}\cdot(\mathbf{v}\otimes\mathbf{u}) = \mathbf{n}\cdot(\rho\,\mathbf{u}\otimes\mathbf{u} - \boldsymbol{\sigma}). \tag{1.12.18}$$

In the absence of a body force, $\dot{\pi} = 0$, the packet momentum remains constant in time.

1.12.6 Accumulation

Using the Reynolds transport equation (1.12.6) to evaluate the left-hand side of (1.12.10), and simplifying, we obtain

$$\iiint_{\text{packet}} \frac{\partial \phi}{\partial t}\,\mathrm{d}V = -\iint_{\text{packet}} \phi\,\mathbf{n}\cdot\mathbf{u}\,\mathrm{d}S + \frac{\mathrm{d}}{\mathrm{d}t}\iiint_{\text{parcel}} \phi\,\mathrm{d}V. \tag{1.12.19}$$

The left-hand side of (1.12.19) is the rate of accumulation of the physical or conceptual entity represented by ϕ inside the coincident packet and parcel.

The packet velocity, $\mathbf{v}$, does not appear explicitly in Eq. (1.12.19). An explicit dependence on $\mathbf{v}$ is mediated through the instantaneous packet position and shape determining the volume and surface integration domains.

Using expression (1.6.5) to evaluate the second term on the right-hand side of (1.12.19), we obtain

$$\iiint_{\text{packet}} \frac{\partial \phi}{\partial t}\,\mathrm{d}V = -\iint_{\text{packet}} \phi(\rho,\mathbf{u})\,\mathbf{n}\cdot\mathbf{u}\,\mathrm{d}S - \iint_{\text{packet}} \mathbf{n}\cdot\boldsymbol{\xi}\,\mathrm{d}S + \iiint_{\text{packet}} \dot{\pi}\,\mathrm{d}V + \iint_{\text{packet}} \dot{\nu}\,\mathrm{d}S. \tag{1.12.20}$$

The left-hand side is the rate of accumulation of the physical or conceptual entity represented by ϕ inside an arbitrary stationary or moving packet. An example is the number of persons in a bus that picks up and discharges passengers at different stops.

1.12.7 Control Volume Occupied by a Packet

A packet may be interpreted as the content of a stationary, moving, or evolving control volume $\mathcal{V}_c$, consisting of the same collection of material or nonmaterial point particles. A tracer at the boundary of the control volume, $\mathcal{S}_c$, moves with the normal component of the velocity of the control volume, $\mathbf{n}\cdot\mathbf{v}$, and with an arbitrary tangential velocity.

Straightforward rearrangement of the transport equation (1.12.20) to introduce the control volume surface velocity, $\mathbf{v}$, provides us with the balance equation

$$\begin{aligned}\iiint_{\mathcal{V}_c} \frac{\partial \phi}{\partial t}\,\mathrm{d}V &= \iint_{\mathcal{S}_c} \phi\,\mathbf{n}^{\text{in}}\cdot\mathbf{v}\,\mathrm{d}S + \iint_{\mathcal{S}_c} \phi\,\mathbf{n}^{\text{in}}\cdot(\mathbf{u}-\mathbf{v})\,\mathrm{d}S \\ &\quad - \iint_{\mathcal{S}_c} \mathbf{n}\cdot\boldsymbol{\xi}\,\mathrm{d}S + \iiint_{\mathcal{V}_c} \dot{\pi}\,\mathrm{d}V + \iint_{\mathcal{V}_c} \dot{\nu}\,\mathrm{d}S,\end{aligned} \tag{1.12.21}$$

where $\mathbf{n}^{\text{in}}$ is the inward normal vector. The first and second terms on the left-hand side combine to yield the first term on the left-hand side of (1.12.20). Note that the boundary distribution but not the volume distribution of $\mathbf{v}$ or $\mathbf{u}-\mathbf{v}$ appears in Eq. (1.12.21).

The first integral on the right-hand side of (1.12.21) expresses the rate of convective entry due to the motion of the control volume. This term does not appear in the case of a stationary control volume or when the packet point-particles exhibit a purely tangential boundary motion.

Table 1.2 Typical balance of a transported quantity over a control volume moving or evolving with velocity **v**

(rate of accumulation inside a stationary or frozen control volume)
= (rate of convective entry due to the motion of the control volume)
+ (rate of inward convection due to the lag velocity $\mathbf{u} - \mathbf{v}$)
+ (rate of short-range surface action or molecular transport)
+ (rate of interior production or loss)
+ (rate of long-range surface production or loss)

A choice on the appropriate term inside each set of parentheses on the right-hand side is made according to context

The difference $\mathbf{u} - \mathbf{v}$ inside the second integral on the right-hand side of (1.12.21) is the local relative velocity, termed the *lag* or *diffusion* velocity. Consequently, the second integral on the right-hand side of (1.12.21) expresses the rate of entry into the control volume due to the lag or diffusion velocity, $\mathbf{u} - \mathbf{v}$.

A typical balance of a transported quantity over an evolving control volume takes the form shown in Table 1.2.

1.12.8 Mixtures

Packets and moving control volumes are employed in the kinematic theory of mixtures discussed in Chaps. 4–8, where $\mathbf{v}$ is the mixture mass velocity, $\mathbf{u}_{\text{mixt}}$, and $\mathbf{u}$ is the species X velocity, $\mathbf{u}_{\text{X}}$. In the context of mixtures, the lag velocity $\mathbf{u}_{\text{X}} - \mathbf{u}_{\text{mixt}}$ is a diffusion velocity expressed in terms of a diffusive mass flux, denoted as $\boldsymbol{j}_{\text{X}}$.

A diffusion law involving the species mass fractions and possibly other driving forces must be introduced for closure. Examples are Fick's law and the fractional diffusion law discussed in Chap. 6.

1.12.9 Exercise

1.12.1 Explain in physical terms why adding an arbitrary component to the point-particle velocity, $\mathbf{v}$, that does not affect the surface normal velocity, $\mathbf{v} \cdot \mathbf{n}$, does not alter the rate of change of a packet's property.

Momentum and Forces

2

In the second chapter, we introduce the concepts of force, traction, and stress developing in a fluid consisting of a single chemical species. By invoking Newton's second law of motion for a material fluid parcel, we derive the Cauchy equation of motion for a compressible or incompressible fluid. By combining the equation of motion with the Reynolds transport equation, we derive a momentum balance and associated equation of momentum transport over a control volume in a way that is rigorous, intuitive, and free of ad-hoc stipulations.

Projecting the equation of motion onto the velocity at a point allows us to develop an expression for the reversible or irreversible conversion of mechanical energy into internal energy, which is then employed in energy balances performed in the context of thermodynamics, as discussed in Chap. 3.

2.1 Traction and the Stress Tensor

Consider the force exerted on an infinitesimal surface element at the boundary of a material fluid parcel, denoted by $\delta \mathbf{F}^{\text{surface}}$. Dividing the force by the element surface area, denoted by δS, we obtain the traction, denoted by

$$\mathbf{f} \equiv \frac{1}{\delta S}\, \delta \mathbf{F}^{\text{surface}}, \tag{2.1.1}$$

as shown in Fig. 2.1. The traction has a normal component, called the normal stress, and a tangential component, called the shear stress.

The traction depends on the location, orientation, and designated side of the infinitesimal surface element. The location is determined by the position vector, $\mathbf{x}$, and the orientation and side are determined by the normal unit vector, $\mathbf{n}$, so that

$$\mathbf{f}(\mathbf{x}, \mathbf{n}), \tag{2.1.2}$$

where the parentheses enclose the arguments of the traction. The orientation of the traction vector, $\mathbf{f}$, with respect to the unit normal vector, $\mathbf{n}$, is arbitrary, as shown in Fig. 2.1.

C. Pozrikidis, *Transport Processes Primer*,
https://doi.org/10.1007/978-1-4939-9909-5_2

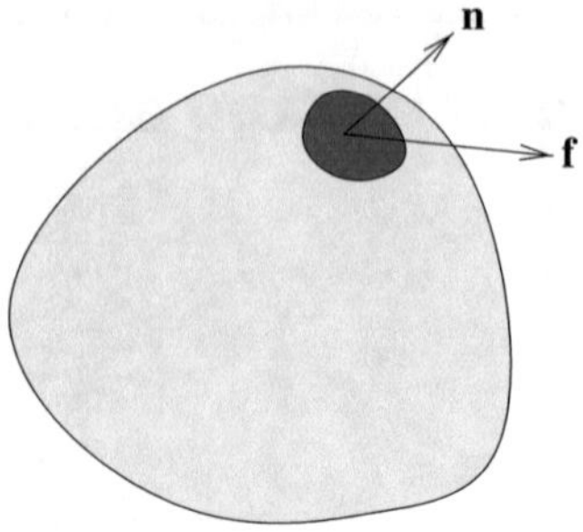

Fig. 2.1 Illustration of the traction, **f**, exerted on a small surface at the boundary of a fluid parcel or control volume; the unit normal unit vector, **n**, is also shown

2.1.1 Force in Terms of the Traction

The traction has dimensions of force divided by squared length. In terms of the traction, the force exerted on the surface of a fluid parcel is given by

$$\mathbf{F}^{\text{surface}} = \iint_{\text{parcel}} \mathbf{f}\, \mathrm{d}S. \tag{2.1.3}$$

The same expression provides us with the force exerted on a boundary, such as the surface of a suspended particle, or on the surface of a control volume occupied by fluid, where **n** is the unit vector normal to the boundary.

2.1.2 Traction in Terms of the Stress Tensor

The traction at a particular position in a fluid derives from the Cauchy stress tensor, $\boldsymbol{\sigma}$, as

$$\mathbf{f} = \mathbf{n} \cdot \boldsymbol{\sigma}. \tag{2.1.4}$$

In a three-dimensional flow, the stress tensor is represented by a 3×3 matrix that encapsulates the stress components in a chosen Cartesian system.

Equation (2.1.4) states that the traction, **f**, is a linear function of the normal unit vector, **n**, with a matrix of proportionality that is equal to the stress tensor, $\boldsymbol{\sigma}$. Setting **n** parallel to the x, y, or z axis reveals that the three *rows* of $\boldsymbol{\sigma}$ contain the components of the traction vector exerted on a small surface element that is perpendicular to the x, y, or z axis.

To derive expression (2.1.4), we apply Newton's second law of motion discussed in Sect. 2.3 for a material parcel having the shape of a tetrahedron, as explained in texts of fluid dynamics.

Unless $\boldsymbol{\sigma}$ is symmetric, the order of multiplication on the right-hand side of (2.1.4) cannot be reversed, that is, $\mathbf{n} \cdot \boldsymbol{\sigma} \neq \boldsymbol{\sigma} \cdot \mathbf{n}$. The condition for the stress tensor to be symmetric is discussed in Sect. 2.8.

2.1.3 Hydrostatics

In the absence of fluid motion, the stress tensor takes the form

$$\boldsymbol{\sigma} = -p\,\mathbf{I}, \tag{2.1.5}$$

where p is the pressure and $\mathbf{I}$ is the identity matrix. The associated traction is given by $\mathbf{f} = -p\,\mathbf{n}$, encapsulating a normal stress in the absence of shear stress.

2.1.4 Pressure in Hydrodynamics

In hydrodynamics, the pressure at a point in a fluid is defined in terms of the trace of the stress tensor as

$$p = -\tfrac{1}{3}\,\mathrm{trace}(\boldsymbol{\sigma}). \tag{2.1.6}$$

We recall that the trace is the sum of the diagonal elements representing the three normal stresses. An equation of state from thermodynamics can be adopted with some trepidation in the case of a gas to evaluate p in terms of the density, ρ, and temperature, T, as discussed in Appendix B.

2.1.5 Deviatoric Stress Tensor

Having defined the pressure, p, we may introduce the deviatoric stress tensor, $\boldsymbol{\sigma}^{\mathrm{dev}}$, sometimes also called the viscous stress tensor, defined by the decomposition

$$\boldsymbol{\sigma} = -p\,\mathbf{I} + \boldsymbol{\sigma}^{\mathrm{dev}}. \tag{2.1.7}$$

Rearranging, we find that

$$\boldsymbol{\sigma}^{\mathrm{dev}} \equiv p\,\mathbf{I} + \boldsymbol{\sigma}. \tag{2.1.8}$$

Since the pressure is the negative of one-third of the stress tensor, the trace of the deviatoric stress tensor is zero.

By definition, the deviatoric stress tensor is identically zero in the absence of observable fluid motion or else for a truly or effectively inviscid fluid, also called an ideal fluid.

We will see that the deviatoric stress tensor is relevant to the irreversible conversion of mechanical energy into internal energy that raises the temperature of the fluid.

2.1.6 Incompressible Newtonian Fluids

In the case of an incompressible Newtonian fluid, $\mathrm{D}\rho/\mathrm{D}t = 0$, the deviatoric stress tensor is given by

$$\boldsymbol{\sigma}^{\mathrm{dev}} = \mu\,(\mathbf{L} + \mathbf{L}^{\mathrm{T}}), \tag{2.1.9}$$

where μ is the fluid viscosity, $\mathbf{L} \equiv \nabla \mathbf{u}$ is the velocity-gradient tensor, and the superscript T denotes the matrix transpose. The sum of the two matrices on the right-hand side of (2.1.9) is twice the rate-of-deformation tensor, 2**E**.

2.1.7 Exercise

2.1.1 Explain in physical terms why $\boldsymbol{\sigma} = -p\,\mathbf{I}$ in a stationary fluid, where p is the pressure and **I** is the identity matrix.

2.2 Parcel Momentum

The momentum of a translating rigid body is the mass of the body multiplied by the velocity. The momentum of a body executing rigid-body motion, a deforming body, or a material fluid parcel engaged in a general type of motion is given by the consequent expression

$$\mathbf{M} \equiv \iiint_{\text{parcel}} \rho\,\mathbf{u}\,\mathrm{d}V. \tag{2.2.1}$$

The rate of change of the parcel momentum is expressed by the ordinary derivative

$$\frac{\mathrm{d}\mathbf{M}}{\mathrm{d}t} = \iiint_{\text{parcel}} \frac{\mathrm{D}(\rho\,\mathbf{u}\,\delta V)}{\mathrm{D}t}, \tag{2.2.2}$$

where δV is a small material volume. Expanding the material derivative of the product, and recalling that $\mathrm{D}\,\delta V/\mathrm{D}t = (\nabla \cdot \mathbf{u})\,\delta V$, we obtain

$$\frac{\mathrm{d}\mathbf{M}}{\mathrm{d}t} = \iiint_{\text{parcel}} \Big(\frac{\mathrm{D}(\rho\,\mathbf{u})}{\mathrm{D}t} + \rho\,\mathbf{u}\,\nabla \cdot \mathbf{u}\Big)\,\mathrm{d}V, \tag{2.2.3}$$

where $\nabla \cdot \mathbf{u}$ is the rate of expansion. Expressing the material derivative inside the integral in terms of Eulerian derivatives, and rearranging, we obtain

$$\frac{\mathrm{d}\mathbf{M}}{\mathrm{d}t} = \iiint_{\text{parcel}} \Big(\frac{\partial(\rho\,\mathbf{u})}{\partial t} + \nabla \cdot (\rho\,\mathbf{u} \otimes \mathbf{u})\Big)\,\mathrm{d}V, \tag{2.2.4}$$

where $\rho\,\mathbf{u} \otimes \mathbf{u}$ is the momentum tensor with components

$$[\,\rho\,\mathbf{u} \otimes \mathbf{u}\,]_{ij} = \rho\,u_i u_j. \tag{2.2.5}$$

The trace of the momentum tensor is the square of the magnitude of the velocity multiplied by the density, $\rho\,|\mathbf{u}|^2$.

2.2.1 Momentum in Terms of the Point-Particle Acceleration

Now expanding the derivatives inside the integral in (2.2.4) and using the continuity equation,

$$\frac{\partial \rho}{\partial t} + \nabla \cdot (\rho\,\mathbf{u}) = 0, \tag{2.2.6}$$

we obtain

$$\frac{\mathrm{d}\mathbf{M}}{\mathrm{d}t} = \iiint_{\text{parcel}} \rho \Big(\frac{\partial \mathbf{u}}{\partial t} + \mathbf{u} \cdot \nabla \mathbf{u} \Big) \, \mathrm{d}V. \tag{2.2.7}$$

Setting the expression inside the parentheses equal to the point-particle acceleration, $\mathrm{D}\mathbf{u}/\mathrm{D}t$, we obtain

$$\frac{\mathrm{d}\mathbf{M}}{\mathrm{d}t} = \iiint_{\text{parcel}} \rho \, \frac{\mathrm{D}\mathbf{u}}{\mathrm{D}t} \, \mathrm{d}V. \tag{2.2.8}$$

We have found that the rate of change of a parcel momentum is the integral of the density multiplied by the point particle acceleration for an incompressible or compressible fluid.

2.2.2 Alternative Derivations

Equation (2.2.8) can be derived directly from (2.2.2) by invoking equation (1.9.6) stating that $\mathrm{D}(\rho \, \delta V)/\mathrm{D}t = 0$, and then expanding the material derivative inside the integral on the right-hand side.

Alternatively, Eq. (2.2.8) arises directly from (1.9.17) by regarding the velocity as the specific momentum and setting $\psi = \mathbf{u}$.

2.2.3 Exercise

2.2.1 Write an expression for the momentum of a translating and rotating rigid body.

2.3 Cauchy's Equation of Motion

Newton's second law of motion for a material fluid parcel stipulates that the rate of change of the parcel momentum is balanced by surface and volume forces,

$$\frac{\mathrm{d}\mathbf{M}}{\mathrm{d}t} = \iint_{\text{parcel}} \mathbf{n} \cdot \boldsymbol{\sigma} \, \mathrm{d}S + \iiint_{\text{parcel}} \rho \, \mathbf{g} \, \mathrm{d}V + \iiint_{\text{parcel}} \iota \, \mathbf{b} \, \mathrm{d}V, \tag{2.3.1}$$

where $\mathbf{g}$ is the acceleration of gravity and $\mathbf{b}$ is an imposed body force field associated with physical constant ι that is a property of the fluid, as in the case of an electrical field acting on an ionic fluid.

We recall that the order of multiplication inside the first integral on the right-hand side, $\mathbf{n} \cdot \boldsymbol{\sigma}$, can be transposed only if the stress tensor $\boldsymbol{\sigma}$ is symmetric.

2.3.1 Instance of a General Form

A general natural law pertinent to the properties of a material parcel was stated in Eq. (1.6.5), repeated below for convenience,

$$\frac{\mathrm{d}}{\mathrm{d}t} \iiint_{\text{parcel}} \phi \, \mathrm{d}V = - \iint_{\text{parcel}} \mathbf{n} \cdot \boldsymbol{\xi} \, \mathrm{d}S + \iiint_{\text{parcel}} \dot{\pi} \, \mathrm{d}V + \iint_{\text{parcel}} \dot{\nu} \, \mathrm{d}S, \tag{2.3.2}$$

where ϕ is a suitable property. Comparing this equation with Newton's law expressed by (2.3.1), we set

$$\phi = \rho\,\mathbf{u}, \qquad \xi = -\boldsymbol{\sigma}, \qquad \dot{\pi} = \rho\,\mathbf{g} + \iota\,\mathbf{b}, \qquad \dot{\nu} = 0. \tag{2.3.3}$$

The last condition, $\dot{\nu} = 0$, ensures that, in classical mechanics, momentum cannot be transmitted by a long-range surface action. In Chap. 5, we will see that species momentum can be generated by molecular interception or transferred by chemical conversion in a binary or multicomponent mixture.

2.3.2 Point-Particle Acceleration

Adopting expression (2.2.8) for the parcel rate of change of momentum, repeated below for convenience,

$$\frac{\mathrm{d}\mathbf{M}}{\mathrm{d}t} = \iiint_{\text{parcel}} \rho\,\frac{\mathrm{D}\mathbf{u}}{\mathrm{D}t}\,\mathrm{d}V, \tag{2.3.4}$$

we obtain

$$\iiint_{\text{parcel}} \rho\,\frac{\mathrm{D}\mathbf{u}}{\mathrm{D}t}\,\mathrm{d}V = \iint_{\text{parcel}} \mathbf{n}\cdot\boldsymbol{\sigma}\,\mathrm{d}S + \iiint_{\text{parcel}} \rho\,\mathbf{g}\,\mathrm{d}V + \iiint_{\text{parcel}} \iota\,\mathbf{b}\,\mathrm{d}V. \tag{2.3.5}$$

Using the divergence theorem to convert the surface integral after the equal sign into a volume integral, and then discarding all volume integrals, we derive the Cauchy equation of motion

$$\rho\,\frac{\mathrm{D}\mathbf{u}}{\mathrm{D}t} = \nabla\cdot\boldsymbol{\sigma} + \rho\,\mathbf{g} + \iota\,\mathbf{b}, \tag{2.3.6}$$

where $\nabla\cdot\boldsymbol{\sigma}$ is the divergence of the stress tensor. The effect of fluid inertia is mediated by the left-hand side involving the point-particle density, ρ, and acceleration, $\mathrm{D}\mathbf{u}/\mathrm{D}t$, for incompressible or compressible fluids.

In index notation, the jth Cartesian component of Cauchy's equation of motion reads

$$\rho\,\frac{\mathrm{D}u_j}{\mathrm{D}t} = \frac{\partial \sigma_{ij}}{\partial x_i} + \rho\,g_j + \iota\,b_j, \tag{2.3.7}$$

where summation is implied over the repeated index, i. We recall that the stress tensor is not necessarily symmetric, that is, σ_{ij} is not necessarily equal to σ_{ji}.

2.3.3 Euler, Bernoulli, and Navier–Stokes Equations

Euler's and the Navier–Stokes equation arise from the Cauchy equation of motion (2.3.6) for special choices of the stress tensor, $\boldsymbol{\sigma}$, in the form of constitutive equations involving the pressure. Bernoulli's equation arises by integrating in space the Cauchy equation of motion under certain conditions.

2.3.4 Eulerian Form

Resolving the material derivative of the velocity, $\mathrm{D}\mathbf{u}/\mathrm{D}t$, into Eulerian derivatives, we obtain the Eulerian form of the equation of motion,

$$\rho\,\big(\frac{\partial \mathbf{u}}{\partial t}+\mathbf{u}\cdot\nabla\mathbf{u}\big)=\nabla\cdot\boldsymbol{\sigma}++\rho\,\mathbf{g}+\iota\,\mathbf{b}. \tag{2.3.8}$$

In index notation, the jth component of this equation reads

$$\rho\,\big(\frac{\partial u_j}{\partial t}+u_i\,\frac{\partial u_j}{\partial x_i}\big)=\frac{\partial\,\sigma_{ij}}{\partial x_i}+\rho\,g_j+\iota\,b_j, \tag{2.3.9}$$

where summation is implied over the repeated index, i. The first term inside the parentheses on the left-hand side does not appear in the case of steady flow. The second term does not appear in the case of unidirectional flow.

The negative of the second term on the left-hand side, $-\rho\,\mathbf{u}\cdot\nabla\mathbf{u}$, can be transferred to the right-hand side and regarded as a fictitious nonlinear inertial force.

2.3.5 Conservative Form

A further statement of Cauchy's equation of motion is given by the conservative form

$$\frac{\partial(\rho\,\mathbf{u})}{\partial t}+\nabla\cdot(\rho\,\mathbf{u}\otimes\mathbf{u})=\nabla\cdot\boldsymbol{\sigma}+\rho\,\mathbf{g}+\iota\,\mathbf{b}, \tag{2.3.10}$$

for compressible or incompressible fluids. We recall that the ij component of the momentum tensor $\rho\,\mathbf{u}\otimes\mathbf{u}$ is equal to $\rho u_i u_j$. In index notation, the jth component of Eq. (2.3.10) reads

$$\frac{\partial(\rho\,u_j)}{\partial t}+\frac{\partial(\rho\,u_i u_j)}{\partial x_i}=\frac{\partial\,\sigma_{ij}}{\partial x_i}+\rho\,g_j+\iota\,b_j. \tag{2.3.11}$$

The qualifier *conservative* refers to the divergence in the second term on the left-hand side involving the momentum tensor.

2.3.6 Stress–Momentum Tensor

The two divergences in (2.3.10) can be combined into a single term, yielding the compact form

$$\frac{\partial(\rho\,\mathbf{u})}{\partial t}=\nabla\cdot\boldsymbol{\tau}+\rho\,\mathbf{g}+\iota\,\mathbf{b}, \tag{2.3.12}$$

where

$$\boldsymbol{\tau}\equiv\boldsymbol{\sigma}-\rho\,\mathbf{u}\otimes\mathbf{u} \tag{2.3.13}$$

is the stress–momentum tensors. In Chap. 7, we will see that the species stress–momentum tensors, but the species stress tensor themselves, are additive in a multicomponent mixture.

2.3.7 Exercise

2.3.1 Derive Eq. (2.3.10) by invoking the continuity equation.

2.4 Rankin–Hugoniot Momentum Condition

The rate of change of the fluid momentum recorded by an observer who moves with arbitrary velocity, **v**, is given by the traveler's derivative,

$$\frac{\mathrm{d}(\rho\,\mathbf{u})}{\mathrm{d}t} = \frac{\partial(\rho\,\mathbf{u})}{\partial t} + \mathbf{v}\cdot\nabla(\rho\,\mathbf{u}). \tag{2.4.1}$$

Combining this equation with the equation of motion (2.3.10), we obtain

$$\frac{\mathrm{d}(\rho\,\mathbf{u})}{\mathrm{d}t} - \mathbf{v}\cdot\nabla(\rho\,\mathbf{u}) + \nabla\cdot(\rho\,\mathbf{u}\otimes\mathbf{u}) = \nabla\cdot\boldsymbol{\sigma} + \rho\,\mathbf{g} + \iota\,\mathbf{b}, \tag{2.4.2}$$

which can be regarded as an evolution equation for the momentum, as recorded by the observer.

Rearranging equation (2.4.2) to transfer all terms but one to the right-hand side, we obtain

$$\frac{\mathrm{d}(\rho\,\mathbf{u})}{\mathrm{d}t} = \mathbf{v}\cdot\nabla(\rho\,\mathbf{u}) - \nabla\cdot(\rho\,\mathbf{u}\otimes\mathbf{u}) + \nabla\cdot\boldsymbol{\sigma} + \rho\,\mathbf{g} + \iota\,\mathbf{b}. \tag{2.4.3}$$

Based on this equation, we find that the rate of change of the momentum jump across a front of momentum discontinuity moving with velocity **v** is given by

$$\frac{\mathrm{d}\big(\Delta(\rho\,\mathbf{u})\big)}{\mathrm{d}t} = \mathbf{v}\cdot\nabla\big(\Delta(\rho\,\mathbf{u})\big) - \nabla\cdot\big(\Delta(\rho\,\mathbf{u}\otimes\mathbf{u})\big) + \nabla\cdot\Delta\boldsymbol{\sigma} + \Delta\rho\,\mathbf{g} + \Delta(\iota\,\mathbf{b}), \tag{2.4.4}$$

where Δ denotes the discontinuity or jump. All terms in this equation are finite (non-infinite), except for three terms on the right-hand side involving spatial derivatives normal to the evolving front.

To ensure regularity, we concentrate on these annoying terms and require that, to leading order,

$$\underline{0} = v_n\,\big(\mathbf{n}\cdot\nabla(\Delta(\rho\mathbf{u})\big) - \mathbf{n}\cdot\nabla\big(\Delta(\rho\,(\mathbf{n}\cdot\mathbf{u})\,\mathbf{u})\big) + \mathbf{n}\cdot\nabla\big(\Delta(\mathbf{n}\cdot\boldsymbol{\sigma})\big), \tag{2.4.5}$$

where **n** is the unit vector normal to the front and $v_n \equiv \mathbf{v}\cdot\mathbf{n}$ is the normal component of the front velocity. Removing from all terms the normal derivative mediated by $\mathbf{n}\cdot\nabla$, we derive the Rankin–Hugoniot momentum condition,

$$v_n\,\Delta(\rho\mathbf{u}) = \Delta(\rho\,(\mathbf{n}\cdot\mathbf{u})\,\mathbf{u}) - \Delta(\mathbf{n}\cdot\boldsymbol{\sigma}). \tag{2.4.6}$$

Consolidating the two terms on the right-hand side, we obtain

$$v_n\,\Delta(\rho\mathbf{u}) = \Delta(\rho\,u_n\,\mathbf{u} - \mathbf{n}\cdot\boldsymbol{\sigma}), \tag{2.4.7}$$

where $u_n \equiv \mathbf{u} \cdot \mathbf{n}$ is the normal component of the fluid velocity. The Rankin–Hugoniot momentum condition relates the jump in momentum to the jump in the rate of transport of momentum to the jump in the traction across a discontinuous front.

2.4.1 One-Dimensional Flow

In the idealized case of one-dimensional flow along the x axis, the Rankin–Hugoniot momentum condition provides us with the expression

$$v_x = \frac{\Delta(\rho\, u_x^2 - \sigma_{xx})}{\Delta(\rho u_x)}, \tag{2.4.8}$$

where u_x is the fluid velocity and σ_{xx} is the in-line stress.

It is instructive to derive Eq. (2.4.8) by performing a momentum balance over a fixed interval $[a, b]$ that includes the moving point of discontinuity, $x_d(t)$, that is, $a < x_d(t) < b$ within a time interval of interest. The momentum contained inside this fixed interval is given by

$$M = \int_a^b \rho\, u_x \,\mathrm{d}x. \tag{2.4.9}$$

A momentum balance requires that

$$\frac{\mathrm{d}M}{\mathrm{d}t} = (\rho\, u_x^2)_{x=a} - (\rho\, u_x^2)_{x=b} + \int_a^b \rho\, g_x \,\mathrm{d}x + \int_a^b \iota\, b_x \,\mathrm{d}x. \tag{2.4.10}$$

Breaking up the integration domain into two pieces, we write

$$M = \int_a^{x_d(t)} \rho\, u_x \,\mathrm{d}x + \int_{x_d(t)}^b \rho\, u_x \,\mathrm{d}x. \tag{2.4.11}$$

Differentiating using the Leibnitz rule, we find that

$$\frac{\mathrm{d}M}{\mathrm{d}t} = -\Delta(\rho\, u_x)\, \frac{\mathrm{d}x_d}{\mathrm{d}t} + \int_a^b \frac{\partial(\rho\, u_x)}{\partial t} \,\mathrm{d}x, \tag{2.4.12}$$

where

$$\Delta(\rho\, u_x) = (\rho\, u_x)_{x_d^+} - (\rho\, u_x)_{x_d^-}, \tag{2.4.13}$$

the superscript $+$ indicates evaluation at the right edge and the superscript $-$ indicates evaluation at the left edge of the discontinuity.

The equation of motion takes the form

$$\frac{\partial(\rho\, u_x)}{\partial t} = -\frac{\partial(\rho\, u_x^2 - \sigma_{xx})}{\partial t} + \rho\, g_x + \iota\, b_x. \tag{2.4.14}$$

Using this equation to evaluate the term inside the integral in (2.4.12), we obtain

$$\frac{\mathrm{d}M}{\mathrm{d}t} = -\Delta(\rho u_x)\,\frac{\mathrm{d}x_d}{\mathrm{d}t} - \int_a^{x_d^-(t)} \frac{\partial(\rho\, u_x^2 - \sigma_{xx})}{\partial x}\,\mathrm{d}x - \int_{x_d^+(t)}^{b} \frac{\partial(\rho\, u_x^2 - \sigma_{xx})}{\partial x}\,\mathrm{d}x$$

$$+ \int_a^b \rho\, g_x\,\mathrm{d}x + \int_a^b \iota b_x\,\mathrm{d}x. \tag{2.4.15}$$

Performing the integrations, we find that

$$\frac{\mathrm{d}M}{\mathrm{d}t} = -\Delta(\rho\, u_x)\,\frac{\mathrm{d}x_d}{\mathrm{d}t} + (\rho\, u_x^2)_{x=a} - (\rho\, u_x^2)_{x=b} + \Delta(\rho u_x^2 - \sigma_{xx})$$

$$+ \int_a^b \rho\, g_x\,\mathrm{d}x + \int_a^b \iota b_x\,\mathrm{d}x, \tag{2.4.16}$$

where

$$\Delta(\rho\, u_x^2 - \sigma_{xx}) = (\rho\, u_x^2 - \sigma_{xx})_{x_d^+} - (\rho\, u_x^2 - \sigma_{xx})_{x_d^-} \quad . \tag{2.4.17}$$

Substituting expression (2.4.16) for $\mathrm{d}M/\mathrm{d}t$ into the left-hand side of (2.4.10), simplifying, rearranging, and setting

$$\frac{\mathrm{d}x_d}{\mathrm{d}t} = v_x, \tag{2.4.18}$$

we recover equation (2.4.8).

2.4.2 Exercise

2.4.1 Derive the Rankin–Hugoniot momentum condition for two-dimensional flow working as discussed in the text for one-dimensional flow.

2.5 Noninertial Frames

Cauchy's equation of motion was derived under the assumption that the frame of reference is inertial, which means that the Cartesian axes are either stationary or translate in space with constant velocity, but neither accelerate nor rotate.

It is sometimes convenient to work with a noninertial frame whose origin translates with respect to an inertial frame with time-dependent velocity, $\mathbf{U}(t)$, while its axes rotate about the instantaneous position of the origin with a time-dependent angular velocity, $\boldsymbol{\Omega}(t)$. An example is a frame of reference attached to a person that carries a cup of coffee.

2.5.1 Position, Velocity, and Acceleration

The position of a point particle in the inertial frame is

$$\mathbf{X} = \mathbf{X}' + \mathbf{x}_0, \tag{2.5.1}$$

where $\mathbf{x}_0$ is the instantaneous position of the origin of the noninertial frame and $\mathbf{X}'$ is the point-particle position in the noninertial frame. By definition,

$$\mathbf{U} = \frac{\mathrm{d}\mathbf{x}_0}{\mathrm{d}t} \tag{2.5.2}$$

is the velocity of the origin of the noninertial system.

The velocity of a point particle in the inertial frame is given by the corresponding expression

$$\mathbf{u} = \mathbf{u}' + \mathbf{U} + \boldsymbol{\Omega} \times \mathbf{X}', \tag{2.5.3}$$

where $\mathbf{u}'$ is the point-particle velocity and $\mathbf{X}'$ is the point-particle position in the noninertial frame.

The acceleration of a point particle in the inertial frame is

$$\mathbf{a} = \mathbf{a}' + \frac{\mathrm{d}\mathbf{U}}{\mathrm{d}t} + 2\,\boldsymbol{\Omega} \times \mathbf{u}' + \frac{\mathrm{d}\boldsymbol{\Omega}}{\mathrm{d}t} \times \mathbf{X}' + \boldsymbol{\Omega} \times (\boldsymbol{\Omega} \times \mathbf{X}'), \tag{2.5.4}$$

where

$$\mathbf{a} = \Big(\frac{\partial \mathbf{u}}{\partial t}\Big)_{\mathbf{x}} + \mathbf{u} \cdot \nabla \mathbf{u}, \qquad \mathbf{a}' = \Big(\frac{\partial \mathbf{u}'}{\partial t}\Big)_{\mathbf{x}'} + \mathbf{u}' \cdot \nabla' \mathbf{u}', \tag{2.5.5}$$

and the gradient ∇' operates with respect to $\mathbf{x}'$.

2.5.2 Equation of Motion

Substituting the right-hand side of (2.5.4) for the point-particle acceleration into Cauchy's equation of motion, and rearranging, we derive a generalized equation of motion in the noninertial frame,

$$\rho\,\frac{\mathrm{D}\mathbf{u}'}{\mathrm{D}t} = \boldsymbol{\Sigma} + \mathbf{f}^{\text{fictitious}} + \rho\,\mathbf{g} + \iota\,\mathbf{b}, \tag{2.5.6}$$

where

$$\mathbf{f}^{\text{fictitious}} \equiv -\rho\,\Big(\frac{\mathrm{d}\mathbf{U}}{\mathrm{d}t} + 2\,\boldsymbol{\Omega} \times \mathbf{u}' + \boldsymbol{\Omega} \times (\boldsymbol{\Omega} \times \mathbf{x}') + \frac{\mathrm{d}\boldsymbol{\Omega}}{\mathrm{d}t} \times \mathbf{x}'\Big) \tag{2.5.7}$$

is a fictitious volumetric inertial force and $\mathbf{x}'$ is the position in the noninertial frame. The fictitious inertial force consists of four terms: (a) the linear acceleration force, $-\rho\,\mathrm{d}\mathbf{U}/\mathrm{d}t$; (b) the Coriolis force, $-2\,\rho\,\boldsymbol{\Omega} \times \mathbf{u}'$; (c) the centrifugal force, $-\rho\,\boldsymbol{\Omega} \times (\boldsymbol{\Omega} \times \mathbf{x}')$; and (d) the angular acceleration force, $-\rho\,(\mathrm{d}\boldsymbol{\Omega}/\mathrm{d}t) \times \mathbf{x}'$. Only one fictitious force survives in the case of steady or unsteady rigid-body rotation.

2.5.3 Exercise

2.5.1 Draw the Coriolis and centrifugal forces on the globe.

2.6 Integral Momentum Balance

To derive an integral momentum transport equation, we integrate the equation of motion (2.3.10) over a control volume $\mathcal{V}_c$ that lies entirely inside the fluid and is bounded by a boundary or a collection of boundaries, denoted by $\mathcal{S}_c$, as shown in Fig. 1.3.

Next, we use the divergence theorem to convert the volume integral of the divergence of the momentum tensor on the left-hand side and the volume integral of the divergence of the stress tensor on the right-hand side into corresponding surface integrals over the boundary $\mathcal{S}_c$, and thereby obtain an integral or macroscopic transport equation,

$$\iiint_{\mathcal{V}_c} \frac{\partial(\rho \mathbf{u})}{\partial t}\, \mathrm{d}V = \iint_{\mathcal{S}_c} (\rho\, \mathbf{u})\, (\mathbf{n}^{\mathrm{in}} \cdot \mathbf{u})\, \mathrm{d}S - \iint_{\mathcal{S}_c} \mathbf{n}^{\mathrm{in}} \cdot \boldsymbol{\sigma}\, \mathrm{d}S + \iiint_{\mathcal{V}_c} \rho\, \mathbf{g}\, \mathrm{d}V + \iiint_{\mathcal{V}_c} \iota\, \mathbf{b}\, \mathrm{d}V, \tag{2.6.1}$$

where $\mathbf{n}^{\mathrm{in}}$ is the inward normal vector. This transport equation can also be derived by applying the generic transport equation (1.8.2) for the fluid momentum, $\phi = \rho\, \mathbf{u}$, combined with Newton's second law of motion for a fluid parcel expressed by (2.3.1) to evaluate the last term on the right-hand side.

2.6.1 Physical Interpretation

The transport equation (2.6.1) states that the rate of accumulation of momentum of the fluid inside a control volume, represented by the left-hand side, is determined by the following surface and volume processes expressed by the four terms on the right-hand side:

(a) The convective flow rate of momentum normal to the boundaries expressed by the first term on the right-hand side.
(b) The force exerted on the boundaries expressed by the second term on the right-hand side.
(c) The gravitational force exerted on the fluid residing inside the control volume expressed by the third term on the right-hand side.
(d) An arbitrary body force exerted on the fluid residing inside the control volume expressed by the fourth term on the right-hand side.

In the absence of accumulation, the four terms on the right-hand side balance to zero.

Recall that the control volume may not be stationary, but may be translate, rotate, or evolve in an arbitrary fashion in time.

2.6.2 Newton's Second Law of Motion for an Open System

In the literature, Eq. (2.6.1) is often accepted as a stand-alone law instead of Newton's second law of motion for a material parcel. Although Newton's second law of motion can be derived from the momentum transport balance (2.6.1) working backward and in hindsight, a great deal of effort must be expended for explaining the merits of this procedure to skeptics.

A piece of advice is given to those who prefer the transport approach: accept or introduce equation (2.6.1) as Newton's second law of motion for an open system, analogous to the first law of thermodynamics for an open system. Then emphasize the duality of the second law.

2.6.3 Engineering Analysis

In practice, the integral momentum balance allows us to develop approximate relations between global properties of a steady or unsteady flow involving velocity and stress. In engineering analysis, we typically derive expressions for boundary forces or pressures in terms of boundary velocities, subject to sensible simplifications for inlet or outlet velocity profiles. The simplifications are guided by experience and physical intuition.

2.6.4 Exercise

2.6.1 Derive Newton's second law of motion for a fluid parcel departing from Eq. (2.6.1).

2.7 Angular Momentum

The angular momentum of a rigid body or fluid parcel engaged in a general type of motion is given by

$$\mathbf{A} \equiv \iiint_{\text{parcel}} \rho\,(\mathbf{X} - \mathbf{x}_0) \times \mathbf{u}\,\mathrm{d}V, \tag{2.7.1}$$

where $\times$ denotes the outer vector product, $\mathbf{X}$ is the position of a material point particle inside the parcel, and $\mathbf{x}_0$ is an arbitrarily specified point that can be set at the origin of the working coordinate system. For the purpose of integration, $\mathbf{X}$ can be replaced with the position vector $\mathbf{x}$ inside the parcel.

2.7.1 Center of Momentum

We see that the angular momentum is an implicit function of the designated center of angular momentum, $\mathbf{x}_0$, in that different choices for $\mathbf{x}_0$ generate different angular momenta, unless the linear momentum defined in (2.2.1) is zero. The center of momentum of the parcel is the unique point, $\mathbf{x}_0$, where the angular momentum is zero.

2.7.2 Rate of Change of Parcel Angular Momentum

The rate of change of a parcel's angular momentum is given by an ordinary time derivative that can be transferred inside the integral as a material derivative,

$$\frac{\mathrm{d}\mathbf{A}}{\mathrm{d}t} = \iiint_{\text{parcel}} \frac{\mathrm{D}(\rho\,(\mathbf{X} - \mathbf{x}_0) \times \mathbf{u}\,\delta V)}{\mathrm{D}t}, \tag{2.7.2}$$

where δV is a small material volume.

Expanding the material derivative inside the integral, and recalling that the rate of change of a small material volume is given by $\mathrm{D}\,\delta V/\mathrm{D}t = (\nabla \cdot \mathbf{u})\,\delta V$, we obtain

$$\frac{\mathrm{d}\,\mathbf{A}}{\mathrm{d}t} = \iiint_{\text{parcel}} \Big(\frac{\mathrm{D}(\rho\,(\mathbf{X}-\mathbf{x}_0)\times\mathbf{u})}{\mathrm{D}t} + \rho\,(\mathbf{X}-\mathbf{x}_0)\times\mathbf{u}\,(\nabla\cdot\mathbf{u}) \Big)\,\mathrm{d}V. \tag{2.7.3}$$

The integrand on the right-hand side can be manipulated further in two ways.

2.7.3 Evolution in Terms of Eulerian Derivatives

First, we express the material derivative inside the integrand in terms of Eulerian derivatives,

$$\frac{\mathrm{D}(\rho\,(\mathbf{X}-\mathbf{x}_0)\times\mathbf{u})}{\mathrm{D}t} = \frac{\partial(\rho\,(\mathbf{X}-\mathbf{x}_0)\times\mathbf{u})}{\partial t} + \mathbf{u}\cdot\nabla\big(\rho\,(\mathbf{X}-\mathbf{x}_0)\times\mathbf{u}\big), \tag{2.7.4}$$

and combining two terms to obtain

$$\frac{\mathrm{d}\,\mathbf{A}}{\mathrm{d}t} = \iiint_{\text{parcel}} \Big(\frac{\partial(\rho\,(\mathbf{X}-\mathbf{x}_0)\times\mathbf{u})}{\partial t} + \nabla\cdot\big(\mathbf{u}\otimes(\rho\,(\mathbf{X}-\mathbf{x}_0)\times\mathbf{u})\big) \Big)\,\mathrm{d}V, \tag{2.7.5}$$

where $\otimes$ denotes the tensor product.

2.7.4 Evolution in Terms of Point-Particle Acceleration

Alternatively, we expand further the first term inside the integral on the right-hand side of (2.7.3), and obtain

$$\frac{\mathrm{d}\mathbf{A}}{\mathrm{d}t} = \iiint_{\text{parcel}} \Big(\frac{\mathrm{D}\mathbf{X}}{\mathrm{D}t}\times(\rho\,\mathbf{u}) + (\mathbf{X}-\mathbf{x}_0)\times\frac{\mathrm{D}(\rho\,\mathbf{u})}{\mathrm{D}t} + \rho\,(\mathbf{X}-\mathbf{x}_0)\times\mathbf{u}\,(\nabla\cdot\mathbf{u}) \Big)\,\mathrm{d}V. \tag{2.7.6}$$

The material derivative of the position of a point particle is the fluid velocity, $\mathrm{D}\mathbf{X}/\mathrm{D}t = \mathbf{u}$, and the cross product of $\mathbf{u}$ and $\rho\mathbf{u}$ is zero. We conclude that

$$\frac{\mathrm{d}\mathbf{A}}{\mathrm{d}t} = \iiint_{\text{parcel}} (\mathbf{X}-\mathbf{x}_0)\times\Big(\frac{\mathrm{D}(\rho\,\mathbf{u})}{\mathrm{D}t} + \rho\,\mathbf{u}\,(\nabla\cdot\mathbf{u})\Big)\,\mathrm{d}V. \tag{2.7.7}$$

Expressing the material derivative inside the integrand in terms of Eulerian derivatives, and combining two terms, we obtain

$$\frac{\mathrm{d}\mathbf{A}}{\mathrm{d}t} = \iiint_{\text{parcel}} (\mathbf{X}-\mathbf{x}_0)\times\Big(\frac{\partial(\rho\,\mathbf{u})}{\partial t} + \nabla\cdot(\rho\,\mathbf{u}\otimes\mathbf{u})\Big)\,\mathrm{d}V. \tag{2.7.8}$$

Now using the continuity equation,

$$\frac{\partial\rho}{\partial t} + \nabla\cdot(\rho\,\mathbf{u}) = 0, \tag{2.7.9}$$

we obtain

$$\frac{\mathrm{d}\mathbf{A}}{\mathrm{d}t} = \iiint_{\text{parcel}} \rho\,(\mathbf{X} - \mathbf{x}_0) \times \Big(\frac{\partial \mathbf{u}}{\partial t} + \mathbf{u} \cdot \nabla \mathbf{u}\Big)\,\mathrm{d}V. \tag{2.7.10}$$

In terms of the material derivative of the velocity expressing the point particle acceleration, $\mathbf{a} \equiv \mathrm{D}\mathbf{u}/\mathrm{D}t$, we obtain

$$\frac{\mathrm{d}\mathbf{A}}{\mathrm{d}t} = \iiint_{\text{parcel}} \rho\,(\mathbf{X} - \mathbf{x}_0) \times \frac{\mathrm{D}\mathbf{u}}{\mathrm{D}t}\,\mathrm{d}V. \tag{2.7.11}$$

At this final stage, the point-particle position, $\mathbf{X}$, can be replaced by the position, $\mathbf{x}$, inside the parcel volume.

In fact, Eq. (2.7.11) arises directly from the general equation (1.9.17) by setting $\psi = (\mathbf{X} - \mathbf{x}_0) \times \mathbf{u}$ and expanding the material derivative of the outer product.

All equations presented in this section, including Eq. (2.7.11), apply for incompressible as well as compressible fluids.

2.7.5 Exercises

2.7.1 Derive Eq. (2.7.5).

2.7.2 Derive Eq. (2.7.11) from (2.7.2) using the differential form of the mass conservation equation, $\mathrm{D}\,(\rho\,\delta V)/\mathrm{D}t = 0$, where δV is a small material volume.

2.8 Newton's Law for the Angular Momentum

Invoking Newton's second law of motion for a material fluid parcel, we balance the rate of change of angular momentum, $\mathbf{A}$, with the torque due to surface and volume forces, by writing

$$\frac{\mathrm{d}\,\mathbf{A}}{\mathrm{d}t} = \iint_{\text{parcel}} (\mathbf{X} - \mathbf{x}_0) \times (\mathbf{n} \cdot \boldsymbol{\sigma})\,\mathrm{d}S + \iiint_{\text{parcel}} (\mathbf{X} - \mathbf{x}_0) \times (\rho\,\mathbf{g} + \iota\,\mathbf{b})\,\mathrm{d}V + \iiint_{\text{parcel}} \lambda\,\mathbf{c}\,\mathrm{d}V. \tag{2.8.1}$$

Note that the torque is computed with respect to the point $\mathbf{x}_0$ employed in the definition of the angular momentum. The vector field $\mathbf{c}$ in the last integral expresses a torque-inducing field associated with a physical constant, λ. A torque-inducing field can be an electrical field applied to a ferrofluid.

2.8.1 Instance of a General Law

A general natural law pertinent to the properties of a material parcel was stated in Eq. (1.6.5), repeated below for convenience,

$$\frac{\mathrm{d}}{\mathrm{d}t} \iiint_{\text{parcel}} \phi\,\mathrm{d}V = -\iint_{\text{parcel}} \mathbf{n} \cdot \boldsymbol{\xi}\,\mathrm{d}S + \iiint_{\text{parcel}} \dot{\pi}\,\mathrm{d}V + \iint_{\text{parcel}} \dot{\nu}\,\mathrm{d}S, \tag{2.8.2}$$

where ϕ is a suitable property field. Comparing this equation with Newton's law for the angular momentum expressed by (2.8.1), we set $\phi = \rho\,(\mathbf{X} - \mathbf{x}_0) \times \mathbf{u}$ and

$$\boldsymbol{\xi} = \boldsymbol{\sigma} \times (\mathbf{X} - \mathbf{x}_0), \qquad \dot{\pi} = (\mathbf{X} - \mathbf{x}_0) \times (\rho\,\mathbf{g} + \iota\,\mathbf{b}) + \lambda\,\mathbf{c}, \qquad \dot{\nu} = 0. \tag{2.8.3}$$

The last condition, $\dot{\nu} = 0$, ensures that angular momentum cannot be generated by long-range surface action. In Chap. 7, we will see that species angular momentum can be generated by molecular interception in a binary or multicomponent mixture.

2.8.2 Stress Tensor Spin

Now considering the first term on the right-hand side of (2.8.1), denoted by

$$\mathcal{F} \equiv \iint_{\text{parcel}} (\mathbf{X} - \mathbf{x}_0) \times (\mathbf{n} \cdot \boldsymbol{\sigma})\, \mathrm{d}S, \tag{2.8.4}$$

we define for convenience $\hat{\mathbf{X}} \equiv \mathbf{X} - \mathbf{x}_0$, and write

$$\mathcal{F}_i \equiv \epsilon_{ijk} \iint_{\text{parcel}} \hat{X}_j\, n_m\, \sigma_{mk}\, \mathrm{d}S. \tag{2.8.5}$$

Applying the divergence theorem to convert the surface integral into a volume integral, we obtain

$$\mathcal{F}_i \equiv \epsilon_{ijk} \iiint_{\text{parcel}} \Big(\frac{\partial X_j}{\partial x_m}\, \sigma_{mk} + \hat{X}_j\, \frac{\partial\, \sigma_{mk}}{\partial x_m} \Big)\, \mathrm{d}V. \tag{2.8.6}$$

Setting $\partial X_j / \partial x_m = \delta_{jm}$, simplifying, and reverting to vector notation, we obtain

$$\mathcal{F} = \iiint_{\text{parcel}} \big(\mathbf{s} + (\mathbf{X} - \mathbf{x}_0) \times \nabla \cdot \boldsymbol{\sigma}\,\big)\, \mathrm{d}V, \tag{2.8.7}$$

where $\mathbf{s}$ is a torque-like vector called the stress tensor spin with components

$$s_i \equiv \epsilon_{ijk}\, \sigma_{jk} = \epsilon_{ijk}\, \widetilde{\sigma}_{jk}, \tag{2.8.8}$$

$\widetilde{\boldsymbol{\sigma}} \equiv \frac{1}{2}\,(\boldsymbol{\sigma} - \boldsymbol{\sigma}^{\mathrm{T}})$ is the skew-symmetric part of the stress tensor, $\boldsymbol{\sigma}$, and the superscript T denotes the matrix transpose. We note that only the skew-symmetric part of $\boldsymbol{\sigma}$ contributes to the spin, $\mathbf{s}$. Conversely,

$$\widetilde{\sigma}_{ij} = \frac{1}{2}\, \epsilon_{ijk}\, s_k. \tag{2.8.9}$$

We have found that

$$\iint_{\text{parcel}} (\mathbf{X} - \mathbf{x}_0) \times (\mathbf{n} \cdot \boldsymbol{\sigma})\, \mathrm{d}S = \iiint_{\text{parcel}} \big(\mathbf{s} + (\mathbf{X} - \mathbf{x}_0) \times \nabla \cdot \boldsymbol{\sigma}\,\big)\, \mathrm{d}V, \tag{2.8.10}$$

where $\mathbf{s} = \nabla \times \boldsymbol{\sigma}$.

2.8.3 Cauchy Angular Momentum Equation

The right-hand side of (2.8.10) can be substituted for the first term on the right-hand side of (2.8.1) to yield a balance equation in terms of volume integrals alone. Using expression (2.7.11) for the parcel rate of change of angular momentum, and discarding the integral signs in (2.8.1), we derive the Cauchy angular-momentum equation

$$\rho\,(\mathbf{x}-\mathbf{x}_0)\times\frac{\mathrm{D}\mathbf{u}}{\mathrm{D}t}=\mathbf{s}+(\mathbf{x}-\mathbf{x}_0)\times\nabla\cdot\boldsymbol{\sigma}+(\mathbf{x}-\mathbf{x}_0)\times(\rho\,\mathbf{g}+\iota\,\mathbf{b})+\lambda\,\mathbf{c}. \tag{2.8.11}$$

Taking the cross product of each term in Cauchy's equation of motion (2.3.6) with the distance $\mathbf{x}-\mathbf{x}_0$, and comparing the result with (2.8.11), we conclude that

$$\mathbf{s}=-\lambda\,\mathbf{c}. \tag{2.8.12}$$

The angular momentum equation (2.8.11) then simplifies to

$$\rho\,(\mathbf{x}-\mathbf{x}_0)\times\frac{\mathrm{D}\mathbf{u}}{\mathrm{D}t}=(\mathbf{x}-\mathbf{x}_0)\times\nabla\cdot\boldsymbol{\sigma}+(\mathbf{x}-\mathbf{x}_0)\times(\rho\,\mathbf{g}+\iota\,\mathbf{b}). \tag{2.8.13}$$

The torque field associated with $\mathbf{c}$ appears implicitly in Eq. (2.8.13) by determining the antisymmetric part of the stress tensor. Substituting (2.8.12) into (2.8.9), we find that

$$\sigma_{ij}-\sigma_{ji}=-\tfrac{1}{2}\,\epsilon_{ijk}\,\lambda\,c_k, \tag{2.8.14}$$

where summation is implied over the repeated index, k.

2.8.4 Symmetry of the Stress Tensor

We have found that, in the absence of a torque-inducing field, the stress tensor is symmetric. In the presence of a torque-inducing field, the stress tensor is not symmetric, that is, $\boldsymbol{\sigma}\neq\boldsymbol{\sigma}^{\mathrm{T}}$.

2.8.5 Eulerian Form

To derive the Eulerian form of the angular momentum equation, we resolve the material of the velocity on the left-hand side of (2.8.13) into Eulerian derivatives, and obtain

$$\rho\,(\mathbf{x}-\mathbf{x}_0)\times\Big(\frac{\partial\mathbf{u}}{\partial t}+\mathbf{u}\cdot\nabla\mathbf{u}\Big)=(\mathbf{x}-\mathbf{x}_0)\times\nabla\cdot\boldsymbol{\sigma}+(\mathbf{x}-\mathbf{x}_0)\times(\rho\,\mathbf{g}+\iota\,\mathbf{b}), \tag{2.8.15}$$

which is merely the cross product of the Cauchy equation of motion with the vectorial distance $\mathbf{x}-\mathbf{x}_0$.

2.8.6 Conservative Form

An alternative statement of the angular momentum equation originating from the expression for the rate of change of a parcel's angular momentum given in (2.7.5) is

$$\frac{\partial(\rho\,(\mathbf{x}-\mathbf{x}_0)\times\mathbf{u})}{\partial t}+\nabla\cdot\big(\mathbf{u}\otimes(\rho\,(\mathbf{x}-\mathbf{x}_0)\times\mathbf{u})\big)$$
$$=(\mathbf{x}-\mathbf{x}_0)\times\nabla\cdot\boldsymbol{\sigma}+(\mathbf{x}-\mathbf{x}_0)\times(\rho\,\mathbf{g}+\iota\,\mathbf{b}). \tag{2.8.16}$$

The qualifier *conservative* refers to the divergence form of the second term on the left-hand side.

2.8.7 Angular Momentum Transport

To derive a transport equation for the angular momentum, we consider a control volume, $\mathcal{V}_c$, that lies entirely inside a fluid, and is bounded by a boundary or a collection of boundaries, $\mathcal{S}_c$, as shown in Fig. 1.3.

Applying the general transport equation (1.8.2) for the fluid angular momentum by setting $\phi = \rho\,(\mathbf{x}-\mathbf{x}_0)\times\mathbf{u}$, and using (2.8.1) to evaluate the last term on the right-hand side, we obtain an angular-momentum transport equation,

$$\iiint_{\mathcal{V}_c}\frac{\partial(\rho(\mathbf{x}-\mathbf{x}_0)\times\mathbf{u})}{\partial t}\,\mathrm{d}V=\iint_{\mathcal{S}_c}(\rho\,(\mathbf{x}-\mathbf{x}_0)\times\mathbf{u})\,(\mathbf{n}^{\mathrm{in}}\cdot\mathbf{u})\,\mathrm{d}S-\iint_{\mathcal{S}_c}(\mathbf{x}-\mathbf{x}_0)\times(\mathbf{n}^{\mathrm{in}}\cdot\boldsymbol{\sigma})\,\mathrm{d}S$$
$$+\iiint_{\mathcal{V}_c}\rho\,(\mathbf{x}-\mathbf{x}_0)\times\mathbf{g}\,\mathrm{d}V+\iiint_{\mathcal{V}_c}\iota\,(\mathbf{x}-\mathbf{x}_0)\times\mathbf{b}\,\mathrm{d}V+\iiint_{\mathcal{V}_c}\lambda\,\mathbf{c}\,\mathrm{d}V. \tag{2.8.17}$$

Note the presence of the external torque term on the right-hand side.

The transport equation (2.8.17) states that the rate of accumulation of angular momentum inside a control volume, expressed by the left-hand side, is determined by the following:

(a) The convective flow rate of angular momentum normal to the boundaries expressed by the fist term on the right-hand side.
(b) The torque exerted on the boundaries expressed by the second term on the right-hand side, including the minus sign in front of the integral.
(c) The torque due to the body force exerted on the fluid residing inside the control volume expressed by the third and fourth terms on the right-hand side.
(d) The torque due to an externally imposed torque field in the case of special fluids.

Thanks to Cauchy's equation of motion for the linear momentum and associated integral balances discussed earlier in this chapter, Eq. (2.8.17) and all previous balance equations derived in this section apply for any arbitrary point, $\mathbf{x}_0$.

2.8.8 Balance on an Open System

In the literature of transport phenomena, Eq. (2.8.17) is often accepted as a self-evident starting point for an open system in place of Newton's second law of motion for the angular momentum of a material parcel.

2.8.9 Exercises

2.8.1 Derive Eq. (2.8.7).

2.8.2 Simplify Eq. (2.8.17) for a fluid in rigid-body motion.

2.9 Mechanical Energy Balance

The mechanical energy of a fluid incorporates (a) the kinetic energy due to the motion of the fluid and (b) the potential energy due to a long-range body force field, such as the gravitational field or an electromagnetic field.

2.9.1 Velocity Projected on Acceleration

A mechanical energy balance arises by projecting Cauchy's equation of motion onto the fluid velocity vector at an arbitrary point in a flow. Integrating this balance over a control volume provides us with an expression for the rate of mechanical energy loss or gain due to the flow and reveals the mechanism of energy cascade and transport.

2.9.2 Cauchy Equation

A convenient starting point for developing the mechanical energy balance is the Eulerian form of Cauchy's equation of motion (2.3.8), repeated below for convenience,

$$\rho\,\Big(\frac{\partial \mathbf{u}}{\partial t} + \mathbf{u}\cdot\nabla\mathbf{u}\Big) = \nabla\cdot\boldsymbol{\sigma} + \rho\,\mathbf{g} + \iota\,\mathbf{b}, \tag{2.9.1}$$

which applies for incompressible or compressible fluids. We recall that **b** is an arbitrary body force field associated with the physical constant ι. The expression inside the parentheses on the left-hand side is the point-particle acceleration.

2.9.3 Velocity Projected on Acceleration

Projecting equation (2.9.1) onto the velocity, **u**, at a point in a flow, we obtain

$$\rho\,\Big(\mathbf{u}\cdot\frac{\partial \mathbf{u}}{\partial t} + (\mathbf{u}\cdot\nabla\mathbf{u})\cdot\mathbf{u}\Big) = \mathbf{u}\cdot(\nabla\cdot\boldsymbol{\sigma}) + (\rho\,\mathbf{g} + \iota\,\mathbf{b})\cdot\mathbf{u}. \tag{2.9.2}$$

The left-hand side is the density multiplied on the projection of the point-particle velocity onto the point-particle acceleration.

Rearranging the left-hand side, we obtain

$$\tfrac{1}{2}\,\rho\,\Big(\frac{\partial\,|\mathbf{u}|^2}{\partial t} + \mathbf{u}\cdot\nabla|\mathbf{u}|^2\Big) = \mathbf{u}\cdot(\nabla\cdot\boldsymbol{\sigma}) + (\rho\,\mathbf{g} + \iota\,\mathbf{b})\cdot\mathbf{u}, \tag{2.9.3}$$

where $|\mathbf{u}|^2 = \mathbf{u} \cdot \mathbf{u}$ is the square of the magnitude of the velocity. Resolving the first term on the right-hand side into two parts,

$$\mathbf{u} \cdot (\nabla \cdot \boldsymbol{\sigma}) = \nabla \cdot (\boldsymbol{\sigma} \cdot \mathbf{u}) - \boldsymbol{\sigma} : \nabla \mathbf{u}, \tag{2.9.4}$$

we obtain

$$\tfrac{1}{2}\,\rho\,\Big(\frac{\partial\,|\mathbf{u}|^2}{\partial t} + \mathbf{u} \cdot \nabla |\mathbf{u}|^2\Big) = \nabla \cdot (\boldsymbol{\sigma} \cdot \mathbf{u}) - \boldsymbol{\sigma} : \nabla \mathbf{u} + (\rho\,\mathbf{g} + \iota\,\mathbf{b}) \cdot \mathbf{u}, \tag{2.9.5}$$

where the double-dot product is explained next.

2.9.4 Double-Dot Product

The double-dot product of $\boldsymbol{\sigma}$ and $\nabla \mathbf{u}$ is a scalar defined as

$$\boldsymbol{\sigma} : \nabla \mathbf{u} \equiv \sigma_{ij}\,\frac{\partial u_j}{\partial x_i}, \tag{2.9.6}$$

where summation is implied over the repeated indices i and j. Using the rules of matrix multiplication, we find that

$$\boldsymbol{\sigma} : \nabla \mathbf{u} = \mathrm{trace}(\boldsymbol{\sigma}^{\mathrm{T}} \cdot \nabla \mathbf{u}) = \mathrm{trace}(\boldsymbol{\sigma} \cdot (\nabla \mathbf{u})^{\mathrm{T}}), \tag{2.9.7}$$

where T denotes the matrix transpose and the trace is the sum of the diagonal elements.

2.9.5 Mechanical Energy Loss or Gain

We will see that the double-dot product, $\boldsymbol{\sigma} : \nabla \mathbf{u}$, expresses the volumetric rate of mechanical energy loss or gain due to stresses developing in the fluid as a result of the motion. Constitutive equations for the stress tensor must be introduced to evaluate this term.

In the simplest circumstances of hydrostatics or ideal fluid flow, $\boldsymbol{\sigma} = -p\,\mathbf{I}$, where p is the pressure, we obtain $\boldsymbol{\sigma} : \nabla \mathbf{u} = -p\,\nabla \cdot \mathbf{u}$, where $\nabla \cdot \mathbf{u} = \mathrm{trace}(\nabla \mathbf{u})$ is the rate of expansion of the fluid.

2.9.6 Lagrangian Form

In terms of the material derivative, $\mathrm{D}/\mathrm{D}t$, Eq. (2.9.5) takes the form

$$\rho\,\frac{\mathrm{D}}{\mathrm{D}t}\big(\tfrac{1}{2}\,|\mathbf{u}|^2\big) = \nabla \cdot (\boldsymbol{\sigma} \cdot \mathbf{u}) - \boldsymbol{\sigma} : \nabla \mathbf{u} + (\rho\,\mathbf{g} + \iota\,\mathbf{b}) \cdot \mathbf{u}. \tag{2.9.8}$$

This equation can also be derived directly by projecting the velocity onto the Lagrangian form of the equation of motion at a point.

2.9.7 Conservative Form

Combining equation (2.9.5) with the continuity equation,

$$\frac{\partial \rho}{\partial t} + \nabla \cdot (\rho\, \mathbf{u}) = 0, \tag{2.9.9}$$

to modify the left-hand side, we obtain the conservative form

$$\frac{\partial}{\partial t}\left(\tfrac{1}{2}\,\rho\,|\mathbf{u}|^2\right) + \nabla \cdot \left(\tfrac{1}{2}\,\rho\,|\mathbf{u}|^2\,\mathbf{u}\right) = \nabla \cdot (\boldsymbol{\sigma} \cdot \mathbf{u}) - \boldsymbol{\sigma} : \nabla \mathbf{u} + (\rho\,\mathbf{g} + \iota\,\mathbf{b}) \cdot \mathbf{u}. \tag{2.9.10}$$

The physical meaning of each term will become evident later in this section when we formulate a corresponding integral balance.

2.9.8 Deviatoric Stress Tensor

The deviatoric stress tensor was defined in (2.1.8) as

$$\boldsymbol{\sigma}^{\text{dev}} \equiv p\,\mathbf{I} + \boldsymbol{\sigma}, \tag{2.9.11}$$

where p is the pressure. By definition, the deviatoric stress tensor is identically zero for an ideal fluid.

In terms of the deviatoric stress tensor, the first term on the right-hand side of (2.9.8) takes the form

$$\nabla \cdot (\boldsymbol{\sigma} \cdot \mathbf{u}) = -\nabla \cdot (p\,\mathbf{u}) + \nabla \cdot (\boldsymbol{\sigma}^{\text{dev}} \cdot \mathbf{u}), \tag{2.9.12}$$

and the second term on the right-hand sides of (2.9.8) takes the form

$$\boldsymbol{\sigma} : \nabla \mathbf{u} = -p\,\nabla \cdot \mathbf{u} + \boldsymbol{\sigma}^{\text{dev}} : \nabla \mathbf{u}. \tag{2.9.13}$$

We recall that, by definition, the pressure is the negative one-third of the trace of the stress tensor.

The negative of the first term on the right-hand side of (2.9.13) is reminiscent of the term $p\,\mathrm{d}v$ term in classical thermodynamics, where $\mathrm{d}v$ is a change in volume. This term is identically zero in the case of an incompressible fluid due to the vanishing rate of expansion, $\nabla \cdot \mathbf{u} = 0$.

Subtracting equation (2.9.13) from Eq. (2.9.12), and simplifying, we obtain

$$\mathbf{u} \cdot (\nabla \cdot \boldsymbol{\sigma}) = -\mathbf{u} \cdot \nabla p + \nabla \cdot (\boldsymbol{\sigma}^{\text{dev}} \cdot \mathbf{u}) - \boldsymbol{\sigma}^{\text{dev}} : \nabla \mathbf{u}. \tag{2.9.14}$$

The first term on the left-hand side is the negative of the directional derivative of the pressure along the velocity, multiplied by the magnitude of the velocity. The rest of the terms do not appear for an ideal fluid.

2.9.9 Integral Mechanical Energy Balance

An integral mechanical energy balance arises by integrating the differential mechanical energy balance (2.9.10) over a control volume, $\mathcal{V}_c$, that is bounded by a surface or a collection of surfaces,

$\mathcal{S}_c$, as shown in Fig. 1.3. Using the divergence theorem to convert the volume integrals involving a divergence into surface integrals, we obtain

$$\begin{aligned}\iiint_{\mathcal{V}_c} \frac{\partial}{\partial t}\left(\tfrac{1}{2}\,\rho\,|\mathbf{u}|^2\right) \mathrm{d}V &= \iint_{\mathcal{S}_c} \left(\tfrac{1}{2}\,\rho\,|\mathbf{u}|^2\right) \mathbf{n}^{\mathrm{in}} \cdot \mathbf{u}\, \mathrm{d}S - \iint_{\mathcal{S}_c} \mathbf{f}^{\mathrm{in}} \cdot \mathbf{u}\, \mathrm{d}S \\ &- \iiint_{\mathcal{V}_c} \boldsymbol{\sigma} : \nabla\mathbf{u}\, \mathrm{d}V + \iiint_{\mathcal{V}_c} \rho\, \mathbf{g} \cdot \mathbf{u}\, \mathrm{d}V + \iiint_{\mathcal{V}_c} \iota\, \mathbf{b} \cdot \mathbf{u}\, \mathrm{d}V,\end{aligned} \tag{2.9.15}$$

where $\mathbf{f}^{\mathrm{in}} = \mathbf{n}^{\mathrm{in}} \cdot \boldsymbol{\sigma}$ is the traction and $\mathbf{n}^{\mathrm{in}}$ is the normal unit vector pointing *into* the control volume.

The left-hand side of (2.9.15) expresses the rate of accumulation of kinetic energy inside the control volume. The five terms on the right-hand side of (2.9.15) represent, respectively,

(a) The rate of convective supply of kinetic energy *into* the control volume by fluid motion.
(b) The rate of working of the traction exerted at the boundary of the control volume.
(c) The rate of recoverable (reversible) or irrecoverable (irreversible) volumetric mechanical energy loss or gain due to stresses developing in the fluid.
(d) The rate of working against the body force. This term is equal to the negative of the rate of change of potential energy due to gravity.
(e) The rate of working against an arbitrary time- or space-dependent body force.

The kinetic-energy transport equation (2.9.15) can be written directly only in hindsight due to the specific form of the third term on the right-hand side. This term can be traced back to the notion of stress and the application of Newton's second law of motion for a material parcel.

2.9.10 Energy Conversion

We have found that the rate of volumetric mechanical energy loss or gain due to the motion of the fluid is given by the double-dot product of the stress tensor and the velocity gradient tensor, $\boldsymbol{\sigma} : \nabla\mathbf{u}$.

In Chap. 3, we will see that dissipated mechanical energy is converted into thermal energy by way of *interior work* that escapes across the boundaries of a fluid parcel or control volume, or raises the temperature of the fluid. This transfer of energy is analogous to that occurring during the sliding of an object on a surface in the presence of friction.

2.9.11 Exercises

2.9.1 Derive identity (2.9.4).

2.9.2 Apply the integral mechanical energy balance for a fluid in time-dependent translation.

3 Energy Balances in a Homogeneous Fluid

The total energy of the fluid residing inside a material fluid parcel is comprised of three constituents: (a) the potential energy due to an external body force, (b) the kinetic energy due to the motion of the fluid, and (c) the internal energy defined in the sense of thermodynamics, as discussed in Appendix B. The sum of the potential and the kinetic energies constitutes the mechanical energy. Although the external force field responsible for the potential energy is typically the gravitational field, other force fields can be encountered.

Expressions for the rate of change of the aforementioned energies can be obtained by combining the equation of motion discussed in Chap. 2 with the laws of thermodynamics under the auspices of the continuum. We will see that mechanical energy is converted into internal energy, ultimately raising the temperature of a viscous fluid and thereby necessitating energy expenditure to drive or sustain a flow. Entropy enters the transport equations by way of Gibbs' fundamental equation of thermodynamics discussed in Appendix B.

3.1 Mechanical Energy of a Fluid Parcel

The mechanical energy of a material fluid parcel consists of the potential energy due to a force field and the kinetic energy due to the motion of the fluid. As point particles move in the domain of a flow, their constituent energies evolve at individual and combined rates.

3.1.1 Potential Energy

The potential energy of a fluid parcel due to the gravitational acceleration field, $\mathbf{g}$, is denoted as $\mathcal{E}_{\text{pot}}^{g}$, and the potential energy due to any other arbitrary body force field, $\mathbf{b}$, is denoted as $\mathcal{E}_{\text{pot}}^{b}$. The total potential energy is

$$\mathcal{E}_{\text{pot}} = \mathcal{E}_{\text{pot}}^{g} + \mathcal{E}_{\text{pot}}^{b}. \tag{3.1.1}$$

The distinction between the gravitational potential energy and some other type of potential energy is motivated by the common approximation that the gravitational acceleration is spatially uniform, whereas an arbitrary body force is allowed to vary over the domain occupied by a fluid.

C. Pozrikidis, *Transport Processes Primer*,
https://doi.org/10.1007/978-1-4939-9909-5_3

3.1.2 Rate of Change of the Potential Energy

The rate of change of the potential energy of a material parcel due to the gravitational field is given by

$$\frac{d\mathcal{E}^g_{\text{pot}}}{dt} = - \iiint_{\text{parcel}} \rho\, \mathbf{g} \cdot \mathbf{u}\, dV, \tag{3.1.2}$$

where ρ is the fluid density and $\mathbf{g}$ is the gravitational acceleration. The right-hand side is the rate of working against the gravitational force field. The potential energy decreases in free gravitational fall, that is, when a point particle moves in the direction of $\mathbf{g}$ so that $\mathbf{g} \cdot \mathbf{u} > 0$.

The rate of change of the potential energy due to an arbitrary distributed force field, $\mathbf{b}$, is given by the corresponding expression

$$\frac{d\mathcal{E}^b_{\text{pot}}}{dt} = - \iiint_{\text{parcel}} \iota\, \mathbf{b} \cdot \mathbf{u}\, dV, \tag{3.1.3}$$

where ι is a density-like physical constant associated with $\mathbf{b}$. The right-hand side is positive when a point particle moves against the body force, that is, when $\mathbf{b} \cdot \mathbf{u} < 0$, and negative otherwise.

Combining the preceding expressions, we find that the rate of change of a parcel's potential energy is

$$\frac{d\mathcal{E}_{\text{pot}}}{dt} = - \iiint_{\text{parcel}} (\rho\, \mathbf{g} + \iota\, \mathbf{b}) \cdot \mathbf{u}\, dV, \tag{3.1.4}$$

where ρ, ι, $\mathbf{b}$, and $\mathbf{u}$ may change in time and over the parcel volume.

3.1.3 Potential Energy Due to Gravity

Assuming that the gravitational acceleration, $\mathbf{g}$, is constant over the parcel volume, we find that the potential energy due to the gravitational field is given by

$$\mathcal{E}^g_{\text{pot}} = - \iiint_{\text{parcel}} \rho\, \mathbf{g} \cdot \mathbf{X}\, dV, \tag{3.1.5}$$

where $\mathbf{X}$ is the position of a point particle inside the parcel, now also regarded as the position vector inside the parcel. The negative sign before the integral of the potential energy takes into consideration that potential energy is gained when a point particle elevates against the body force (hiking to the top of a mountain). An arbitrary constant can be added to the right-hand side of (3.1.5).

To confirm expression (3.1.5), we take the derivative of the potential energy integral, transfer the derivative inside the integral as a material derivative, recall that, because of mass conservation, $D(\rho\, \delta V)/Dt = 0$, where δV is a small material volume, set $D\mathbf{X}/Dt = \mathbf{u}$, and treat $\mathbf{g}$ as a constant to recover (3.1.2).

Substituting (3.1.5) into (3.1.4), we obtain

$$\frac{d\mathcal{E}_{\text{pot}}}{dt} = - \frac{d}{dt}\Big(\iiint_{\text{parcel}} \rho\, \mathbf{g} \cdot \mathbf{X}\, dV \Big) - \iiint_{\text{parcel}} \iota\, \mathbf{b} \cdot \mathbf{u}\, dV. \tag{3.1.6}$$

This expression will be used in Sect. 3.2 in conjunction with the first law of thermodynamics.

3.1.4 Potential Energy Due to an Arbitrary Force Field

The potential energy due to the arbitrary force field is given by

$$\mathcal{E}_{\text{pot}}^{b} = -\iiint_{\text{parcel}} \iota f_{\text{pot}}^{b}(\mathbf{X}, t)\, \mathrm{d}V, \tag{3.1.7}$$

where the function $f_{\text{pot}}^{b}(\mathbf{x})$ evolves according to the differential equation

$$\frac{\mathrm{D}}{\mathrm{D}t}\Big(\frac{\iota}{\rho}\, f_{\text{pot}}^{b}(\mathbf{x}, t)\Big) = \frac{\iota}{\rho}\, \mathbf{b} \cdot \mathbf{u}. \tag{3.1.8}$$

When $\iota = \rho$ and $\mathbf{b} = \mathbf{g}$, we obtain $f_{\text{pot}}^{g}(\mathbf{X}, t) = \mathbf{g} \cdot \mathbf{X}$.

Given the additional complication of extracting the potential energy associated with $\mathbf{b}$ in its general form, we prefer to treat it indirectly in terms of its rate of change.

3.1.5 Kinetic Energy

The instantaneous kinetic energy of a material parcel is given by

$$\mathcal{E}_{\text{kin}} \equiv \frac{1}{2} \iiint_{\text{parcel}} \rho\, |\mathbf{u}|^2\, \mathrm{d}V, \tag{3.1.9}$$

where $|\mathbf{u}|^2 = \mathbf{u} \cdot \mathbf{u}$ is the square of the magnitude of the velocity. Taking the time derivative of this expression, transferring the derivative inside the integral as a material derivative, and recalling once again that, because of mass conservation, $\mathrm{D}(\rho\, \delta V)/\mathrm{D}t = 0$, we obtain

$$\frac{\mathrm{d}\mathcal{E}_{\text{kin}}}{\mathrm{d}t} = \frac{1}{2} \iiint_{\text{parcel}} \rho\, \frac{\mathrm{D}(\mathbf{u} \cdot \mathbf{u})}{\mathrm{D}t}\, \mathrm{d}V. \tag{3.1.10}$$

Carrying out the differentiation inside the integral, we obtain

$$\frac{\mathrm{d}\mathcal{E}_{\text{kin}}}{\mathrm{d}t} = \iiint_{\text{parcel}} \rho\, \mathbf{u} \cdot \frac{\mathrm{D}\mathbf{u}}{\mathrm{D}t}\, \mathrm{d}V. \tag{3.1.11}$$

The integrand involves the projection of the velocity, $\mathbf{u}$, onto the point-particle acceleration, $\mathrm{D}\mathbf{u}/\mathrm{D}t$. The kinetic energy does not change when the acceleration is perpendicular to the velocity.

Now we use the Cauchy equation of motion to express the acceleration, $\mathrm{D}\mathbf{u}/\mathrm{D}t$, in terms of the divergence of the stress tensor and the body force, obtaining

$$\frac{\mathrm{d}\mathcal{E}_{\text{kin}}}{\mathrm{d}t} = \iiint_{\text{parcel}} \mathbf{u} \cdot (\nabla \cdot \boldsymbol{\sigma})\, \mathrm{d}V + \iiint_{\text{parcel}} (\rho\, \mathbf{g} + \iota \mathbf{b}) \cdot \mathbf{u}\, \mathrm{d}V, \tag{3.1.12}$$

for a compressible or incompressible fluid.

3.1.6 Mechanical Energy

The last integral in (3.1.12) is equal to $-\mathrm{d}\mathcal{E}_{\mathrm{pot}}/\mathrm{d}t$ according to (3.1.4), yielding the rate of chance of the mechanical energy,

$$\mathcal{E}_{\mathrm{mech}} = \mathcal{E}_{\mathrm{kin}} + \mathcal{E}_{\mathrm{pot}}, \tag{3.1.13}$$

as a single integral,

$$\frac{\mathrm{d}\mathcal{E}_{\mathrm{mech}}}{\mathrm{d}t} = \iiint_{\mathrm{parcel}} \mathbf{u} \cdot (\boldsymbol{\nabla} \cdot \boldsymbol{\sigma})\, \mathrm{d}V. \tag{3.1.14}$$

Splitting the integrand on the right-hand side into two parts, and rearranging, we obtain

$$\frac{\mathrm{d}\mathcal{E}_{\mathrm{mech}}}{\mathrm{d}t} = \iiint_{\mathrm{parcel}} \boldsymbol{\nabla} \cdot (\boldsymbol{\sigma} \cdot \mathbf{u})\, \mathrm{d}V - \iiint_{\mathrm{parcel}} \boldsymbol{\sigma} : \boldsymbol{\nabla}\mathbf{u}\, \mathrm{d}V, \tag{3.1.15}$$

where the double-dot product is a scalar given by $\boldsymbol{\sigma} : \boldsymbol{\nabla}\mathbf{u} \equiv \sigma_{ij}\, \partial u_j/\partial x_i$.

Next, we use the divergence theorem to convert the first integral on the right-hand side of (3.1.15) into a surface integral. Rearranging the resulting expression, we derive a mechanical energy balance expressed by the equation

$$\frac{\mathrm{d}\mathcal{E}_{\mathrm{mech}}}{\mathrm{d}t} = \iint_{\mathrm{parcel}} \mathbf{f} \cdot \mathbf{u}\, \mathrm{d}S - \iiint_{\mathrm{parcel}} \boldsymbol{\sigma} : \boldsymbol{\nabla}\mathbf{u}\, \mathrm{d}V, \tag{3.1.16}$$

where $\mathbf{f} = \mathbf{n} \cdot \boldsymbol{\sigma}$ is the traction and $\mathbf{n}$ is the normal unit vector pointing *outward* from the parcel.

3.1.7 Mechanical Energy Loss or Gain

The first integral on the right-hand side of (3.1.16) is the rate of working of the traction on the parcel surface. It then follows that the second integral on the right-hand side expresses the rate of mechanical energy loss or gain,

$$\dot{\mathcal{E}}_{\mathrm{mech\ loss}} = \iiint_{\mathrm{parcel}} \boldsymbol{\sigma} : \boldsymbol{\nabla}\mathbf{u}\, \mathrm{d}V, \tag{3.1.17}$$

consistent with the third term on the right-hand side of (2.9.15) pertaining to a control volume. The gain is the negative of the loss.

Subject to this definition, the balance equation (3.1.16) takes the simple form

$$\frac{\mathrm{d}\mathcal{E}_{\mathrm{mech}}}{\mathrm{d}t} = \iint_{\mathrm{parcel}} \mathbf{f} \cdot \mathbf{u}\, \mathrm{d}S - \dot{\mathcal{E}}_{\mathrm{mech\ loss}}. \tag{3.1.18}$$

We recall that this equation applies to incompressible as well as compressible fluids and for symmetric as well as asymmetric stress tensors.

3.1.8 Energy Conversion

In Sect. 3.2, we will see that part of the mechanical energy dissipated inside a parcel escapes by conduction across the parcel surface. The remainder of the dissipated energy raises the internal energy of the fluid residing inside the parcel.

3.1.9 Exercises

3.1.1 Confirm that the units of the three terms in (3.1.18) are the same.

3.1.2 Confirm that Eqs. (3.1.7) and (3.1.8) are consistent with (3.1.3).

3.2 First Law of Thermodynamics

The first law of thermodynamics prescribes that any change in the internal energy of a material body, $\mathcal{E}_{\text{int}}$, is given by

$$\Delta\mathcal{E}_{\text{int}} = Q_{\text{added}} - W_{\text{performed}}, \tag{3.2.1}$$

where Δ denotes the change, Q_{added} is heat added to the body, and $W_{\text{performed}}$ is work performed by the body. The heat added to the body is the negative of the heat released by the body.

3.2.1 Work Performed and Work Received

Work performed by a body is the negative of work received by the body from the ambience,

$$W_{\text{performed}} = -W_{\text{received}}. \tag{3.2.2}$$

The first law of thermodynamics then takes the form

$$\Delta\mathcal{E}_{\text{int}} = Q_{\text{added}} + W_{\text{received}}. \tag{3.2.3}$$

Not surprisingly, heat added to a body and work received by the body act to increase the internal energy of the body.

3.2.2 First Law of Thermodynamics for a Fluid Parcel

The internal energy of a homogeneous fluid accounts for (a) the kinetic energy associated with molecular vibrations and rotational motions in a frame of reference moving with the local macroscopic fluid velocity, **u**, and (b) the energy associated with a force interaction potential between the atoms or molecules comprising the fluid. An intermolecular force interaction potential is absent in the case of an ideal gas.

In the present context of a fluid in motion, the first law of thermodynamics prescribes that the rate of change of the internal energy of a material fluid parcel is given by

$$\frac{\mathrm{d}\mathcal{E}_{\text{int}}}{\mathrm{d}t} = \dot{Q}_{\text{cond}} + \dot{Q}_{\text{int gen}} + \dot{Q}_{\text{surf gen}} + \dot{\mathcal{E}}_{\text{mech loss}}, \tag{3.2.4}$$

where a dot over a variable should be read as *rate of*, subject to following definitions:

- $\mathcal{E}_{\text{int}}$ is the parcel internal energy.
- $\dot{Q}_{\text{cond}}$ is the rate of inward heat conduction across the parcel surface.

- $\dot{\mathcal{Q}}_{\text{int gen}}$ is the rate of interior energy generation incorporating types of energy that are not included in the nominal internal energy of the fluid. Examples are energies due to electromagnetic exposure, nuclear processes, volume radiation, or a chemical reaction.
- $\dot{\mathcal{Q}}_{\text{surf gen}}$ is the rate of surface heat generation due, for example, to boundary radiation.
- $\dot{\mathcal{E}}_{\text{mech loss}}$ is the rate of mechanical energy loss defined in (3.1.17).

Note that the first and third terms on the right-hand side of (3.2.4) pertain to the surface of the parcel, whereas the second and fourth terms pertain to the volume of the parcel.

3.2.3 Heat Added and Work Received

Comparing (3.2.1) with (3.2.4), we deduce that the rate of heat addition, $\dot{Q}_{\text{added}}$, is mediated by the first three terms on the right-hand side of (3.2.4),

$$\dot{Q}_{\text{added}} = \dot{Q}_{\text{cond}} + \dot{\mathcal{Q}}_{\text{int gen}} + \dot{\mathcal{Q}}_{\text{surf gen}}, \tag{3.2.5}$$

while the rate of work received is mediated by the last term on the right-hand side of (3.2.4), that is,

$$\dot{W}_{\text{received}} = -\dot{W}_{\text{performed}} = \dot{\mathcal{E}}_{\text{mech loss}}. \tag{3.2.6}$$

Work is transferred from the macroscopic or integral (observable) to the microscopic or differential (molecular) level by way of the mechanical energy loss term, $\dot{\mathcal{E}}_{\text{mech loss}}$.

3.2.4 Conductive Heat Flux

With regard to heat conduction, we introduce the conductive heat or thermal flux, $\mathbf{q}$, and write

$$\dot{Q}_{\text{cond}} = -\iint_{\text{parcel}} \mathbf{n} \cdot \mathbf{q} \, \mathrm{d}S, \tag{3.2.7}$$

where $\mathbf{n}$ is the normal unit vector pointing outward from the parcel. The heat flux can be regarded as the counterpart of the velocity for heat transport.

3.2.5 Interior Heat Generation

With regard to the rate of interior heat generation, we write

$$\dot{\mathcal{Q}}_{\text{int gen}} = \iiint_{\text{parcel}} \dot{q} \, \mathrm{d}V, \tag{3.2.8}$$

where $\dot{q}$ is the local volumetric rate of interior heat generation with units of energy over volume divided by time elapsed.

3.2.6 Surface Heat Generation

With regard to the rate of surface heat generation, we write

$$\dot{\mathcal{Q}}_{\text{surf gen}} = \iint_{\text{parcel}} \dot{q}_{\text{surf}} \, \mathrm{d}S, \tag{3.2.9}$$

where $\dot{q}_{\text{surf}}$ is the local surface rate of heat generation with units of energy over surface divided by time elapsed.

3.2.7 Specific Internal Energy

The internal energy of a fluid parcel can be expressed in terms of the specific internal energy, u, as

$$\mathcal{E}_{\text{int}} = \iiint_{\text{parcel}} \rho \, u \, \mathrm{d}V, \tag{3.2.10}$$

where ρ is the density. The specific internal energy is the internal energy divided by the mass of an infinitesimal fluid parcel.

3.2.8 Energy Balance

Adding Eq. (3.2.4) to the parcel mechanical energy balance (3.1.18), repeated below for convenience,

$$\frac{\mathrm{d}\mathcal{E}_{\text{mech}}}{\mathrm{d}t} = \iint_{\text{parcel}} \mathbf{f} \cdot \mathbf{u} \, \mathrm{d}S - \dot{\mathcal{E}}_{\text{mech loss}}, \tag{3.2.11}$$

and recalling that the mechanical energy is the sum of the kinetic and potential energies, we obtain an evolution equation for the total energy of the parcel consisting of the kinetic energy, the potential energy due to all external body forces, and the internal energy,

$$\mathcal{E}_{\text{tot}} = \mathcal{E}_{\text{kin}} + \mathcal{E}_{\text{pot}} + \mathcal{E}_{\text{int}}, \tag{3.2.12}$$

in the form

$$\frac{\mathrm{d}\mathcal{E}_{\text{tot}}}{\mathrm{d}t} = \dot{\mathcal{Q}}_{\text{cond}} + \iint_{\text{parcel}} \mathbf{f} \cdot \mathbf{u} \, \mathrm{d}S + \dot{\mathcal{Q}}_{\text{int gen}} + \dot{\mathcal{Q}}_{\text{surf gen}}. \tag{3.2.13}$$

Note that the mechanical loss term disappears in the addition.

In the absence of interior heat generation associated with an external action, $\dot{\mathcal{Q}}_{\text{int gen}} = 0$, only surface integrals appear on the right-hand side of the total-energy balance equation (3.2.13). Consequently, the total energy of the parcel may change only by way of transport across the parcel surface or work done by stresses over the parcel surface.

Table 3.1 Equations determining the rate of change of the energy of a fluid parcel; the third equation can be derived from the first two equations; the second equation can be derived from the first and second equations

Cauchy equation of motion	$\frac{d\mathcal{E}_{mech}}{dt} = \iint_{parcel} \mathbf{f} \cdot \mathbf{u} \, dS - \dot{\mathcal{E}}_{mech\ loss}$
First law of thermodynamics	$\frac{d\mathcal{E}_{int}}{dt} = \dot{\mathcal{Q}}_{cond} + \dot{\mathcal{Q}}_{int\ gen} + \dot{\mathcal{Q}}_{surf\ gen} + \dot{\mathcal{E}}_{mech\ loss}$
First law of thermodynamics	$\frac{d\mathcal{E}_{tot}}{dt} = \dot{\mathcal{Q}}_{cond} + \iint_{parcel} \mathbf{f} \cdot \mathbf{u} \, dS + \dot{\mathcal{Q}}_{int\ gen} + \dot{\mathcal{Q}}_{surf\ gen}$

3.2.9 First Law of Thermodynamics for a Closed System

The total energy balance (3.2.13) was derived from the internal energy balance (3.2.4) by way of the parcel mechanical energy balance (3.1.18). The inverse derivation could also have been carried out. Specifically, Eq. (3.2.13) could have been written at the outset based on the first law of thermodynamics for a closed system that does not allow mass to enter across the boundaries, as discussed in the next paragraph. The parcel mechanical energy balance (3.1.18) could then be invoked to derive internal energy balance (3.2.4).

The first law of thermodynamics for a closed system prescribes that the total energy of a material body or parcel—consisting of the kinetic, the potential, and the internal energy—changes because of heat entering through the surface of the body or parcel, mechanical work done on the surface of the body or parcel, and interior or surface heat generation or removal due to an external action.

Three fundamental equations of energy transport are summarized in Table 3.1 with reference to a material parcel.

3.2.10 Specific Form

Substituting into the total energy balance (3.2.13) expression (3.2.7) for the rate of surface conduction, $\dot{\mathcal{Q}}_{cond}$, expression (3.2.8) for the rate of interior heat generation, $\dot{\mathcal{Q}}_{int\ gen}$, and expression (3.2.9) for the rate of surface heat generation, $\dot{\mathcal{Q}}_{surf\ gen}$, we obtain

$$\frac{d\mathcal{E}_{tot}}{dt} = -\iint_{parcel} \mathbf{n} \cdot \mathbf{q} \, dS + \iint_{parcel} \mathbf{f} \cdot \mathbf{u} \, dS + \iiint_{parcel} \dot{q} \, dV + \iint_{parcel} \dot{q}_{surf} \, dS. \tag{3.2.14}$$

Expressing the traction in terms of the stress tensor, $\mathbf{f} = \mathbf{n} \cdot \boldsymbol{\sigma}$, and rearranging the first two terms on the right-hand side, we obtain

$$\frac{d\mathcal{E}_{tot}}{dt} = -\iint_{parcel} \mathbf{n} \cdot (\mathbf{q} - \boldsymbol{\sigma} \cdot \mathbf{u}) \, dS + \iiint_{parcel} \dot{q} \, dV + \iint_{parcel} \dot{q}_{surf} \, dS, \tag{3.2.15}$$

where the unit normal vector, $\mathbf{n}$, points outward from the parcel. The term $\mathbf{q} - \boldsymbol{\sigma} \cdot \mathbf{u}$ inside the first integral on the right-hand side expresses a compound thermal–hydrodynamic energy flux.

3.2.11 Instance of a General Form

A general natural law pertinent to the properties of a material parcel was stated in Eq. (1.6.5), repeated below for convenience,

$$\frac{\mathrm{d}}{\mathrm{d}t}\iiint_{\text{parcel}} \phi \,\mathrm{d}V = -\iint_{\text{parcel}} \mathbf{n}\cdot\boldsymbol{\xi}\,\mathrm{d}V + \iiint_{\text{parcel}} \dot{\pi}\,\mathrm{d}V + \iint_{\text{parcel}} \dot{\nu}\,\mathrm{d}S. \tag{3.2.16}$$

Comparing this equation with the total energy balance expressed by (3.2.15), we set

$$\phi = \rho\,(\tfrac{1}{2}\,|\mathbf{u}|^2 - \mathbf{g}\cdot\mathbf{X} - \frac{\iota}{\rho}\,f^b_{\text{pot}} + u) \tag{3.2.17}$$

and

$$\boldsymbol{\xi} = \mathbf{q} - \boldsymbol{\sigma}\cdot\mathbf{u}, \qquad \dot{\pi} = \dot{q}, \qquad \dot{\nu} = \dot{q}_{\text{surf}}. \tag{3.2.18}$$

The expression for ϕ arises by using expression (3.2.10) for the internal energy, expressions (3.1.6) and (3.1.7) for the potential energy, and expression (3.1.9) for the kinetic energy encapsulated in $\mathcal{E}_{\text{tot}}$.

3.2.12 Exercise

3.2.1 Discuss whether nuclear disintegration can be a source of interior heat generation.

3.3 Total-Energy Transport

The compound specific energy of the fluid, including the kinetic energy, the potential energy associated with gravity, and the internal energy, is defined as

$$\vartheta \equiv \tfrac{1}{2}\,|\mathbf{u}|^2 - \mathbf{g}\cdot\mathbf{X} + u, \tag{3.3.1}$$

where $\mathbf{X}$ is the position of a point particle inside a material parcel and u is the specific internal energy. In the absence of an arbitrary body force other than gravity, $\mathbf{b} = \underline{0}$, the compound specific energy is the total specific energy.

The exclusion of an arbitrary body force $\mathbf{b}$ from ϑ is due to the general unavailability of an explicit expression for the associated potential energy. If such an expression was available, ϑ could be redefined to include the effect of the arbitrary body force associated with the physical constant ι.

3.3.1 Parcel Balance Equation

To derive a parcel balance equation for ϑ, we substitute into the total energy balance (3.2.13) the following expressions: expression (3.2.7) for the rate of surface conduction, $\dot{\mathcal{Q}}_{\text{cond}}$; expression (3.2.8) for the rate of interior heat generation, $\dot{\mathcal{Q}}_{\text{int gen}}$; expression (3.2.9) for the rate of surface heat generation, $\dot{\mathcal{Q}}_{\text{surf gen}}$; expression (3.2.10) for the internal energy encapsulated in $\mathcal{E}_{\text{tot}}$; expression (3.1.6) for the potential energy encapsulated in $\mathcal{E}_{\text{tot}}$; and expression (3.1.9) for the kinetic energy encapsulated in $\mathcal{E}_{\text{tot}}$. Rearranging the resulting equation, we obtain an explicit parcel balance equation,

$$\frac{\mathrm{d}}{\mathrm{d}t}\iiint_{\text{parcel}} \rho\,\vartheta\,\mathrm{d}V = -\iint_{\text{parcel}} \mathbf{n}\cdot\mathbf{q}\,\mathrm{d}S + \iint_{\text{parcel}} \mathbf{f}\cdot\mathbf{u}\,\mathrm{d}S + \iiint_{\text{parcel}} \iota\,\mathbf{b}\cdot\mathbf{u}\,\mathrm{d}V + \iiint_{\text{parcel}} \dot{q}\,\mathrm{d}V + \iint_{\text{parcel}} \dot{q}_{\text{surf}}\,\mathrm{d}S, \tag{3.3.2}$$

where $\mathbf{f} = \mathbf{n}\cdot\boldsymbol{\sigma}$ is the traction over the parcel surface and $\mathbf{n}$ is the unit normal vector pointing outward from the parcel. Note that the mechanical loss term does *not* appear in this equation.

3.3.2 Eulerian Form

The associated Eulerian form arises by using the Reynolds transport equation (1.7.7) for the property $\phi = \rho\vartheta$ to restate the left-hand side of (3.3.2), obtaining the transport equation

$$\begin{aligned}\iiint_{\text{parcel}} \frac{\partial(\rho\,\vartheta)}{\partial t}\,\mathrm{d}V + \iint_{\text{parcel}} \rho\,\vartheta\,\mathbf{n}\cdot\mathbf{u}\,\mathrm{d}S = -\iint_{\text{parcel}} \mathbf{n}\cdot\mathbf{q}\,\mathrm{d}S \\ + \iint_{\text{parcel}} \mathbf{f}\cdot\mathbf{u}\,\mathrm{d}S + \iiint_{\text{parcel}} \iota\,\mathbf{b}\cdot\mathbf{u}\,\mathrm{d}V + \iiint_{\text{parcel}} \dot{q}\,\mathrm{d}V + \iint_{\text{parcel}} \dot{q}_{\text{surf}}\,\mathrm{d}S.\end{aligned} \tag{3.3.3}$$

The effect of surface radiation, expressed by the last integral on the right-hand side, will be disregarded in the forthcoming analysis.

3.3.3 Evolution Equations

Applying the divergence theorem to convert the two surface integrals into volume integrals over the parcel on the right-hand side of (3.3.3), and discarding the volume integration signs over the arbitrary parcel in all terms, we obtain the conservative form of an energy evolution equation,

$$\frac{\partial(\rho\,\vartheta)}{\partial t} + \nabla\cdot\big(\rho\,\vartheta\,\mathbf{u}\big) = -\nabla\cdot\mathbf{q} + \nabla\cdot(\boldsymbol{\sigma}\cdot\mathbf{u}) + \iota\,\mathbf{b}\cdot\mathbf{u} + \dot{q}, \tag{3.3.4}$$

where

$$\vartheta \equiv \frac{1}{2}\,|\mathbf{u}|^2 - \mathbf{g}\cdot\mathbf{x} + u \tag{3.3.5}$$

is a specific energy and $\mathbf{x}$ is the position vector.

The first two terms on the right-hand side of (3.3.4) can be transferred inside the divergence of the second term on the left-hand side, yielding

$$\frac{\partial(\rho\,\vartheta)}{\partial t} + \nabla\cdot\big(\rho\,\vartheta\,\mathbf{u} + \mathbf{q} - \boldsymbol{\sigma}\cdot\mathbf{u}\big) = \iota\,\mathbf{b}\cdot\mathbf{u} + \dot{q}. \tag{3.3.6}$$

The expression inside the second set of the parentheses on the left-hand side defines a compound energy flux.

3.3.4 Lagrangian Form

Using the continuity equation, we obtain the associated Lagrangian form involving the material derivative, $\mathrm{D}/\mathrm{D}t$,

$$\rho\,\frac{\mathrm{D}\vartheta}{\mathrm{D}t} = -\nabla\cdot\mathbf{q} + \nabla\cdot(\boldsymbol{\sigma}\cdot\mathbf{u}) + \iota\,\mathbf{b}\cdot\mathbf{u} + \dot{q}. \tag{3.3.7}$$

Both the conservative form expressed by Eq. (3.3.4) or (3.3.6) and the Lagrangian form expressed by (3.3.7) apply for incompressible as well as incompressible fluids.

Although the three constituents of ϑ shown in (3.3.5) may change during the motion of a point particle, the sum is determined by the evolution equation (3.3.7).

3.3.5 Interpretation

The first, second, and fourth terms on the right-hand side of (3.3.4) or (3.3.7) express the effect of heat conduction, rate of working of the traction, and rate of interior energy generation. The third term on the right-hand side is the negative of the rate of working against an arbitrary body force, **b**.

3.3.6 Deviatoric Stress Tensor

In terms of the deviatoric stress tensor defined as $\boldsymbol{\sigma}^{\text{dev}} \equiv p\,\mathbf{I}+\boldsymbol{\sigma}$, the energy evolution equation (3.3.4) takes the form

$$\frac{\partial(\rho\,\vartheta)}{\partial t} + \nabla\cdot\big(\rho\,\vartheta\,\mathbf{u}\big) = -\nabla\cdot\mathbf{q} - \nabla\cdot(p\,\mathbf{u}) + \nabla\cdot(\boldsymbol{\sigma}^{\text{dev}}\cdot\mathbf{u}) + \iota\,\mathbf{b}\cdot\mathbf{u} + \dot{q}, \tag{3.3.8}$$

and the associated Lagrangian equation (3.3.7) takes the form

$$\rho\,\frac{\mathrm{D}\vartheta}{\mathrm{D}t} = -\nabla\cdot\mathbf{q} - \nabla\cdot(p\,\mathbf{u}) + \nabla\cdot(\boldsymbol{\sigma}^{\text{dev}}\cdot\mathbf{u}) + \iota\,\mathbf{b}\cdot\mathbf{u} + \dot{q}, \tag{3.3.9}$$

where p is the pressure. The third term on the right-hand side of either equation does not appear in the case of ideal-fluid flow.

3.3.7 Integral Energy Balance

An integral energy balance arises by integrating the differential energy balance expressed by (3.3.4) over a control volume, $\mathcal{V}_c$, that is bounded by a surface or a collection of surfaces, $\mathcal{S}_c$, as shown in Fig. 1.3.

Applying the divergence theorem to convert volume integrals into surface integrals, we obtain the balance equation

$$\begin{aligned}\iiint_{\mathcal{V}_c}\frac{\partial(\rho\,\vartheta)}{\partial t}\,\mathrm{d}V &= \iiint_{\mathcal{S}_c}\rho\,\vartheta\,\mathbf{n}^{\text{in}}\cdot\mathbf{u}\,\mathrm{d}S + \iint_{\mathcal{S}_c}\mathbf{n}^{\text{in}}\cdot\mathbf{q}\,\mathrm{d}S\\ &\quad - \iint_{\mathcal{S}_c}\mathbf{f}^{\text{in}}\cdot\mathbf{u}\,\mathrm{d}S + \iiint_{\mathcal{V}_c}\iota\,\mathbf{b}\cdot\mathbf{u}\,\mathrm{d}V + \iiint_{\mathcal{V}_c}\dot{q}\,\mathrm{d}V,\end{aligned} \tag{3.3.10}$$

where $\mathbf{f}^{\text{in}} = \mathbf{n}^{\text{in}}\cdot\boldsymbol{\sigma}$ is the boundary traction and $\mathbf{n}^{\text{in}}$ is the normal unit vector pointing *into* the control volume.

The left-hand side of (3.3.10) expresses the rate of accumulation of the compound energy encapsulated in ϑ inside the control volume, including the internal, kinetic, and potential energy associated with gravity. The five terms on the right-hand side represent, respectively:

(a) The rate of supply of compound energy *into* the control volume by convection.
(b) The rate of supply of thermal energy *into* the control volume by conduction.

(c) The rate of working of the traction over the boundary against the fluid velocity.
(d) The rate of working against a distributed body force throughout the control volume.
(e) The rate of interior heat generation due to an external agent, such as laser heating.

Other sources or sinks of energy, such as surface radiation, can be added to the right-hand side.

3.3.8 Energy Balance for an Open System

The transport equation (3.3.10) could have been written at the outset based on the first law of thermodynamics for an open system that allows matter to enter or exit through the boundaries, and thus carry energy across the boundaries. A mechanical dissipation term is not included in this balance. Working backward from Eq. (3.3.10), we may derive the evolution equation (3.3.4) or (3.3.7) for the compound energy represented by ϑ.

3.3.9 Exercises

3.3.1 Derive the evolution equations (3.3.4) and (3.3.7).

3.3.2 Add a heat radiation term to the integral energy balance expressed by Eq. (3.3.10).

3.4 Internal-Energy Transport

Substituting expressions (3.1.17), (3.2.10), (3.2.8), and (3.2.7) into (3.2.4) expressing the first law of thermodynamics for a fluid parcel in the absence of surface heat generation, we obtain

$$\frac{\mathrm{d}}{\mathrm{d}t}\iiint_{\text{parcel}} \rho\, u \,\mathrm{d}V = -\iint_{\text{parcel}} \mathbf{n}\cdot\mathbf{q}\,\mathrm{d}S + \iiint_{\text{parcel}} \boldsymbol{\sigma} : \nabla\mathbf{u}\,\mathrm{d}V + \iiint_{\text{parcel}} \dot{q}\,\mathrm{d}V, \tag{3.4.1}$$

where u is the specific internal energy and $\dot{q}$ is the volumetric rate of interior energy generation. We recall that the unit normal vector, $\mathbf{n}$, points outward from the parcel.

Equation (3.4.1) shows that the internal energy of a parcel or material point particle changes for three reasons represented by each term on the right-hand side:

(a) Heat conduction across the parcel surface with associated heat flux $\mathbf{q}$.
(b) Mechanical energy loss or gain due to the fluid motion inside the parcel.
(c) Interior heat generation.

A body force is unable to directly raise or lower the internal energy of a fluid.

3.4.1 Evolution Equations

The time derivative on the left-hand side of (3.4.1) can be manipulated by applying the Reynolds transport equation (1.7.7) for the property $\phi = \rho u$. Using the divergence theorem to convert the surface integral involving the heat flux into a volume integral on the right-hand side of (3.4.1), and then discarding the integral signs in all terms of the resulting equation, we obtain the conservative form of an internal-energy evolution equation,

$$\frac{\partial(\rho u)}{\partial t} + \nabla \cdot \big(\rho\, u\, \mathbf{u}\big) = -\nabla \cdot \mathbf{q} + \boldsymbol{\sigma} : \nabla\mathbf{u} + \dot{q}. \tag{3.4.2}$$

The divergence in the first term on the right-hand side can be combined with that of the second term on the left-hand side.

Combining equation (3.4.2) with the continuity equation to simplify the left-hand side, we obtain an expression for the rate of change of the specific internal energy following a point particle in the Lagrangian form

$$\rho\,\frac{\mathrm{D}u}{\mathrm{D}t} = -\nabla \cdot \mathbf{q} + \boldsymbol{\sigma} : \nabla\mathbf{u} + \dot{q}, \tag{3.4.3}$$

where $\mathrm{D}/\mathrm{D}t$ is the material derivative. This Lagrangian form and the conservative form expressed by (3.4.2) apply for incompressible as well as compressible fluids.

3.4.2 Deviatoric Stress Tensor

In terms of the deviatoric stress tensor defined as $\boldsymbol{\sigma}^{\mathrm{dev}} \equiv p\,\mathbf{I} + \boldsymbol{\sigma}$, the internal-energy evolution equation (3.4.3) takes the form

$$\rho\,\frac{\mathrm{D}u}{\mathrm{D}t} = -\nabla \cdot \mathbf{q} - p\,\nabla \cdot \mathbf{u} + \boldsymbol{\sigma}^{\mathrm{dev}} : \nabla\mathbf{u} + \dot{q}, \tag{3.4.4}$$

where p is the pressure. The second term on the right-hand side expresses the rate of working against pressure due to expansion or contraction. Since the pressure is a positive physical entity in the context of thermodynamics, expansion causes an expected decrease in the internal energy of a fluid.

Now invoking the continuity equation to set

$$\nabla \cdot \mathbf{u} = \rho\,\frac{\mathrm{D}}{\mathrm{D}t}\Big(\frac{1}{\rho}\Big), \tag{3.4.5}$$

and rearranging, we obtain another version of the internal-energy equation,

$$\frac{\mathrm{D}u}{\mathrm{D}t} + p\,\frac{\mathrm{D}}{\mathrm{D}t}\Big(\frac{1}{\rho}\Big) = \frac{1}{\rho}\big(-\nabla \cdot \mathbf{q} + \boldsymbol{\sigma}^{\mathrm{dev}} : \nabla\mathbf{u} + \dot{q}\,\big). \tag{3.4.6}$$

The second term on the left-hand side is reminiscent of the $p\,\mathrm{d}v$ term in classical thermodynamics, where $\mathrm{d}v$ is the change in specific volume, as discussed in Appendix B. Equation (3.4.6) can be regarded as an alternative statement of the first law of thermodynamics.

3.4.3 Integral Internal-Energy Balance

An integral internal-energy balance arises by integrating the conservative form of the internal-energy balance (3.4.2) over a control volume, $\mathcal{V}_c$, that is bounded by a surface or a collection of surfaces, denoted by $\mathcal{S}_c$. Applying the divergence theorem, we obtain

$$\iiint_{\mathcal{V}_c} \frac{\partial(\rho u)}{\partial t} = \iint_{\mathcal{S}_c} \rho u\, \mathbf{n}^{\text{in}} \cdot \mathbf{u}\, \mathrm{d}S + \iint_{\mathcal{S}_c} \mathbf{n}^{\text{in}} \cdot \mathbf{q}\, \mathrm{d}S + \iiint_{\mathcal{V}_c} \boldsymbol{\sigma} : \nabla \mathbf{u}\, \mathrm{d}V + \iiint_{\mathcal{V}_c} \dot{q}\, \mathrm{d}V, \tag{3.4.7}$$

where $\mathbf{n}^{\text{in}}$ is the normal unit vector pointing *into* the control volume, as shown in Fig. 1.3.

The left-hand side of (3.4.7) expresses the rate of accumulation of internal energy inside the control volume. The four terms on the right-hand side represent the effects of (a) convection of internal energy *into* the control volume, (b) thermal energy transport *into* the control volume by conduction, (c) mechanical energy loss or gain, and (d) internal heat generation. We recall that the expression for the mechanical energy loss or gain arises directly from Cauchy's equation of motion without further consideration.

3.4.4 Thermodynamics

To this end, equations describing the behavior of the fluid and relations from thermodynamics must be introduced to relate the specific internal energy to the fluid temperature and pressure, as discussed in Sect. 3.7.

3.4.5 Exercises

3.4.1 Derive Eq. (3.4.2) working as discussed in the text.

3.4.2 Express the integral balance (3.4.7) in terms of the deviatoric stress tensor.

3.4.3 Discuss and explain the differences between the transport equations (3.3.2) and (3.4.1).

3.5 Kinetic–Internal Energy Transport

It is useful to derive transport and balance equations for the sum of the kinetic and internal energies, $\mathcal{E}_{\text{kin}}$ and $\mathcal{E}_{\text{int}}$, excluding effect of the potential energy associated with gravity or another body force. Introducing the specific kinetic–integral energy,

$$\varepsilon \equiv \tfrac{1}{2}\, |\mathbf{u}|^2 + u, \tag{3.5.1}$$

and applying the Reynolds transport equation (1.7.8) for the property field $\phi = \rho\, \varepsilon$, we obtain

$$\frac{\mathrm{d}(\mathcal{E}_{\text{kin}} + \mathcal{E}_{\text{int}})}{\mathrm{d}t} = \iiint_{\text{parcel}} \frac{\partial(\rho\, \varepsilon)}{\partial t}\, \mathrm{d}V + \iint_{\text{parcel}} \rho\, \varepsilon\, \mathbf{n} \cdot \mathbf{u}\, \mathrm{d}S, \tag{3.5.2}$$

where $\mathbf{n}$ is the outward normal unit vector. Substituting (3.5.2) together with expression (3.1.4) for the rate of change of the potential energy into the total energy balance (3.2.13), and rearranging, we obtain

$$\begin{aligned} \iiint_{\text{parcel}} \frac{\partial(\rho\, \varepsilon)}{\partial t}\, \mathrm{d}V + \iint_{\text{parcel}} \rho\, \varepsilon\, \mathbf{n} \cdot \mathbf{u}\, \mathrm{d}S &= - \iint_{\text{parcel}} \mathbf{n} \cdot \mathbf{q}\, \mathrm{d}S \\ + \iint_{\text{parcel}} \mathbf{f} \cdot \mathbf{u}\, \mathrm{d}S + \iiint_{\text{parcel}} (\rho\, \mathbf{g} + \iota \mathbf{b}) \cdot \mathbf{u}\, \mathrm{d}V &+ \iiint_{\text{parcel}} \dot{q}\, \mathrm{d}V, \end{aligned} \tag{3.5.3}$$

where $\mathbf{f} = \mathbf{n} \cdot \boldsymbol{\sigma}$ is the traction exerted on the parcel surface. The effect of the potential energy due to gravity or another body force is expressed by the third term on the right-hand side.

3.5.1 Evolution Equation

To derive an associated evolution equation, we apply the divergence theorem to convert the surface integrals into volume integrals over the parcel. Since the volume of the parcel is arbitrary, we may discard the volume integral signs to obtain

$$\frac{\partial(\rho\,\varepsilon)}{\partial t} + \nabla \cdot \big(\rho\,\varepsilon\,\mathbf{u}\big) = -\nabla \cdot \mathbf{q} + \nabla \cdot (\boldsymbol{\sigma} \cdot \mathbf{u}) + (\rho\,\mathbf{g} + \iota\mathbf{b}) \cdot \mathbf{u} + \dot{q}. \tag{3.5.4}$$

The three divergences could be combined into a single term.

Using the continuity equation to manipulate the left-hand side, we obtain the corresponding Lagrangian form

$$\rho\,\frac{\mathrm{D}\varepsilon}{\mathrm{D}t} = -\nabla \cdot \mathbf{q} + \nabla \cdot (\boldsymbol{\sigma} \cdot \mathbf{u}) + (\rho\,\mathbf{g} + \iota\mathbf{b}) \cdot \mathbf{u} + \dot{q}, \tag{3.5.5}$$

where $\mathrm{D}/\mathrm{D}t$ is the material derivative.

3.5.2 Deviatoric Stress Tensor

In terms of the deviatoric stress tensor defined as $\boldsymbol{\sigma}^{\mathrm{dev}} \equiv p\,\mathbf{I} + \boldsymbol{\sigma}$, the transport equation (3.5.4) becomes

$$\frac{\partial(\rho\,\varepsilon)}{\partial t} + \nabla \cdot \big(\rho\,(\varepsilon + \frac{p}{\rho})\,\mathbf{u}\big) = -\nabla \cdot \mathbf{q} + \nabla \cdot (\boldsymbol{\sigma}^{\mathrm{dev}} \cdot \mathbf{u}) + (\rho\,\mathbf{g} + \iota\mathbf{b}) \cdot \mathbf{u} + \dot{q}. \tag{3.5.6}$$

Note that the pressure has been transferred inside the divergence on the second term on the left-hand side. The Lagrangian transport equation (3.5.5) takes the corresponding form

$$\rho\,\frac{\mathrm{D}\varepsilon}{\mathrm{D}t} = -\nabla \cdot \mathbf{q} - \nabla \cdot (p\mathbf{u}) + \nabla \cdot (\boldsymbol{\sigma}^{\mathrm{dev}} \cdot \mathbf{u}) + (\rho\,\mathbf{g} + \iota\mathbf{b}) \cdot \mathbf{u} + \dot{q}. \tag{3.5.7}$$

3.5.3 Integral Kinetic–Internal Energy Balance

An integral kinetic–internal energy balance arises by integrating the conservative form (3.5.4) over a control volume, $\mathcal{V}_c$, that is bounded by a surface or a collection of surfaces, $\mathcal{S}_c$, as shown in Fig. 1.3. Applying the divergence theorem to convert volume integrals of divergences into surface integrals, we obtain

$$\begin{aligned} \iiint_{\mathcal{V}_c} \frac{\partial(\rho\varepsilon)}{\partial t} &= \iint_{\mathcal{S}_c} \rho\varepsilon\,\mathbf{n}^{\mathrm{in}} \cdot \mathbf{u}\,\mathrm{d}S + \iint_{\mathcal{S}_c} \mathbf{n}^{\mathrm{in}} \cdot \mathbf{q}\,\mathrm{d}S \\ &- \iint_{\mathcal{S}_c} (\mathbf{n}^{\mathrm{in}} \cdot \boldsymbol{\sigma}) \cdot \mathbf{u}\,\mathrm{d}S + \iiint_{\mathcal{V}_c} (\rho\,\mathbf{g} + \iota\mathbf{b}) \cdot \mathbf{u}\,\mathrm{d}V + \iiint_{\mathcal{V}_c} \dot{q}\,\mathrm{d}V, \end{aligned} \tag{3.5.8}$$

where $\mathbf{n}^{\text{in}}$ is the normal unit vector pointing *into* the control volume. The interpretation of the various terms is similar to that for the total energy discussed in Sect. 3.3.

3.5.4 Exercise

3.5.1 Derive an integral balance associated with (3.5.4).

3.6 Enthalpy Transport

The specific enthalpy, h, is defined with respect to the specific internal energy, u, pressure, p, and specific volume, $v \equiv 1/\rho$, by the equivalent equations

$$h = u + p\,v, \qquad u = h - p\,v, \tag{3.6.1}$$

as discussed in Appendix B. In terms of the density, ρ,

$$h = u + \frac{p}{\rho}, \qquad u = h - \frac{p}{\rho}. \tag{3.6.2}$$

The significance of the specific enthalpy will become evident in hindsight.

3.6.1 Parcel Enthalpy

The enthalpy of a material fluid parcel is found by integrating over the parcel volume,

$$H = \iiint_{\text{parcel}} \rho\, h\, \mathrm{d}V. \tag{3.6.3}$$

We see that the specific enthalpy is the enthalpy divided by the mass of an infinitesimal fluid parcel.

3.6.2 Evolution of the specific enthalpy

Substituting the second expression in (3.6.2) into the specific internal-energy evolution equation (3.4.3), repeated below for convenience,

$$\rho\,\frac{\mathrm{D}u}{\mathrm{D}t} = -\boldsymbol{\nabla}\cdot\mathbf{q} + \boldsymbol{\sigma} : \boldsymbol{\nabla}\mathbf{u} + \dot{q}, \tag{3.6.4}$$

we obtain an evolution equation for the specific enthalpy following a point particle,

$$\rho\,\frac{\mathrm{D}h}{\mathrm{D}t} - \rho\,\frac{\mathrm{D}}{\mathrm{D}t}\Big(\frac{p}{\rho}\Big) = -\boldsymbol{\nabla}\cdot\mathbf{q} + \boldsymbol{\sigma} : \boldsymbol{\nabla}\mathbf{u} + \dot{q}, \tag{3.6.5}$$

where $\mathrm{D}/\mathrm{D}t$ is the material derivative. The second term on the left-hand side is subject to simplification.

3.6.3 Simplification

Let $f(\rho)$ be an arbitrary function of the density, ρ, and ϕ be an arbitrary scalar field. Invoking the definition of the material derivative, we write

$$\frac{\mathrm{D}(f\phi)}{\mathrm{D}t} \equiv \frac{\partial(f\phi)}{\partial t} + \mathbf{u}\cdot\nabla(f\phi). \tag{3.6.6}$$

Expanding the product, we obtain

$$\frac{\mathrm{D}(f\phi)}{\mathrm{D}t} = f(\rho)\,\frac{\mathrm{D}\phi}{\mathrm{D}t} + \phi\, f'(\rho)\,\frac{\mathrm{D}\rho}{\mathrm{D}t}. \tag{3.6.7}$$

Using the continuity equation, we set

$$\frac{\mathrm{D}\rho}{\mathrm{D}t} = -\rho\,\nabla\cdot\mathbf{u} \tag{3.6.8}$$

in the last term on the right-hand side, and obtain

$$\frac{\mathrm{D}(f\phi)}{\mathrm{D}t} = f(\rho)\,\frac{\mathrm{D}\phi}{\mathrm{D}t} - \phi\,\rho\, f'(\rho)\,\nabla\cdot\mathbf{u}. \tag{3.6.9}$$

Now making the specific choice $f(\rho) = 1/\rho$ and setting $f'(\rho) = -1/\rho^2$, we obtain

$$\rho\,\frac{\mathrm{D}}{\mathrm{D}t}\Big(\frac{\phi}{\rho}\Big) = \frac{\mathrm{D}\phi}{\mathrm{D}t} + \phi\,\nabla\cdot\mathbf{u}. \tag{3.6.10}$$

For $\phi = p$, we obtain

$$\rho\,\frac{\mathrm{D}}{\mathrm{D}t}\Big(\frac{p}{\rho}\Big) = \frac{\mathrm{D}p}{\mathrm{D}t} + p\,\nabla\cdot\mathbf{u}. \tag{3.6.11}$$

This equation can be used to simplify the second term on the left-hand side of (3.6.5).

3.6.4 Final Act

Substituting the right-hand side of (3.6.11) into the second term on the left-hand side of (3.6.5), and rearranging the resulting equation, we obtain

$$\rho\,\frac{\mathrm{D}h}{\mathrm{D}t} = -\nabla\cdot\mathbf{q} + \boldsymbol{\sigma} : \nabla\mathbf{u} + p\,\nabla\cdot\mathbf{u} + \frac{\mathrm{D}p}{\mathrm{D}t} + \dot{q}. \tag{3.6.12}$$

The second and third terms on the right-hand side combine to yield the compact expression

$$\rho\,\frac{\mathrm{D}h}{\mathrm{D}t} = -\nabla\cdot\mathbf{q} + \boldsymbol{\sigma}^{\mathrm{dev}} : \nabla\mathbf{u} + \frac{\mathrm{D}p}{\mathrm{D}t} + \dot{q}, \tag{3.6.13}$$

where $\boldsymbol{\sigma}^{\text{dev}} \equiv p\,\mathbf{I} + \boldsymbol{\sigma}$ is the deviatoric stress tensor. We see that the deviatoric stress tensor arises naturally in the enthalpy evolution equation together with the material derivative of the pressure. A body force is unable to alter the enthalpy of the fluid.

Equations describing the behavior of the fluid and relations from thermodynamics will be introduced to relate the specific enthalpy to the fluid temperature and pressure, as discussed in Sect. 3.7.

3.6.5 Eulerian Conservative Form

Expressing the material derivative of h in terms of Eulerian derivatives, and then using the continuity equation, we obtain the conservative Eulerian form of the enthalpy transport equation,

$$\frac{\partial(\rho\, h)}{\partial t} + \nabla \cdot \big(\rho\, h\, \mathbf{u}\big) = -\nabla \cdot \mathbf{q} + \boldsymbol{\sigma}^{\text{dev}} : \nabla \mathbf{u} + \frac{\mathrm{D}p}{\mathrm{D}t} + \dot{q}. \tag{3.6.14}$$

This equation would be hard to write directly based on a differential enthalpy balance in the absence of hindsight.

3.6.6 Integral Enthalpy Balance

An integral enthalpy balance arises by integrating equation (3.6.14) over a control volume, $\mathcal{V}_c$, that is bounded by one surface or a collection of surfaces, $\mathcal{S}_c$, as shown in Fig. 1.3. Applying the divergence theorem, we obtain

$$\begin{aligned} \iiint_{\mathcal{V}_c} \frac{\partial(\rho h)}{\partial t} = & \iint_{\mathcal{S}_c} \rho\, h\, \mathbf{n}^{\text{in}} \cdot \mathbf{u}\, \mathrm{d}S + \iint_{\mathcal{S}_c} \mathbf{n}^{\text{in}} \cdot \mathbf{q}\, \mathrm{d}S \\ & + \iiint_{\mathcal{V}_c} \boldsymbol{\sigma}^{\text{dev}} : \nabla \mathbf{u}\, \mathrm{d}V + \iiint_{\mathcal{V}_c} \frac{\mathrm{D}p}{\mathrm{D}t}\, \mathrm{d}V + \iiint_{\mathcal{V}_c} \dot{q}\, \mathrm{d}V, \end{aligned} \tag{3.6.15}$$

where $\mathbf{n}^{\text{in}}$ is the normal unit vector pointing *into* the control volume. The material derivative of the pressure must be available throughout the control volume.

3.6.7 Exercise

3.6.1 Discuss the physical interpretation of the enthalpy.

3.7 Temperature Evolution

The specific internal energy, u, can be considered a function of the absolute temperature, T, and specific volume, $v \equiv 1/\rho$,

$$u(T, v), \tag{3.7.1}$$

where ρ is the density. The rate of change of the specific internal energy following a point particle is given by the evolution equation

$$\frac{\mathrm{D}u}{\mathrm{D}t} = c_v \frac{\mathrm{D}T}{\mathrm{D}t} + \left(\frac{\partial u}{\partial v}\right)_T \frac{\mathrm{D}v}{\mathrm{D}t}, \tag{3.7.2}$$

where $\mathrm{D}/\mathrm{D}t$ is the material derivative and

$$c_v \equiv \left(\frac{\partial u}{\partial T}\right)_v \tag{3.7.3}$$

is the heat capacity under constant volume or density.

3.7.1 In Terms of c_v

Now using the version of the continuity equation stated in (1.9.10), repeated below for convenience,

$$\frac{1}{v}\frac{\mathrm{D}v}{\mathrm{D}t} = \boldsymbol{\nabla} \cdot \mathbf{u}, \tag{3.7.4}$$

and recalling that $v = 1/\rho$, we obtain

$$\frac{\mathrm{D}u}{\mathrm{D}t} = c_v \frac{\mathrm{D}T}{\mathrm{D}t} + \frac{1}{\rho}\left(\frac{\partial u}{\partial v}\right)_T \boldsymbol{\nabla} \cdot \mathbf{u}, \tag{3.7.5}$$

where $\boldsymbol{\nabla} \cdot \mathbf{u}$ is the rate of expansion. Since

$$\left(\frac{\partial u}{\partial v}\right)_T = T\left(\frac{\partial p}{\partial T}\right)_v - p, \tag{3.7.6}$$

as shown in Eq. (B.7.6), Appendix B, we obtain

$$\rho \frac{\mathrm{D}u}{\mathrm{D}t} = \rho\, c_v \frac{\mathrm{D}T}{\mathrm{D}t} + \left(T\left(\frac{\partial p}{\partial T}\right)_v - p\right) \boldsymbol{\nabla} \cdot \mathbf{u}, \tag{3.7.7}$$

which relates a small change in the internal energy to the corresponding change in temperature, to the rate of expansion.

In the case of an ideal gas or strictly incompressible fluid, because $u = c_v T$ and thus $(\partial u/\partial v)_T = 0$, the last term on the right-hand side of (3.7.2) and all other terms originating from this term do not appear, yielding $\mathrm{D}u/\mathrm{D}t = c_v\, \mathrm{D}T/\mathrm{D}t$.

Setting the right-hand side of (3.7.7) equal to that of the specific internal-energy evolution equation (3.4.3), and rearranging, we obtain an evolution equation for the temperature involving c_v,

$$\rho\, c_v \frac{\mathrm{D}T}{\mathrm{D}t} = -\boldsymbol{\nabla} \cdot \mathbf{q} + \boldsymbol{\sigma}^{\mathrm{dev}} : \boldsymbol{\nabla}\mathbf{u} - T\left(\frac{\partial p}{\partial T}\right)_v \boldsymbol{\nabla} \cdot \mathbf{u} + \dot{q}, \tag{3.7.8}$$

where $\boldsymbol{\sigma}^{\mathrm{dev}}$ is the deviatoric stress tensor. An equivalent evolution equation in terms of the heat capacity under constant volume, c_p, will be derived next based on the specific enthalpy.

3.7.2 In Terms of c_p

The specific enthalpy, h, can be regarded a function of the absolute temperature, T, and pressure, p,

$$h(T, p). \tag{3.7.9}$$

The rate of change of the specific enthalpy following a point particle is given by the evolution equation

$$\frac{\mathrm{D}h}{\mathrm{D}t} = c_p \frac{\mathrm{D}T}{\mathrm{D}t} + \Big(\frac{\partial h}{\partial p}\Big)_T \frac{\mathrm{D}p}{\mathrm{D}t}, \tag{3.7.10}$$

where the partial derivative

$$c_p \equiv \Big(\frac{\partial h}{\partial T}\Big)_p \tag{3.7.11}$$

is the heat capacity under constant pressure.

Referring to Eq. (B.6.25), Appendix B, we find that the second coefficient on the right-hand side of (3.7.10) is given by

$$\Big(\frac{\partial h}{\partial p}\Big)_T = \frac{1}{\rho}(1 - \widehat{\alpha}\, T), \tag{3.7.12}$$

where

$$\widehat{\alpha} \equiv \frac{1}{v}\Big(\frac{\partial v}{\partial T}\Big)_p = -\frac{1}{\rho}\Big(\frac{\partial \rho}{\partial T}\Big)_p = -\Big(\frac{\partial \ln \rho/\rho_0}{\partial T}\Big)_p \tag{3.7.13}$$

is the coefficient of thermal expansion with units of inverse temperature, $v \equiv 1/\rho$ is the specific volume, and ρ_0 is a reference density. We can write

$$\widehat{\alpha}\, T = -\Big(\frac{\partial \ln \rho/\rho_0}{\partial \ln T/T_0}\Big)_p, \tag{3.7.14}$$

where T_0 is a chosen reference temperature.

Substituting (3.7.12) into (3.7.10), we obtain an expression that relates a small change in the specific enthalpy to corresponding changes in temperature and pressure,

$$\rho \frac{\mathrm{D}h}{\mathrm{D}t} = \rho\, c_p \frac{\mathrm{D}T}{\mathrm{D}t} + (1 - \widehat{\alpha}\, T) \frac{\mathrm{D}p}{\mathrm{D}t}. \tag{3.7.15}$$

In the case of an ideal gas or strictly incompressible liquid, $\widehat{\alpha} = 1/T$; consequently, the last term on the right-hand side does not appear.

Now combining equations (3.7.15) and (3.6.13), and rearranging, we obtain an alternative evolution equation for the temperature,

$$\rho\, c_p \frac{\mathrm{D}T}{\mathrm{D}t} = -\nabla \cdot \mathbf{q} + \boldsymbol{\sigma}^{\mathrm{dev}} : \nabla \mathbf{u} + \widehat{\alpha}\, T \frac{\mathrm{D}p}{\mathrm{D}t} + \dot{q}. \tag{3.7.16}$$

The rate of change of the pressure following a point particle, $\mathrm{D}p/\mathrm{D}t$, can be computed simultaneously, specified, or otherwise approximated. In the case of an ideal gas or strictly incompressible liquid, $\widehat{\alpha}\, T = 1$ in the penultimate term on the right-hand side of (3.7.16).

3.7.3 Reconciliation

To reconcile the evolution equations (3.7.8) and (3.7.16) for the temperature derived in this section, we subtract the former from the latter and obtain

$$\rho\,(c_p - c_v)\,\frac{\mathrm{D}T}{\mathrm{D}t} = \widehat{\alpha}\, T\,\frac{\mathrm{D}p}{\mathrm{D}t} + T\left(\frac{\partial p}{\partial T}\right)_v \nabla\cdot\mathbf{u}. \tag{3.7.17}$$

Substituting the definition of $\widehat{\alpha}$ from the first expression in (3.7.13), using the continuity equation (3.7.4), and recalling that $v\rho = 1$, we obtain

$$(c_p - c_v)\,\frac{\mathrm{D}T}{\mathrm{D}t} = T\left(\frac{\partial v}{\partial T}\right)_p \frac{\mathrm{D}p}{\mathrm{D}t} + T\left(\frac{\partial p}{\partial T}\right)_v \frac{\mathrm{D}v}{\mathrm{D}t}, \tag{3.7.18}$$

which implies that

$$(c_p - c_v)\,\mathrm{d}T = T\left(\frac{\partial v}{\partial T}\right)_p \mathrm{d}p + T\left(\frac{\partial p}{\partial T}\right)_v \mathrm{d}v. \tag{3.7.19}$$

Substituting into the left-hand side the expression

$$\mathrm{d}T = \left(\frac{\partial T}{\partial p}\right)_v \mathrm{d}p + \left(\frac{\partial T}{\partial v}\right)_p \mathrm{d}v, \tag{3.7.20}$$

we find that

$$(c_p - c_v)\left(\frac{\partial T}{\partial p}\right)_v = T\left(\frac{\partial v}{\partial T}\right)_p, \qquad (c_p - c_v)\left(\frac{\partial T}{\partial v}\right)_p = T\left(\frac{\partial p}{\partial T}\right)_v. \tag{3.7.21}$$

The satisfaction of these equations is guaranteed by the equation

$$c_p - c_v = T\left(\frac{\partial v}{\partial T}\right)_p \left(\frac{\partial p}{\partial T}\right)_v, \tag{3.7.22}$$

as shown in Eq. (B.6.15), Appendix B. An expression for the ratio of the heat capacities, c_p/c_v, may also be derived, as discussed in Sect. B.6, Appendix B.

3.7.4 Integral Thermal Energy Balance

An integral thermal energy balance arises by integrating equation (3.7.16) over a control volume, $\mathcal{V}_c$, that is bounded by one surface or a collection of surfaces, $\mathcal{S}_c$, as shown in Fig. 1.3. Resolving the material derivative into Eulerian derivatives, using the continuity equation, treating c_p as a constant, and applying the divergence theorem to convert volume integrals into surface integrals, we obtain the balance equation

$$\iiint_{\mathcal{V}_c} \frac{\partial(\rho\, c_p (T - T_r))}{\partial t} = \iint_{\mathcal{S}_c} \rho\, c_p (T - T_r)\, \mathbf{n}^{\text{in}} \cdot \mathbf{u}\, \mathrm{d}S + \iint_{\mathcal{S}_c} \mathbf{n}^{\text{in}} \cdot \mathbf{q}\, \mathrm{d}S$$
$$+ \iiint_{\mathcal{V}_c} \boldsymbol{\sigma}^{\text{dev}} : \boldsymbol{\nabla}\mathbf{u}\, \mathrm{d}V + \iiint_{\mathcal{V}_c} \widehat{\alpha}\, T \frac{\mathrm{D}p}{\mathrm{D}t}\, \mathrm{d}V + \iiint_{\mathcal{V}_c} \dot{q}\, \mathrm{d}V, \tag{3.7.23}$$

where T_r is a reference temperature and $\mathbf{n}^{\text{in}}$ is the normal unit vector pointing *into* the control volume.

A similar thermal energy balance involving c_v can be written based on Eq. (3.7.8).

3.7.5 Exercise

3.7.1 Derive the evolution equation (3.7.16) directly from (3.7.8) for an ideal gas.

3.8 Entropy Transport

The concept of entropy is discussed in Appendix B with reference to Gibbs' fundamental equation of thermodynamics. The entropy of a material parcel is given by

$$S = \iiint_{\text{parcel}} \rho\, s\, \mathrm{d}V, \tag{3.8.1}$$

where ρ is the density and s is the specific entropy defined as the entropy divided by the mass of an infinitesimal parcel.

3.8.1 Gibbs Fundamental Equation

Gibbs' fundamental equation of thermodynamics relates small changes in the specific entropy, s, to corresponding changes in the specific internal energy, u, and specific volume, $v = 1/\rho$. For a point particle, Gibbs' equation takes the form

$$\frac{\mathrm{D}u}{\mathrm{D}t} = T \frac{\mathrm{D}s}{\mathrm{D}t} - p \frac{\mathrm{D}v}{\mathrm{D}t}, \tag{3.8.2}$$

where T is the absolute temperature, p is the pressure, and ρ is the density, as discussed in Appendix B. In fact, Eq. (3.8.2) can be regarded as the definition of the specific entropy.

Substituting into (3.8.2) $v = 1/\rho$ and carrying out the differentiation in the last term, we obtain

$$\frac{\mathrm{D}u}{\mathrm{D}t} = T \frac{\mathrm{D}s}{\mathrm{D}t} + \frac{p}{\rho^2} \frac{\mathrm{D}\rho}{\mathrm{D}t}. \tag{3.8.3}$$

The fraction p/ρ^2 in front of the derivative of the density can be regarded as a chemical potential pertaining to the density of a single-component fluid. We note that if the density and specific entropy of a point particle remain constant in time, the specific internal energy also remains constant in time.

3.8.2 Gibbs Fundamental Equation in Terms of the Enthalpy

Substituting into (3.8.3) the definition $u = \eta - p/\rho$, we obtain a companion evolution equation for the specific enthalpy, η, in terms of the specific entropy,

$$\frac{D\eta}{Dt} = T\,\frac{Ds}{Dt} + \frac{1}{\rho}\frac{Dp}{Dt}. \tag{3.8.4}$$

We note that if the pressure and specific entropy of a point particle remain constant in time, the specific enthalpy also remains constant in time.

3.8.3 Entropy Evolution Equation

Substituting the right-hand side of (3.8.4) into the left-hand side of the enthalpy evolution equation (3.6.12), invoking the continuity equation to set $D\rho/Dt = -\rho\,\nabla\cdot\mathbf{u}$, and rearranging, we obtain a specific-entropy evolution equation,

$$\rho\,\frac{Ds}{Dt} = \frac{1}{T}\big(-\nabla\cdot\mathbf{q} + \boldsymbol{\sigma}^{\rm dev} : \nabla\mathbf{u} + \dot{q}\big). \tag{3.8.5}$$

It is important to remember that this equation is merely a restatement of total energy conservation for a compressible or incompressible fluid, subject to Gibbs' fundamental equation of thermodynamics. Equation (3.8.5) shows that, in the absence of heat conduction, viscous dissipation, and interior heat generation, point particles retain their entropy as they move in the available domain of flow, $Ds/Dt = 0$.

Combining equation (3.8.5) with the corresponding internal-energy evolution equation (3.4.4) or with the corresponding enthalpy evolution equation (3.6.13), we obtain

$$p\,\nabla\cdot\mathbf{u} = \rho\,\big(T\frac{Ds}{Dt} - \frac{Du}{Dt}\big), \qquad \frac{Dp}{Dt} = -\rho\,\big(T\frac{Ds}{Dt} - \frac{Dh}{Dt}\big). \tag{3.8.6}$$

Resolving the material derivative on the left-hand side of (3.8.5) into Eulerian derivatives and using the continuity equation, we derive the Eulerian conservative form of the entropy evolution equation,

$$\frac{\partial(\rho s)}{\partial t} + \nabla\cdot(\rho\,s\,\mathbf{u}) = \frac{1}{T}\big(-\nabla\cdot\mathbf{q} + \boldsymbol{\sigma}^{\rm dev} : \nabla\mathbf{u} + \dot{q}\big). \tag{3.8.7}$$

Rearranging the first term on the right-hand side of (3.8.5), we obtain

$$\rho\,\frac{Ds}{Dt} = -\nabla\cdot\big(\frac{1}{T}\,\mathbf{q}\big) - \frac{1}{T^2}\,\mathbf{q}\cdot\nabla T + \frac{1}{T}\,\boldsymbol{\sigma}^{\rm dev} : \nabla\mathbf{u} + \frac{1}{T}\,\dot{q}. \tag{3.8.8}$$

The Eulerian conservative form of the left-hand side is the same as that shown in (3.8.7).

3.8.4 Evolution of Parcel Entropy

Integrating equation (3.8.8) over a parcel volume, applying the divergence theorem to the first term on the right-hand side, and using the continuity equation to manipulate the left-hand side, we obtain the parcel entropy transport equation

$$\frac{\mathrm{d}}{\mathrm{d}t}\iiint_{\text{parcel}} \rho\, s\, \mathrm{d}V = -\iint_{\text{parcel}} \frac{1}{T}\mathbf{q}\cdot\mathbf{n}\,\mathrm{d}S - \iiint_{\text{parcel}} \frac{1}{T^2}\mathbf{q}\cdot\nabla T\,\mathrm{d}V$$
$$+\iiint_{\text{parcel}} \frac{1}{T}\boldsymbol{\sigma}^{\text{dev}} : \nabla\mathbf{u}\,\mathrm{d}V + \iiint_{\text{parcel}} \frac{1}{T}\dot{q}\,\mathrm{d}V, \tag{3.8.9}$$

where $\mathbf{n}$ is the outward unit vector. The penultimate integral on the right-hand side expresses the effect of fluid motion by way of the deviatoric stress tensor, excluding the pressure. Since the integrand is nonnegative, viscous stresses cause an increase in the entropy of a fluid parcel. Conversely, the evolution equation (3.8.8) can be derived from (3.8.9) by reversing the aforementioned steps.

3.8.5 Why Entropy?

Introducing the notion of entropy in terms of the Gibbs fundamental equation requires an explanation, which is provided in Sect. 3.9 with regard to the second and third terms on the right-hand side of (3.8.9) and with reference to the second law of thermodynamics.

3.8.6 Integral Entropy Balance

An integral entropy balance arises by integrating (3.8.7) over a control volume, $\mathcal{V}_c$, bounded by a surface or a collection of surfaces, $\mathcal{S}_c$, as shown in Fig. 1.3. Applying the divergence theorem, we obtain the integral balance equation

$$\iiint_{\mathcal{V}_c} \frac{\partial(\rho s)}{\partial t}\,\mathrm{d}V = \iint_{\mathcal{S}_c} \rho s\,\mathbf{n}^{\text{in}}\cdot\mathbf{u}\,\mathrm{d}S + \iint_{\mathcal{S}_c} \frac{1}{T}\mathbf{n}^{\text{in}}\cdot\mathbf{q}\,\mathrm{d}S$$
$$-\iiint_{\mathcal{V}_c} \frac{1}{T^2}\mathbf{q}\cdot\nabla T\,\mathrm{d}V + \iiint_{\mathcal{V}_c} \frac{1}{T}\boldsymbol{\sigma}^{\text{dev}} : \nabla\mathbf{u}\,\mathrm{d}V + \iiint_{\mathcal{V}_c} \frac{1}{T}\dot{q}\,\mathrm{d}V, \tag{3.8.10}$$

where $\mathbf{n}^{\text{in}}$ is the normal unit vector pointing *into* the control volume. The first term on the right-hand side expresses the effect of convection, the second and third terms express the effect of thermal conduction, the fourth term expresses the effect of mechanical energy dissipation, and the fifth term expresses the effect of homogeneous heat production.

3.8.7 Exercise

3.8.1 Derive the balance equation (3.8.10).

3.9 Second Law of Thermodynamics

One version of the second law of thermodynamics stipulates that the entropy of a system that is closed with respect to mass, such as a material fluid parcel, increases at a rate that is no less than the rate of heat supply divided by the temperature. In the case of a material fluid parcel,

$$\frac{d}{dt}\iiint_{\text{parcel}} \rho\, s\, dV \geq -\iint_{\text{parcel}} \frac{1}{T}\,\mathbf{q}\cdot\mathbf{n}\, dS + \iiint_{\text{parcel}} \frac{1}{T}\,\dot{q}\, dV, \tag{3.9.1}$$

where $\mathbf{n}$ is the outward unit normal vector. The equality applies in the case of an idealized reversible process. Transferring the time derivative into the integral as a material derivative, as shown in (1.9.17), we obtain

$$\iiint_{\text{parcel}} \rho\,\frac{Ds}{Dt}\, dV \geq -\iint_{\text{parcel}} \frac{1}{T}\,\mathbf{q}\cdot\mathbf{n}\, dS + \iiint_{\text{parcel}} \frac{1}{T}\,\dot{q}\, dV. \tag{3.9.2}$$

Using the divergence theorem to transform the surface integral into a volume integral, and then discarding the integral signs on both sides, we obtain the entropy inequality

$$\rho\,\frac{Ds}{Dt} \geq -\nabla\cdot\left(\frac{1}{T}\,\mathbf{q}\right) + \frac{1}{T}\,\dot{q}. \tag{3.9.3}$$

The notion and significance of the entropy are connected intimately with this inequality.

3.9.1 Clausius–Duhem Inequality

Combining inequality (3.9.3) with the parcel entropy evolution equation (3.8.8), we obtain the Clausius–Duhem inequality

$$-\frac{1}{T}\,\mathbf{q}\cdot\nabla T + \boldsymbol{\sigma}^{\text{dev}} : \nabla\mathbf{u} \geq 0, \tag{3.9.4}$$

which applies at every point in a compressible or incompressible fluid.

3.9.2 Clausius–Duhem Inequality for a Symmetric Stress Tensor

In the event that the deviatoric stress tensor is symmetric, the Clausius–Duhem inequality takes the form

$$-\frac{1}{T}\,\mathbf{q}\cdot\nabla T + \boldsymbol{\sigma}^{\text{dev}} : [\nabla\mathbf{u}] \geq 0, \tag{3.9.5}$$

where $[\nabla\mathbf{u}]$ is the symmetric part of the velocity-gradient tensor given by

$$[\nabla\mathbf{u}] = \tfrac{1}{2}\,(\nabla\mathbf{u} + (\nabla\mathbf{u})^{\text{T}}) = \mathbf{E} + \tfrac{1}{3}\,(\nabla\cdot\mathbf{u})\,\mathbf{I}, \tag{3.9.6}$$

the superscript T denotes the matrix transpose, $\mathbf{E}$ is the rate-of-deformation tensor, $\nabla\cdot\mathbf{u}$ is the rate of expansion, and $\mathbf{I}$ is the identity matrix. Since the trace of the deviatoric stress tensor is zero by construction, we may write

$$-\frac{1}{T}\,\mathbf{q}\cdot\nabla T + \boldsymbol{\sigma}^{\text{dev}} : \mathbf{E} \geq 0 \tag{3.9.7}$$

for a symmetric stress tensor.

When the motion of the fluid does not affect the conductive flux, and *vice versa*, we obtain a pair of inequalities pertinent to heat transport or hydrodynamic energy production

$$\mathbf{q} \cdot \nabla T \leq 0, \qquad \boldsymbol{\sigma}^{\text{dev}} : \nabla \mathbf{u} \geq 0. \tag{3.9.8}$$

The first inequality requires that the heat flux due to a temperature gradient may not have a component in the direction of the temperature gradient. Consequently, heat can only flow from a region of high temperature to a region of low temperature. This statement can be accepted as another version of the second law of thermodynamics.

The second inequality in (3.9.8) requires that the deviatoric stress tensor due to fluid motion always acts to dissipate mechanical energy, converting it into internal energy.

3.9.3 Exercises

3.9.1 Discuss a physical system where $\mathbf{q} \cdot \nabla T$ is nearly zero.

3.10 Fourier's Law of Conduction

Fourier's law of conduction is a semi-empirical relation stipulating that the heat flux at a point in a conductive medium, $\mathbf{q}$, is related to the local temperature gradient, ∇T, by a linear relationship,

$$\mathbf{q} = -\mathbf{K} \cdot \nabla T, \tag{3.10.1}$$

where $\mathbf{K}$ is the thermal conductivity tensor. In medical imaging technology (MRI), the counterpart of $\mathbf{K}$ is known as the diffusion tensor.

Equation (3.10.1) allows for the possibility that a temperature gradient in a certain direction may induce a flux in some other direction, that is, $\mathbf{q}$ and ∇T are not necessarily aligned. The physical reason is that a conductive medium is not necessarily isotropic. An example is a medium hosting an array of parallel fibers or wires.

3.10.1 Divergence of the Flux

In the case of a spatially uniform conductivity tensor, the divergence of the heat flux is

$$\nabla \cdot \mathbf{q} = \mathbf{K} : \nabla\nabla T = [\mathbf{K}] : \nabla\nabla T, \tag{3.10.2}$$

where $\nabla\nabla T$ is the matrix of second derivatives whose ij element is $\partial^2 T/\partial x_i \partial x_j$, the square brackets denote the symmetric part,

$$[\mathbf{K}] = \tfrac{1}{2}\,(\mathbf{K} + \mathbf{K}^{\mathrm{T}}), \tag{3.10.3}$$

and the superscript T denotes the matrix transpose. We see that the antisymmetric part of $\mathbf{K}$ has no effect $\nabla \cdot \mathbf{q}$, and therefore on the energy equations discussed previously in this chapter.

3.10.2 Isotropic Materials

In the case of an isotropic medium, the conductivity tensor is diagonal, $\mathbf{K} = k\,\mathbf{I}$, where k is the scalar thermal conductivity and $\mathbf{I}$ is the identity matrix. Fourier's law prescribes that

$$\mathbf{q} = -k\,\nabla T, \tag{3.10.4}$$

so that a temperature gradient in a certain direction induces a heat flux in the opposite direction. If it did not, the orientation of the flux would be indeterminate.

3.10.3 Clausius–Duhem Inequality

The first inequality in (3.9.8) requires that

$$\nabla T \cdot \mathbf{K} \cdot \nabla T \geq 0, \tag{3.10.5}$$

which can be restated as

$$\mathbf{K} : \mathcal{T} \geq 0 \qquad \text{or} \qquad [\mathbf{K}] : \mathcal{T} \geq 0, \tag{3.10.6}$$

where $\mathcal{T} \equiv \nabla T \otimes \nabla T$ is a symmetric matrix with components

$$\mathcal{T}_{ij} = \frac{\partial T}{\partial x_i}\,\frac{\partial T}{\partial x_j}. \tag{3.10.7}$$

The second inequality in (3.10.6) requires that $[\mathbf{K}]$ must be positive semi-definite, that is, it should have real and nonnegative eigenvalues.

3.10.4 Onsager Principle

In the literature, it is often mentioned that Onsager's principle implies that the thermal conductivity tensor $\mathbf{K}$ is symmetric and positive-definite. An consequence is that three mutually orthogonal directions can always be found where the local flux is parallel to the local temperature gradient.

3.10.5 Exercise

3.10.1 Compare the conductivity of copper to that of aluminum.

3.11 Helmholtz Free Energy

The specific Helmholtz free energy, a, is defined with respect to the specific internal energy, u, and specific entropy, s, as

$$a = u - Ts, \qquad\qquad u = a + Ts, \tag{3.11.1}$$

where T is the absolute temperature. Physically, the Helmholtz free energy is the work-available part of the internal energy, $a < u$.

The Helmholtz free energy of a material fluid parcel is found by integrating the specific free energy,

$$A = \iiint_{\text{parcel}} \rho\, a\, \mathrm{d}V, \tag{3.11.2}$$

where ρ is the density. We see that the specific Helmholtz free energy is the Helmholtz free energy divided by the mass of an infinitesimal fluid parcel.

3.11.1 Evolution Equations

Taking the material derivative of the first equation in (3.11.1), and expanding the derivative of the product Ts, we obtain

$$\frac{\mathrm{D}a}{\mathrm{D}t} = \frac{\mathrm{D}u}{\mathrm{D}t} - T\,\frac{\mathrm{D}s}{\mathrm{D}t} - s\,\frac{\mathrm{D}T}{\mathrm{D}t}. \tag{3.11.3}$$

Combining this equation with the fundamental equation of thermodynamics expressed by (3.8.3), repeated below for convenience,

$$\frac{\mathrm{D}u}{\mathrm{D}t} = T\,\frac{\mathrm{D}s}{\mathrm{D}t} + \frac{p}{\rho^2}\,\frac{\mathrm{D}\rho}{\mathrm{D}t}, \tag{3.11.4}$$

we obtain

$$\frac{\mathrm{D}a}{\mathrm{D}t} = \frac{p}{\rho^2}\,\frac{\mathrm{D}\rho}{\mathrm{D}t} - s\,\frac{\mathrm{D}T}{\mathrm{D}t}. \tag{3.11.5}$$

If the density and temperature of a point particle remain constant in time, the specific Gibbs energy also remains constant in time. The condition of constant density is satisfied for an incompressible fluid. An evolution equation for a can be obtained by substituting the evolution equation for T shown in (3.7.8) or (3.7.16).

3.11.2 Exercise

3.11.1 Derive and discuss an integral balance for the Helmholtz free energy.

3.12 Gibbs Free Energy

The specific Gibbs free energy, g, is defined with respect to the fluid density, ρ, the specific internal energy, u, the specific enthalpy, η, the specific entropy, s, the pressure, p, and the specific volume, $v \equiv 1/\rho$, by the equations

$$u = g - pv + Ts = \eta - pv, \qquad u = g - \frac{p}{\rho} + Ts = \eta - \frac{p}{\rho}. \tag{3.12.1}$$

Explicitly,

$$g = u + pv - Ts = \eta - Ts, \qquad g = u + \frac{p}{\rho} - Ts = \eta - Ts. \tag{3.12.2}$$

The Gibbs free energy of a material fluid parcel is given by

$$G = \iiint_{\text{parcel}} \rho\, g \, \mathrm{d}V. \tag{3.12.3}$$

The specific Gibbs energy is the Gibbs energy divided by the mass of an infinitesimal fluid parcel.

3.12.1 Evolution Equations

Taking the material derivative of the definition $g = \eta - Ts$, and expanding the derivative of the product, we obtain

$$\frac{\mathrm{D}g}{\mathrm{D}t} = \frac{\mathrm{D}\eta}{\mathrm{D}t} - T\,\frac{\mathrm{D}s}{\mathrm{D}t} - s\,\frac{\mathrm{D}T}{\mathrm{D}t}. \tag{3.12.4}$$

Combining this equation with (3.8.4), we obtain

$$\frac{\mathrm{D}g}{\mathrm{D}t} = \frac{1}{\rho}\frac{\mathrm{D}p}{\mathrm{D}t} - s\,\frac{\mathrm{D}T}{\mathrm{D}t}. \tag{3.12.5}$$

If the pressure and temperature of a point particle remain constant in time, the specific Gibbs energy also remains constant in time. An evolution equations for g can be obtained by substituting into (3.12.5) the evolution equation for T shown in (3.7.8) or (3.7.16).

3.12.2 Exercise

3.12.1 Derive and discuss an integral balance for the Gibbs free energy.

Solutions and Mixtures

4

We have discussed the flow and transport properties of homogeneous fluids consisting of a single chemical species, allowing for spatial density variations. In the remainder of this book, we turn out attention to inhomogeneous fluids consisting of two or more chemical species, including the effect of a chemical reaction. Preliminary concepts and the notion of diffusion are introduced in this chapter and then discussed further and elaborated in subsequent chapters.

4.1 Molecular, Mass, and Other Fractions

Consider a small region of a two-component (binary) fluid at rest or in motion containing red and blue molecules with generally different molecular weights, as shown in Fig. 4.1. The red and blue species are assumed to be miscible in any proportion, that is, spontaneous phase separation does not occur.

4.1.1 Molecular (Molar) Fractions

The fraction of red molecules inside a specified small volume is denoted by f_R, and the corresponding fraction of blue molecules is denoted by f_B, subject to the normalization condition

$$f_R + f_B = 1. \tag{4.1.1}$$

For example, the fraction of red molecules in the region depicted in Fig. 4.1 is $f_R = 5/9$, and the fraction of blue molecules is $f_B = 4/9$. The two molecular fractions add to unity.

We refer to f_R and f_B as the *molecular* or *molar fractions*. In some physical applications, the molecules can be replaced by particles, bubbles, or drops, and the molar fractions can be regarded as particle fractions.

C. Pozrikidis, *Transport Processes Primer*,
https://doi.org/10.1007/978-1-4939-9909-5_4

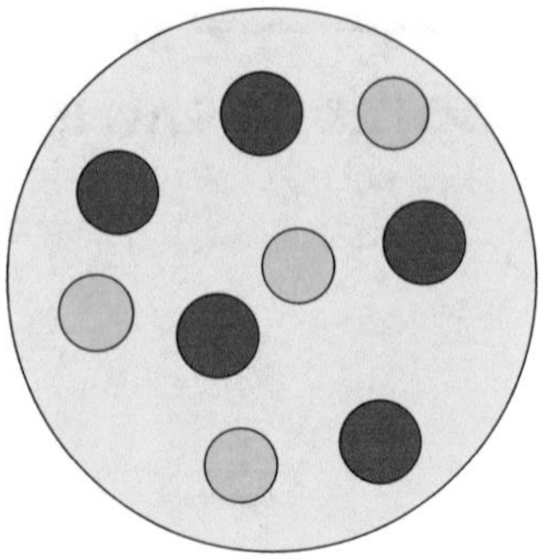

Fig. 4.1 Illustration of a small region in a binary fluid containing red and blue molecules. Molecular fractions and mass fractions can be defined

4.1.2 Probabilities

The molecular fractions may also be interpreted as probabilities of finding a red or blue molecule inside a small region in the binary fluid. In this viewpoint, and assuming independence, the probability of collision of two red molecules is f_{R}^2, the probability of collision of two blue molecules is f_{B}^2, the probability of collision of a red molecule with a blue molecule is $f_{\mathrm{R}} f_{\mathrm{B}}$, and the probability of collision of a blue molecule with a red molecule is $f_{\mathrm{B}} f_{\mathrm{R}}$. These four probabilities add to unity, as required under the condition that a collision takes place.

4.1.3 Continuum Approximation

In the context of the continuum approximation discussed in Sect. 1.1, the molecular fractions depend on position, $\mathbf{x}$, and time, t,

$$f_{\mathrm{R}}(\mathbf{x}, t), \qquad f_{\mathrm{B}}(\mathbf{x}, t). \tag{4.1.2}$$

We will assume that f_{R} and f_{B} are differentiable functions of $\mathbf{x}$ and t. However, artificial discontinuities can be accepted as mathematical idealizations.

An implied assumption is that the blue and red species are mixed all the way down to the molecular level, as discussed in Sect. 4.5. This assumption makes an important distinction between chemical mixtures and multicomponent materials; the second category includes granular sugar and gravel.

4.1.4 Mole and Molar Mass

One mole of a homogeneous substance, denoted by the standard unit symbol *mol*, is defined as a collection of N_{A} molecules, where

$$N_{\mathrm{A}} = 6.022140857 \times 10^{23} \tag{4.1.3}$$

is the Avogadro number. The molar mass of a species, M, is the mass of one mole.

For example, the molar mass of water is 18 g/mol. This means that a two-liter bottle of water contains n moles, where

$$n \times 18 \,\frac{\mathrm{g}}{\mathrm{mol}} = 2\,\mathrm{lt} \times 1\,\frac{\mathrm{kg}}{\mathrm{lt}}, \tag{4.1.4}$$

yielding $n = 2,000/18$ mol $\simeq 111$ mol.

4.1.5 Mixture Molar Mass

Let M_R and M_B be the molar masses of the red and blue molecules. The nominal molar mass of the mixture is defined as

$$M(\mathbf{x}, t) \equiv f_R M_R + f_B M_B, \tag{4.1.5}$$

where f_R and f_B are the species molar fractions. Note that $M(\mathbf{x}, t)$ is a function of position and time according to the continuum approximation by way of the species molar fractions, f_R and f_B.

In the case of an arbitrary number of species, the molar mass of the mixture is defined as

$$M(\mathbf{x}, t) \equiv \sum_X f_X M_X, \tag{4.1.6}$$

where the sum is over all species.

We can think of a mixture as though it consisted of an effective species with a continuous spatial molar mass distribution. One mole of a mixture is comprised on N_A molecules of all constituent species proportioned according to the molar fractions.

4.1.6 Molar-Mass Concentrations

Dividing equation (4.1.5) by the mixture molar mass, M, we obtain

$$\phi_R M_R + \phi_B M_B = 1, \tag{4.1.7}$$

where

$$\phi_R \equiv \frac{1}{M} f_R, \qquad \phi_B \equiv \frac{1}{M} f_B \tag{4.1.8}$$

are the red and blue *molar-mass concentrations* with units of mol/mass. Spatial and time dependencies enter the numerators and the common denominator. By definition,

$$\phi_R + \phi_B = \frac{1}{M}. \tag{4.1.9}$$

We recall that the units of M are mass/mol. In the case of a multicomponent mixture,

$$\phi_X \equiv \frac{1}{M} f_X, \qquad \sum_X \phi_X M_X = 1, \qquad \sum_X \phi_X = \frac{1}{M}, \tag{4.1.10}$$

where the sum is over all species.

4.1.7 Mass Fractions

It is useful to introduce the red and blue *mass fractions*

$$\omega_R \equiv f_R \frac{M_R}{M} = \phi_R M_R, \qquad \omega_B \equiv f_B \frac{M_B}{M} = \phi_B M_B. \tag{4.1.11}$$

By definition, the mass fractions add to unity,

$$\omega_R + \omega_B = 1. \tag{4.1.12}$$

The mass fractions, ω_R and ω_B, are equal to the molecular fractions, f_R and f_B, only in the case of precisely or virtually equal molar masses, as in the case of isotopes. The mass fractions of all species in a multicomponent mixture add to unity by construction,

$$\sum_X \omega_X = 1, \tag{4.1.13}$$

where the sum is over all species.

4.1.8 Mass Densities

The mass densities of the red and blue species, also called the species densities, mass concentrations, or partial densities, are given by

$$\rho_R = \omega_R\, \rho, \qquad \rho_B = \omega_B\, \rho, \tag{4.1.14}$$

where ρ is the density of the mixture defined as the mass divided by the volume of an infinitesimal mixture parcel containing red and blue molecules at a particular location, as shown in Fig. 4.1. By definition,

$$\rho = \rho_R + \rho_B \tag{4.1.15}$$

at every instant and each point in the fluid. Conversely, the mass fractions are given by

$$\omega_R = \frac{\rho_R}{\rho}, \qquad \omega_B = \frac{\rho_B}{\rho}. \tag{4.1.16}$$

Equality of species densities, ρ_R and ρ_B, implies equality of molecular fractions, f_R and f_B, and *vice versa*, only in the case of equal molar masses.

The mixture density of a multicomponent mixture consisting of an arbitrary number of species is the sum of all species densities,

$$\rho = \sum_X \rho_X, \tag{4.1.17}$$

where $\rho_X = \omega_X \rho$.

4.1.9 Molar Concentrations

The red or blue molar concentrations, c_R and c_B, are the numbers of red or blue moles contained in a small mixture parcel, divided by the parcel volume. The mixture molar concentration, c, is the total number of red and blue moles contained in a small mixture parcel, divided by the parcel volume,

$$c \equiv c_R + c_B. \tag{4.1.18}$$

In terms of the mixture molar mass, $M \equiv f_R M_R + f_B M_B$, the mixture concentration is given by

$$c = \frac{\rho}{M}, \tag{4.1.19}$$

where ρ is the mixture density.

By virtue of these definitions, the molar (molecular) fractions are related to the molar concentrations by

$$f_R = \frac{c_R}{c}, \qquad f_B = \frac{c_B}{c}. \tag{4.1.20}$$

We see that the ratio of the molar concentrations is equal to the ratio of the molar fractions,

$$\frac{f_R}{f_B} = \frac{c_R}{c_B}. \tag{4.1.21}$$

Using the preceding definitions, we find that

$$c_R = f_R\, c = f_R\, \frac{\rho}{M}, \qquad c_B = f_B\, c = f_B\, \frac{\rho}{M}, \tag{4.1.22}$$

and also

$$c_R = \frac{\rho_R}{M_R} = \omega_R\, \frac{\rho}{M_R} = \phi_R\, \rho, \qquad c_B = \frac{\rho_B}{M_B} = \omega_B\, \frac{\rho}{M_B} = \phi_B\, \rho, \tag{4.1.23}$$

where $\phi_R = f_R/M$ and $\phi_B = f_B/M$ are the species molar-mass concentrations.

Similar definitions apply for a multicomponent mixture whose molar concentration is

$$c = \sum_X c_X \tag{4.1.24}$$

and the sum is over all species.

4.1.10 Partial Volumes

The mixture mass contained inside a small volume, $\mathrm{d}V$, is $\rho\, \mathrm{d}V$, and the species X mass contained in this volume is

$$\rho_X\, \mathrm{d}V = \omega_X\, \rho\, \mathrm{d}V. \tag{4.1.25}$$

Integrating this equation over a region provides us with the corresponding X mass content. We may define a differential species X partial volume by the equation

$$\mathrm{d}V_X \equiv \omega_X\, \mathrm{d}V, \tag{4.1.26}$$

so that $\rho\, \mathrm{d}V_X = \rho_X\, \mathrm{d}V$, where $\mathrm{d}V = \sum_X \mathrm{d}V_X$ by construction.

Similar partial volumes can be defined with respect to the molar fractions, f_X, or with respect to any other species fractions that add to unity over all species. In a generalization, we introduce a set of positive dimensionless coefficients, φ_X, that sum to unity, define $\mathrm{d}V_X \equiv \varphi_X \,\mathrm{d}V$, and write $\varrho_X \,\mathrm{d}V_X = \rho_X \,\mathrm{d}V$, where

$$\varrho_X \equiv \frac{1}{\varphi_X}\,\rho_X = \frac{\omega_X}{\varphi_X}\,\rho \tag{4.1.27}$$

is a configurational density. If a bucket of sand and a bucket of cement are mixed, then the configurational densities pertain to each bucket.

4.1.11 Why So Many?

The foregoing and other forthcoming notions and definitions are summarized in Table 4.1. The pluralism of symbols and diversity of notation regarding material constitution can be distracting. One reason for the proliferation of symbols is that mixture and multi-fluid theories have been developed by chemists, physical chemists, physicists, engineers, and others with specific preferences, objectives, and points of departure regarding the description of multicomponent materials.

A more essential reason is that thermodynamics and chemical reaction kinetics are best described in terms of species molar fractions, whereas mass conservation and Newton's second law of motion are best described in terms of species mass fractions or mass densities. For example, the ideal-gas law specifies that $c = p/RT$, where R is the ideal-gas constant and T is the absolute temperature, which makes the use of the concentration, c, instead of the density, ρ, appropriate when p and T are specified.

4.1.12 Exercise

4.1.1 How many moles of air are there inside an empty 1 l bottle at 25° C? Air can be treated as an ideal gas.

4.2 Gradients in a Binary Mixture

The gradient of a scalar function of position, $\psi(\mathbf{x})$, is a vector whose Cartesian components are the corresponding partial derivatives,

$$\nabla\psi = \left(\frac{\partial\psi}{\partial x},\ \frac{\partial\psi}{\partial y},\ \frac{\partial\psi}{\partial z}\right). \tag{4.2.1}$$

Physically, the gradient $\nabla\psi$ points in the direction of maximum rate of change of ψ with respect to directional arc length.

4.2.1 Molecular Fractions, Molar-Mass Fractions, and Mass Concentrations

Relations between gradients of functions of interest pertaining to mixtures can be developed by straightforward differentiation. We will demonstrate that the red gradients in a binary mixture,

$$\nabla f_R, \qquad \nabla\phi_R, \qquad \nabla\omega_R, \tag{4.2.2}$$

Table 4.1 Definitions pertaining to a binary or multicomponent mixture

		Units
f_X	Molecular (molar) fraction of species X	Dimensionless
	$\sum f_X = 1$	
M_X	Molar mass of species X	Mass/mol
M	Mixture molar mass	Mass/mol
	$M = \sum f_X M_X$	
ϕ_X	Molar-mass concentration of species X	Mol/mass
	$\phi_X = f_X/M = \omega_X/M_X,$	
	$\sum \phi_X = \frac{1}{M}, \quad \sum \phi_X M_X = 1$	
ω_X	Mass fraction of species X	Dimensionless
	$\omega_X = f_X M_X/M = \phi_X M_X, \quad \sum \omega_X = 1$	
	$f_X = \omega_X M/M_X, \quad \sum \omega_X/M_X = 1/M$	
ρ	Mixture density	Mass/volume
	$\sum \rho_X = \rho$	
ρ_X	Density, mass concentration,	Mass/volume
	partial density of species X	
	$\rho_X = \omega_X \rho = c M_R f_R$	
c	Mixture molar concentration	Mol/volume
	$c = \rho/M$	
c_X	Species X molar concentration	Mol/volume
	$c_X = c f_X, \quad \sum c_X = c$	
$\mathbf{u}_X$	Species X velocity	Velocity
$\mathbf{u}$	Mixture mass, barycentric, momentum velocity	Velocity
	$\mathbf{u} = \sum \omega_X \mathbf{u}_X, \quad \rho\,\mathbf{u} = \sum \rho_X \mathbf{u}_X$	
$\mathbf{u}_X^{\text{diff}}$	Species X diffusion mass velocity	Velocity
	$\mathbf{u}_X^{\text{diff}} \equiv \mathbf{u}_X - \mathbf{u} = \frac{1}{\rho_X}\,\boldsymbol{j}_X, \quad \mathbf{u}_X = \mathbf{u} + \mathbf{u}_X^{\text{diff}}$	
$\mathbf{m}_X$	Species X mass flux	Mass /(area time)
	$\mathbf{m}_X = \rho_X \mathbf{u}_X = \rho_X \mathbf{u} + \boldsymbol{j}_X$	
$\mathbf{u}^*$	Mixture molar velocity	Velocity
	$\mathbf{u}^* = \sum f_X \mathbf{u}_X$	
$\mathbf{m}_X^*$	Species X molar flux	Mol/(area time)
	$\mathbf{m}_X^* = c_X \mathbf{u}_X = c_X \mathbf{u} + \boldsymbol{j}_X^*$	
$\boldsymbol{j}_X$	Species X diffusive mass flux	Mass/(area time)
	$\boldsymbol{j}_X = \rho_X (\mathbf{u}_X - \mathbf{u}), \quad \sum \boldsymbol{j}_X = \underline{0}$	
$\boldsymbol{j}_X^*$	Species X diffusive molar flux	Mol/(area time)
	$\boldsymbol{j}_X^* = c_X (\mathbf{u}_X - \mathbf{u}^*), \quad \sum \boldsymbol{j}_X^* = \underline{0}$	
$\dot{r}_X^*$	Rate of species X mole generation	Mol/(volume time)
$\dot{r}_X$	Rate of species X mass generation	Mass/(volume time)
	$\sum r_X = 0, \quad \dot{r}_X = M_X \dot{r}_X^*$	
σ_X	Species X stress tensor	Force/area
σ	Mixture stress tensor	Force/area

The summation symbol is over red and blue, or any number of species

point in the same direction of maximum rate of change of the red species with respect to directional arc length. Similarly, the blue gradients in a binary mixture,

$$\nabla f_{\rm B}, \qquad \nabla \phi_{\rm B}, \qquad \nabla \omega_{\rm B}, \tag{4.2.3}$$

point in the same direction of maximum rate of change of the blue species with respect to directional arc length.

These observations allow us to employ any of these gradients in phenomenological equations providing us with expressions for diffusive fluxes according to Fick's law, as discussed in Sect. 6.1. The gradients are multiplied by corresponding diffusivities that are related by coefficients involving the species molar masses. However, this is not the case for multicomponent mixtures, as discussed in Sect. 4.3.

4.2.2 Relation Between $\nabla\phi_{\rm R}$ and $\nabla f_{\rm R}$

To develop a relation between the gradient of the molar-mass concentration, $\nabla\phi_{\rm R}$, and the gradient of the molecular fraction, $\nabla f_{\rm R}$, in a binary mixture, we differentiate the definition $\phi_{\rm R} \equiv f_{\rm R}/M$ and recall that the molar mass of the mixture, M, depends on $f_{\rm R}$ and $f_{\rm B}$ by the definition $M = M_{\rm R} f_{\rm R} + M_{\rm B} f_{\rm B}$, to find that

$$\nabla\phi_{\rm R} = \frac{1}{M}\nabla f_{\rm R} + f_{\rm R}\,\nabla\frac{1}{M}. \tag{4.2.4}$$

Carrying out the second differentiation on the right-hand side, we find that

$$\nabla\phi_{\rm R} = \frac{1}{M}\nabla f_{\rm R} - f_{\rm R}\,\frac{1}{M^2}\nabla M. \tag{4.2.5}$$

Since $f_{\rm R} + f_{\rm B} = 1$ in a binary mixture,

$$\nabla M = M_{\rm R}\nabla f_{\rm R} + M_{\rm B}\nabla f_{\rm B} = (M_{\rm R} - M_{\rm B})\,\nabla f_{\rm R}. \tag{4.2.6}$$

Substituting the right-hand side into the last term of (4.2.5), we obtain

$$\nabla\phi_{\rm R}=\frac{1}{M}\,\Big(1-f_{\rm R}\,\frac{M_{\rm R}-M_{\rm B}}{M}\Big)\,\nabla f_{\rm R}. \tag{4.2.7}$$

Further simplification provides us with the final expression

$$\nabla\phi_{\rm R} = \frac{M_{\rm B}}{M^2}\,\nabla f_{\rm R}. \tag{4.2.8}$$

A similar expression can be derived for the blue species,

$$\nabla\phi_{\rm B} = \frac{M_{\rm R}}{M^2}\,\nabla f_{\rm B}. \tag{4.2.9}$$

Note that $\nabla f_{\rm B} + \nabla f_{\rm R} = \underline{0}$, but the sum $\nabla\phi_{\rm B} + \nabla\phi_{\rm R}$ is not necessarily zero. Instead, the preceding two relations show that

$$M_{\rm R}\nabla\phi_{\rm R} + M_{\rm B}\nabla\phi_{\rm B} = \underline{0}, \tag{4.2.10}$$

as required from the identity $\phi_R M_R + \phi_B M_B = 1$. Equations (4.2.8) and (4.2.9) can be inverted readily to give f_R in terms of ϕ_R and f_B in terms of ϕ_B.

Equation (4.2.8) confirms that the gradient $\nabla\phi_R$ is aligned with the gradient ∇f_R. Correspondingly, Eq. (4.2.9) confirms that the gradient $\nabla\phi_B$ is aligned with the gradient ∇f_B. However, we will see that ∇f_R is aligned with $\nabla\phi_R$ only in the case of a binary mixture.

4.2.3 Relation Between $\nabla\omega_R$ and ∇f_R

A relation between the gradients ∇f_R and $\nabla\omega_R$, and *vice versa*, can be derived by substituting into (4.2.8) and (4.2.9) the expressions $\phi_R = \omega_R/M_R$ and $\phi_B = \omega_B/M_B$, finding that

$$\nabla\omega_R = \frac{M_R M_B}{M^2}\,\nabla f_R, \qquad \nabla\omega_B = \frac{M_R M_B}{M^2}\,\nabla f_B. \tag{4.2.11}$$

Adding these two equations, we obtain zeros on both sides. These relations can be inverted readily to give ∇f_R in terms of $\nabla\omega_R$ and ∇f_B in terms of $\nabla\omega_B$.

The relations in (4.2.11) confirm that the gradient $\nabla\omega_R$ is aligned with the gradient ∇f_R, and the gradient $\nabla\omega_B$ is aligned with the gradient ∇f_B. However, alignment occurs only in the case of a binary mixture.

4.2.4 Relation Between ∇f_R and ∇c_R

The red species concentration, c_R, is related to the red molar fraction, f_R, by $c_R = c\, f_R$, where c is the mixture concentration. Using this definition, we find that

$$\nabla c_R = c\,\nabla f_R + f_R\,\nabla c, \tag{4.2.12}$$

which shows that ∇c_R is aligned with ∇f_R only in the case of a mixture with uniform concentration, c.

4.2.5 Exercise

4.2.1 Derive a relation between the matrices of second derivatives, $\nabla\nabla\omega_R$ and $\nabla\nabla f_R$.

4.3 Gradients in a Multicomponent Mixture

Working as in Sect. 4.2, we derive generalized relations for a multicomponent mixture that reduce to those presented in Sect. 4.2 for a binary mixture.

4.3.1 First Relation Between $\nabla\phi_R$ and ∇f_X

Departing from the definition $\phi_R = f_R/M$, we derive the counterpart of Eq. (4.2.7),

$$\nabla\phi_R = \frac{1}{M}\,\Big(\nabla f_R - f_R\,\frac{1}{M}\sum M_X \nabla f_X\Big), \tag{4.3.1}$$

where the sum is over all species, X. Rearranging, we find that

$$\nabla \phi_{\rm R} = \frac{1}{M^2} \sum M_{\rm X}(f_{\rm X} \nabla f_{\rm R} - f_{\rm R} \nabla f_{\rm X}). \tag{4.3.2}$$

The summed term is zero when X = R. We have found that

$$\nabla \phi_i = A_{ij} \, \nabla \, f_j \tag{4.3.3}$$

for species i and j, where summation is implied over the repeated index j, and the elements of the matrix **A** are given by

$$A_{ii} = \frac{1}{M} \left(1 - \frac{1}{M} f_i \, M_i \right), \qquad A_{ij} = -\frac{1}{M^2} f_i \, M_j \quad \text{for} \quad i \neq j. \tag{4.3.4}$$

In fact, the matrix **A** is singular.

4.3.2 Second Relation Between $\nabla \phi_{\rm R}$ and $\nabla f_{\rm X}$

Since $\sum f_{\rm X} = 1$, the first term on the right-hand side of Eq. (4.3.1) can be rearranged to give

$$\nabla \phi_{\rm R} = -\frac{1}{M^2} \left(M {\sum}' \, \nabla f_{\rm X} + f_{\rm R} \sum M_X \nabla f_{\rm X}\right), \tag{4.3.5}$$

where the prime after the sum indicates omission of the term X = R. Rearranging, we find that

$$\nabla \phi_{\rm R} = -\frac{1}{M^2} \left({\sum}' \, (M + f_{\rm R} M_X) \nabla f_{\rm X} + f_{\rm R} \, M_{\rm R} \nabla f_{\rm R}\right), \tag{4.3.6}$$

which is equivalent to

$$\nabla \phi_{\rm R} = -\frac{1}{M^2} {\sum}' \left(M + f_{\rm R} (M_{\rm X} - M_{\rm R}) \right) \nabla f_{\rm X}. \tag{4.3.7}$$

We have found that

$$\nabla \phi_i = B_{ij} \, \nabla \, f_j \tag{4.3.8}$$

for species i and j, where summation is implied over the repeated index j, and **B** is a matrix with zero diagonal elements, $B_{ii} = 0$, and off-diagonal elements given by

$$B_{ij} = -\frac{1}{M^2} \left(M + f_i \, (M_j - M_i) \right) \quad \text{for} \quad i \neq j. \tag{4.3.9}$$

The matrix **B** is distinct and different from the matrix **A**. In fact,

$$B_{ij} = A_{ij} - A_{ii}, \tag{4.3.10}$$

that is, the matrix **B** arises from the matrix **A** by subtracting the diagonal element in each row.

4.3.3 Relation Between ∇f_R and $\nabla \phi_X$

To find the inverse relation between ∇f_R and $\nabla \phi_X$, we depart from the definition $f_R = M\phi_R$, recall that $\sum \phi_X = 1/M$, and set

$$f_R = \frac{\phi_R}{\sum \phi_X}. \tag{4.3.11}$$

Taking the gradient of both sides and using the rules of quotient differentiation, we obtain

$$\nabla f_R = M\,(\,\nabla\phi_R - M\phi_R \sum \nabla\phi_X\,), \tag{4.3.12}$$

which can be restated as

$$\nabla f_R = \frac{M}{M_R}\big(\,(1-f_R)\,M_R\nabla\phi_R - M_R\,f_R \sum{}'\nabla\phi_X\big), \tag{4.3.13}$$

where a prime after the sum indicates omission of the term X = R. Now recalling that $\sum \phi_X M_X = 1$, we find that

$$M_R\nabla\phi_R = -\sum{}' M_X\nabla\phi_X. \tag{4.3.14}$$

Substituting this expression into the first term on the right-hand side of (4.3.13), and rearranging, we obtain

$$\nabla f_R = -\frac{M}{M_R}\sum{}'\big(\,M_X - f_R\,(M_X - M_R)\big)\,\nabla\phi_X, \tag{4.3.15}$$

which shows that

$$\nabla f_i = C_{ij}\,\nabla\,\phi_j \tag{4.3.16}$$

for species i and j, where summation is implied over the repeated index i, and **C** is a matrix with zero diagonal elements, $C_{ii} = 0$, and off-diagonal elements given by

$$C_{ij} = -\frac{M}{M_i}\big(\,M_j - f_i\,(M_j - M_i)\big) \quad \text{for} \quad i \neq j. \tag{4.3.17}$$

The matrix **C** is *not* the inverse of the matrix **B** defined in (4.3.9).

4.3.4 Relation Between $\nabla\omega_R$ and ∇f_X

Substituting into expressions (4.3.2), (4.3.7), and (4.3.15) the definition $\phi_X = \omega_X/M_X$ and rearranging, we obtain the expression

$$\nabla\omega_R = \frac{M_R}{M^2}\sum M_X(f_X\nabla f_R - f_R\nabla f_X), \tag{4.3.18}$$

the alternative expression

$$\nabla \omega_{\rm R} = -\frac{M_{\rm R}}{M^2} \sum{}' \big(M + f_{\rm R}(M_X - M_{\rm R}) \big) \nabla f_{\rm X}, \tag{4.3.19}$$

and the inverse expression

$$\nabla f_{\rm R} = -\frac{M}{M_{\rm R}} \sum{}' \big(1 - f_{\rm R} \,(1 - \frac{M_{\rm R}}{M_{\rm X}}) \big) \nabla \omega_{\rm X}, \tag{4.3.20}$$

where a prime after the sum indicates omission of the term X = R.

4.3.5 Exercise

4.3.1 Demonstrate that the matrix **A** is singular.

4.4 Species Velocity and Momentum

Let $\mathbf{u}_{\rm R}(\mathbf{x}, t)$ be the average (arithmetic mean) velocity of the red molecules, and let $\mathbf{u}_{\rm B}(\mathbf{x}, t)$ be the average velocity of the blue molecules at a certain location, $\mathbf{x}$, and time, t, defined in the context of the continuum approximation, as discussed in Sect. 1.1 for homogeneous fluids. In general, the local and instantaneous red and blue velocities will be different.

4.4.1 Mass (Barycentric) Velocity

The mass velocity of the mixture, also called the *barycentric velocity*, is defined in terms of the species velocities, $\mathbf{u}_{\rm R}$ and $\mathbf{u}_{\rm B}$, and species mass fractions, $\omega_{\rm R}$ and $\omega_{\rm B}$, as

$$\mathbf{u}(\mathbf{x}, t) \equiv \omega_{\rm R} \mathbf{u}_{\rm R} + \omega_{\rm B} \mathbf{u}_{\rm B}. \tag{4.4.1}$$

It is important to note that the mass fluid velocity is weighed with respect to the species *mass fractions,* $\omega_{\rm R}$ and $\omega_{\rm B}$, and *not* with respect to the molecular (molar) fractions, $f_{\rm R}$ and $f_{\rm B}$. The mixture velocity $\mathbf{u}$ can be interpreted as the velocity of the local center of mass of the mixture species.

In the case of a multicomponent mixture, the mixture mass velocity is given by

$$\mathbf{u} = \sum_{\rm X} \omega_{\rm X} \mathbf{u}_X, \tag{4.4.2}$$

where the sum is over all species and $\sum \omega_X = 1$. Depending on physical context, some of the species velocities or the mixture velocity itself may be zero.

4.4.2 Significance of the Mixture Mass Velocity

To demonstrate the significance of the mixture mass velocity, we observe that the rate of convective mixture mass transport across a small surface δS in the fluid is

$$\rho \, \mathbf{n} \cdot \mathbf{u} \, \delta S = \rho \, (\omega_{\rm R} \, \mathbf{n} \cdot \mathbf{u}_{\rm R} + \omega_{\rm B} \, \mathbf{n} \cdot \mathbf{u}_{\rm B}) \, \delta S, \tag{4.4.3}$$

where $\mathbf{n}$ is the unit vector normal to the surface. Invoking the definition of the species densities, $\rho_R = \omega_R \rho$ and $\rho_B = \omega_B \rho$, we find that

$$\rho\, \mathbf{n} \cdot \mathbf{u}\, \delta S = (\rho_R\, \mathbf{n} \cdot \mathbf{u}_R + \rho_B\, \mathbf{n} \cdot \mathbf{u}_B)\, \delta S, \tag{4.4.4}$$

which is the sum of the rates of mass transport of the red and blue species. In contrast, if we had defined by velocity $\mathbf{u}$ in terms of the molar fractions, species mass additivity would not be satisfied.

4.4.3 Species Mass Flux

It is sometimes convenient to employ the mass flux of a mixture,

$$\mathbf{m} \equiv \rho\, \mathbf{u} = \mathbf{m}_R + \mathbf{m}_B, \tag{4.4.5}$$

where

$$\mathbf{m}_R \equiv \rho_R \mathbf{u}_R, \qquad \mathbf{m}_B \equiv \rho_B \mathbf{u}_B \tag{4.4.6}$$

are species mass fluxes. The rate of mass transport introduced in (4.4.3) is

$$\rho\, \mathbf{n} \cdot \mathbf{u}\, \delta S = \mathbf{n} \cdot \mathbf{m}\, \delta S = \mathbf{n} \cdot (\mathbf{m}_R + \mathbf{m}_B)\, \delta S. \tag{4.4.7}$$

The rate of mass transport of the red species is $\mathbf{n} \cdot \mathbf{m}_R\, \delta S$, and the rate of mass transport of the blue species is $\mathbf{n} \cdot \mathbf{m}_B\, \delta S$.

4.4.4 Momentum Velocity

The momentum of an infinitesimal parcel with density ρ and volume δV is

$$\rho\, \mathbf{u}\, \delta V = \rho\, (\omega_R \mathbf{u}_R + \omega_B \mathbf{u}_B)\, \delta V. \tag{4.4.8}$$

Invoking the definition of the species densities, ρ_R and ρ_B, we obtain

$$\rho\, \mathbf{u}\, \delta V = (\rho_R \mathbf{u}_R + \rho_B \mathbf{u}_B)\, \delta V, \tag{4.4.9}$$

where the right-hand side is the sum of the momenta of the red and blue species. In contrast, if we had defined by velocity $\mathbf{u}$ in terms of the molar fractions, species momentum additivity would not be satisfied.

4.4.5 Mixture Governing Equations

We conclude and will confirm that the mixture velocity $\mathbf{u}$ defined in (4.4.1) is both the mass velocity appearing in the continuity equation for the mixture,

$$\frac{\partial \rho}{\partial t} + \nabla \cdot (\rho \mathbf{u}) = 0, \tag{4.4.10}$$

and the momentum velocity appearing in Cauchy's equation of motion for the mixture,

$$\frac{\partial(\rho\,\mathbf{u})}{\partial t} + \nabla\cdot(\rho\,\mathbf{u}\otimes\mathbf{u}) = \nabla\cdot\boldsymbol{\sigma} + \rho\,\mathbf{g} + \iota_R\,\mathbf{b}_R + \iota_B\,\mathbf{b}_B, \tag{4.4.11}$$

where ρ is the mixture density, $\boldsymbol{\sigma}$ is a mixture stress tensor defined as the case of a homogeneous fluid in terms of the traction, as discussed in Sect. 2.1, $\mathbf{g}$ is the acceleration of gravity, and $\mathbf{b}_R$ and $\mathbf{b}_B$ are force fields specific to the red or blue species associated with physical constants ι_R and ι_B.

4.4.6 Significance of the Species Velocities

Equations (4.4.10) and (4.4.11) are identical to those governing the motion of a homogeneous fluid with $\iota\mathbf{b} = \iota_R\,\mathbf{b}_R + \iota_B\,\mathbf{b}_B$. The red and blue species velocity fields, $\mathbf{u}_R$ and $\mathbf{u}_B$, do not appear explicitly in these macroscopic governing equations. Instead, the species velocities influence the motion of the mixture by way of boundary conditions, stipulations on the species velocities, and in terms of the mixture stress tensor, $\boldsymbol{\sigma}$.

4.4.7 Mixture Lag Velocities

Consideration of the individual species velocities with respect to the mixture velocity is a key to understanding the physical mechanism of diffusion. The mixture velocity, $\mathbf{u}$, lags behind the red and blue velocities by

$$\Delta\mathbf{u}_R \equiv \mathbf{u}_R - \mathbf{u}, \qquad \Delta\mathbf{u}_B \equiv \mathbf{u}_B - \mathbf{u}. \tag{4.4.12}$$

Substituting $\mathbf{u} \equiv \omega_R\mathbf{u}_R + \omega_B\mathbf{u}_B$, we find that

$$\Delta\mathbf{u}_R = \omega_B\,(\mathbf{u}_R - \mathbf{u}_B), \qquad \Delta\mathbf{u}_B = \omega_R\,(\mathbf{u}_B - \mathbf{u}_R) \tag{4.4.13}$$

in a binary mixture. Note the juxtaposition of subscripts on the right-hand sides. Combining these equations, we obtain

$$\omega_R\Delta\mathbf{u}_R + \omega_B\Delta\mathbf{u}_B = \underline{0}. \tag{4.4.14}$$

Multiplying both sides of this equation by the mixture density, ρ, converts ω_R to ρ_R and ω_B to ρ_B, yielding

$$\rho_R\Delta\mathbf{u}_R + \rho_B\Delta\mathbf{u}_B = \underline{0}. \tag{4.4.15}$$

The lag velocities in a binary mixture point into opposite directions.

The difference between the average velocity of the red and blue species and the mixture velocity is given by

$$\tfrac{1}{2}\,(\mathbf{u}_R+\mathbf{u}_B)-\mathbf{u}=\tfrac{1}{2}\,(\omega_B-\omega_R)\,(\mathbf{u}_R-\mathbf{u}_B). \tag{4.4.16}$$

We observe that the difference is zero only in the case of equal mass fractions independent of the species velocities, or else in the case of equal species velocities independent of the species mass fractions.

4.4.8 Multicomponent Mixtures

The lag velocity of the mixture relative to the species X velocity in a multicomponent mixture is defined as

$$\Delta \mathbf{u}_X \equiv \mathbf{u}_X - \mathbf{u}, \tag{4.4.17}$$

where $\mathbf{u} = \sum_X \omega_X \mathbf{u}_X$ is the mixture barycentric velocity. By definition,

$$\sum_X \omega_X \Delta \mathbf{u}_X = \mathbf{0}. \tag{4.4.18}$$

Multiplying both sides of this equation by the mixture density, ρ, converts ω_X to ρ_X inside the sum.

4.4.9 Diffusion Velocities

In the context of mass transfer for molecular mixtures discussed in this book, the lag velocities defined in (4.4.17) are due to molecular diffusion associated with random walks, and are therefore appropriately called *diffusion velocities*. Different physical mechanics may be responsible for lag velocities in nonmolecular mixtures and granular materials, such as blood, washed gravel, modified gravel, rip-rap, and sand.

4.4.10 Exercises

4.4.1 Describe a set of physical circumstances or engineering applications where the force fields $\mathbf{b}_R$ and $\mathbf{b}_B$ are different.

4.4.2 Derive expression (4.4.16).

4.5 Solutions vs. Multicomponent Materials

A suspension of small particles, an emulsion of droplets or deformable biological cells, a cup of coffee with added and stirred cream, and a scoop of unwashed contractor's gravel are all mixtures. However, these materials typically respond differently under flow conditions, essentially due to the length scale where mixing of the constituent particles takes place. When mixing occurs all the way down to the molecular level, the mixture is truly a molecular mixture, also known as a *chemical solution.*

4.5.1 Coffee

The flow of pure coffee through a straight circular tube is described by Poiseuille law involving the coffee viscosity. The flow of pure cream is described by Poiseuille law involving the cream viscosity. The flow of coffee with added and well-mixed cream is also described by Poiseuille law involving a viscosity that is intermediate between that of coffee and cream.

Coffee with added and well-stirred cream is a molecular mixture. When the beverage is well mixed, coffee and cream point particles move with the *same* velocity at any location, despite their distinct chemical constitutions.

4.5.2 Blood

Consider blood flow through a tube, such as a blood vessel. The red blood cells (red species) exhibit a distinct hydrodynamics that is significantly different from that of the ambient plasma (blue species.) Even when blood is well mixed, plasma and red-blood-cell particles move with different velocities at any location. Treating blood as an effectively homogeneous medium (molecular mixture) is sensible only under restricted conditions.

Sometimes, nonmolecular mixtures are called *heterogeneous, multicomponent*, or *multiphase* mixtures to distinguish them from genuine chemical solutions.

4.5.3 Diffusion in Molecular Mixtures

The standard theory of diffusion discussed in this book is pertinent to molecular mixtures or chemical solutions that behave like a homogeneous fluid under uniform (spatially homogeneous) molecular fractions, as in the case of coffee with well-mixed cream. This means that the mixture lag velocity for species X, $\Delta \mathbf{u}_X \equiv \mathbf{u}_X - \mathbf{u}$, is due exclusively to diffusion mediated by random molecular motions, as discussed in Sect. 4.9. Conversely, because of diffusion, the species velocity fields are generally different from each other, even when the individual species are isotopes.

4.5.4 Kinetic Energy

A further distinction between solutions and multicomponent mixtures can be made with regard to the kinetic energy. The nominal volumetric kinetic energy of a molecular mixture is equal to $\frac{1}{2}\,\rho\,|\mathbf{u}|^2$, the volumetric red kinetic energy is $\frac{1}{2}\,\rho_R|\mathbf{u}_R|^2$, and the volumetric blue kinetic energy is $\frac{1}{2}\,\rho_B|\mathbf{u}_B|^2$. The term *volumetric* conveys the notion of energy contained inside a small volume, divided by the small volume.

Invoking the definition of the mixture mass velocity, $\mathbf{u} \equiv \omega_R \mathbf{u}_R + \omega_B \mathbf{u}_B$, we find that the difference between the sum of the species kinetic energies and the mixture kinetic energy in a binary mixture is

$$\frac{1}{2}\,\rho_R\,|\mathbf{u}_R|^2 + \frac{1}{2}\,\rho_B\,|\mathbf{u}_B|^2 - \frac{1}{2}\,\rho\,|\mathbf{u}|^2 = \frac{1}{2}\,\rho\,\varepsilon^2, \tag{4.5.1}$$

where

$$\varepsilon^2 \equiv \omega_R\,|\mathbf{u}_R - \mathbf{u}|^2 + \omega_B\,|\mathbf{u}_B - \mathbf{u}|^2, \tag{4.5.2}$$

which can be restated as

$$\varepsilon^2 = \omega_R\,\omega_B\,|\mathbf{u}_R - \mathbf{u}_B|^2. \tag{4.5.3}$$

Thus,

$$\frac{1}{2}\,\rho_R\,|\mathbf{u}_R|^2 + \frac{1}{2}\,\rho_B\,|\mathbf{u}_B|^2 - \frac{1}{2}\,\rho\,|\mathbf{u}|^2 = \frac{1}{2}\,\rho\,\omega_R\,\omega_B\,|\mathbf{u}_R - \mathbf{u}_B|^2. \tag{4.5.4}$$

We see that the sum of the kinetic energies of the species is always higher than that of the mixture.

Moreover, the kinetic energy of a molecular mixture is approximately equal to the sum of the kinetic energies of the species only when

$$|\mathbf{u}_R - \mathbf{u}_B|^2 \ll |\mathbf{u}|^2. \tag{4.5.5}$$

This inequality is significant in the context of energetics and thermodynamics. If this inequality is not satisfied, a mixture cannot be regarded a solution and must be treated as a multicomponent material in the context of energetics.

4.5.5 Exercises

4.5.1 Consider a system consisting of two translating rigid bodies. Define the mass velocity of the system and show that the corresponding kinetic energy is not necessarily the sum of the kinetic energies of the bodies.

4.5.2 Derive (4.5.2) and (4.5.3).

4.6 Fick–Stefan Framework

Consider an arbitrary parcel of a binary mixture consisting of red and blue molecules at a particular instant in time. In the Fick–Stefan theory of mixtures, a mixture parcel is regarded as an instantaneous superposition of monochromatic parcels consisting of red or blue molecules, as shown in Fig. 4.2, or any number of species parcels consisting of identical molecules in a multicomponent solution. The superposition is analogous to that of red–green–blue (RGB) images comprising arbitrary color or that of three fundamental images on a Kodak color film.

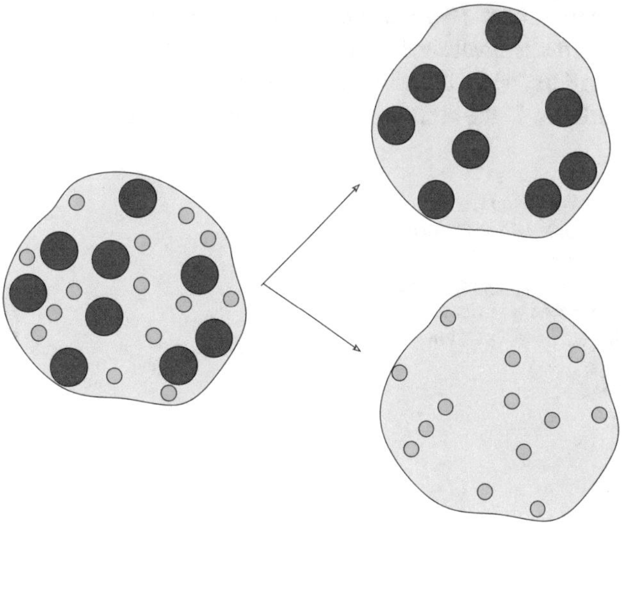

Fig. 4.2 A fluid parcel consisting of red and blue molecules can be regarded as an instantaneous superposition of monochromatic parcels consisting of red or blue molecules. The red and blue parcels move through each other in a ghost-like fashion as mutual apparitions

4.6.1 Point Particles

Red and blue point particles can be defined as in the case of a homogeneous fluid, as discussed in Sect. 1.4, and are found *simultaneously* at any point in space. Red point particles move with the red velocity, $\mathbf{u}_R$, and blue point particles move with the blue velocity, $\mathbf{u}_B$. In the context of continuum mechanics, species point particles consist of a fixed collection of molecules.

Monochromatic parcels are comprised of the same permanent collection of red or blue point particles. Mixture point particles are nonphysical entities in that they do *not* consist of a fixed collection of red and blue molecules, except in the absence of diffusion. By definition, mixture point particles move with the mixture mass velocity, $\mathbf{u}$.

4.6.2 The Boundaries of Red and Blue Parcels Are Sharp and Well Defined

Under the auspices of the continuum approximation, and in the absence of singularities, red point particles at the boundary of a red parcel remain at the boundary of the red parcel, and blue point particles at the boundary of a blue parcel remain at the boundary of a blue parcel. Diffusion neither smears the boundary of monochromatic parcels nor defies the fundamental principles of kinematics. Mixture point particles at the boundary of a mixture parcel moving with the mixture mass velocity also remain at the boundary of a mixed parcel during the motion.

4.6.3 Red and Blue Parcel Separation

In the absence of diffusion, the red and blue parcels constituting a mixture parcel remain attached on top of one another as they are convected with a common velocity field in a flow. By contrast, when diffusion occurs, the constituent red and blue parcels separate at subsequent times. Parcel separation is a key to understanding the essence and kinematics of diffusion and multiphase flow.

Assume that a red and a blue parcel coincide at a certain time, as shown in Fig. 4.3a. The red contour indicates the red parcel moving with velocity $\mathbf{u}_R$, the blue contour indicates the blue parcel moving with velocity $\mathbf{u}_B$, and the shaded area indicates the mixture parcel moving with the mixture mass velocity, $\mathbf{u}$. The three velocities are different even if the red and blue molecules are chemically identical (isotopes), and are the same only in the absence of diffusion.

After a small time interval has elapsed, the monochromatic parcels have separated, as shown in Fig. 4.3b, while the associated mixture parcel lies between the two deformed monochromatic parcel surfaces. In the absence of global motion, the mixture parcel is stationary.

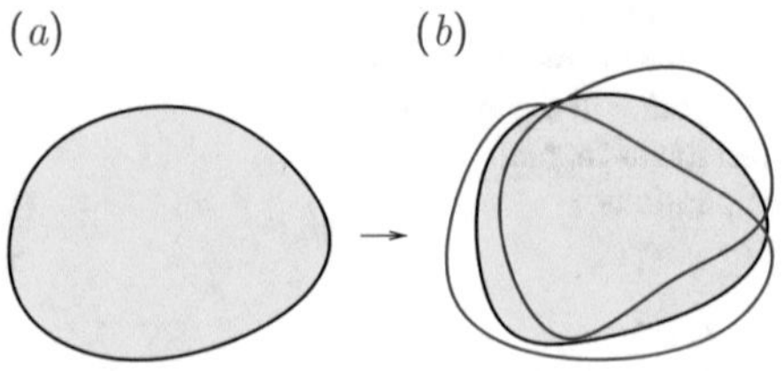

Fig. 4.3 After a small time interval has elapsed, two coincident red and blue parcels separate. The separation is attributed to the action of diffusion

4.6.4 Diffusion as the Result of Red and Blue Parcel Separation

Red and blue parcel separation in a molecular mixture is attributed exclusively to diffusion. Conversely, diffusion can be attributed to red and blue parcel separation, that is, the physical process of diffusion can be associated with the mechanical process of separation, in that red and blue parcels do not separate in the absence of diffusion.

In this interpretation, diffusion appears as a device that implements our empiricism regarding the relative motion of red and blue parcels in a binary fluid. This empiricism is combined with our phenomenological observation that, unless energy is continuously expended, physical systems tend to increase their entropy by becoming homogeneous in time.

4.6.5 Diffusion in a Box

Consider the one-dimensional diffusion experiment in a box described in Fig. 4.4. A red rectangular material parcel is confined between the left boundary and the white vertical line, as shown in Fig. 4.4a. Diffusion will cause the binary mixture to obtain a uniform color in time. Consequently, the white vertical line marking the boundary of the red parcel will shift to the right, as shown in Fig. 4.4b.

The mass of the aforementioned red parcel is conserved, while the volume of the parcel increases during the evolution. This means that the red parcel behaves like a compressible medium by undergoing dilation. If m_R^l is the red mass on the left and m_R^r is the red mass on the right of the vertical white line in Fig. 4.2a, then

$$\frac{m_\text{R}^l}{W} = \frac{m_\text{R}^l + m_\text{R}^r}{L}, \tag{4.6.1}$$

where L is the length of the box and W is the length of the parcel in Fig. 4.2b.

Conversely, consider a rectangular blue material parcel confined between the right boundary and the yellow vertical line shown in Fig. 4.2c. Diffusion will cause the yellow vertical line to shift to the left, as shown in Fig. 4.2d. The mass in the blue parcel is conserved, while the parcel volume increases during the evolution.

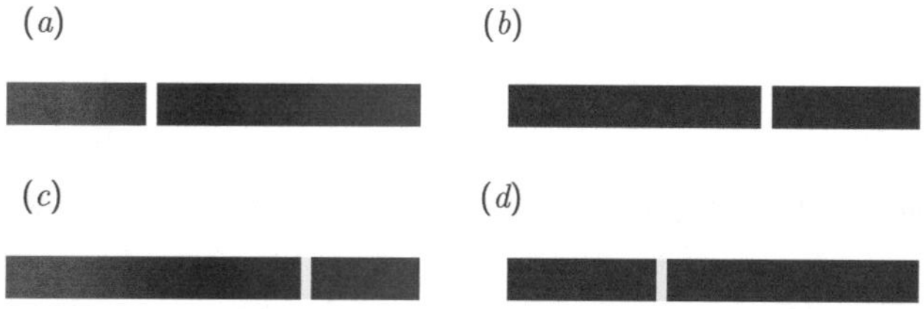

Fig. 4.4 Diffusion causes a binary mixture in a box to obtain a uniform color. The white vertical lines in (**a**) marking the boundary of a red parcel shifts to the right, as shown in (**b**), while the yellow vertical line in (**c**) marking the boundary of a blue parcel shifts to the left, as shown in (**d**). Depending on the initial condition and molar masses of the red and blue species, a net transport of mass to the right or left may take place

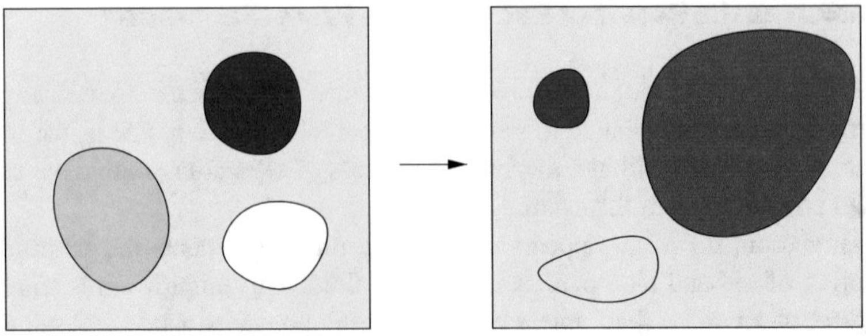

Fig. 4.5 Illustration of a red parcel expanding and dilating and a blue parcel condensing and contracting under the influence of the velocity associated with the diffusive flux. A mixed parcel of an effectively incompressible fluid moving with the fluid velocity retains its volume

4.6.6 Parcel Expansion

We saw that monochromatic red or blue parcels do not necessarily retain their volume as they move in the available domain of flow, even when the mixture is perfectly or nearly incompressible. The presence of the blue species affects the rate of expansion of the red species, and *vice versa.*

4.6.7 Parcel Advection

In summary, a monochromatic parcel advected in an arbitrary flow with its own velocity field, $\mathbf{u}_R$, will expand or contract as it moves about the domain of flow until the red mass fraction becomes uniform over the parcel area or volume, equal to the total red mass divided by the total area or volume of flow. By contrast, a mixture parcel of an incompressible fluid will retain its area or volume as it is convected by the flow.

The process is illustrated in Fig. 4.5 for flow in a square domain where a red parcel expands and dilates as its contour moves under the partial influence of the velocity associated with the red diffusive flux, while a blue parcel is compressed and condenses as its contour moves under the partial influence of the velocity associated with the blue diffusive flux. A mixture parcel of an effectively incompressible fluid moving with the fluid velocity, $\mathbf{u}$, retains its volume.

4.6.8 Exercises

4.6.1 Discuss a nonmaterial conceptual mixture where two species move through each other with different velocities.

4.6.2 Draw the counterpart of Fig. 4.3 for an idealized one-dimensional flow.

4.7 Species Kinematics

Red and blue velocity fields enjoy their own kinematics, which is distinct and different from the overall kinematics of the mixture associated with the mass velocity, $\mathbf{u}$.

4.7.1 Red and Blue Lagrangian Frameworks

A red or blue Lagrangian framework pertinent to red or blue material point particles can be introduced, as discussed in Sect. 1.4 for homogeneous fluids. Let $\boldsymbol{\alpha}_{\rm R}$ be the Lagrangian labeling vector for red particles and $\boldsymbol{\alpha}_{\rm B}$ be the Lagrangian labeling vector for blue particles. The instantaneous positions of red and blue point particles are denoted by

$$\mathbf{x} = \mathbf{X}_{\rm R}(\boldsymbol{\alpha}_{\rm R}, t), \qquad \mathbf{x} = \mathbf{X}_{\rm B}(\boldsymbol{\alpha}_{\rm B}, t). \tag{4.7.1}$$

Partial derivatives with respect to time, t, or with respect to each one of the three scalar components of $\boldsymbol{\alpha}_{\rm R}$ or $\boldsymbol{\alpha}_{\rm B}$ are well defined. The mapping of labeling to the physical space is mediated by the generally time-dependent functions

$$\mathbf{X}_{\rm R}(t) = \mathcal{C}_t(\boldsymbol{\alpha}_{\rm R}), \qquad \mathbf{X}_{\rm B}(t) = \mathcal{C}_t(\boldsymbol{\alpha}_{\rm B}). \tag{4.7.2}$$

Similar species mapping functions are defined in a multicomponent mixture.

A red differential volume in labeling space, $\mathrm{d}V_{\rm R}(\boldsymbol{\alpha}_{\rm R})$, is mapped to a red differential material volume in physical space, $\mathrm{d}V_{\rm R}(\mathbf{X}_{\rm R})$, as discussed in Sect. 1.4. The two volumes are related by the equation

$$\mathrm{d}V_{\rm R}(\mathbf{X}_{\rm R}) = \mathcal{J}_{\rm R}(\boldsymbol{\alpha}_{\rm R})\,\mathrm{d}V(\boldsymbol{\alpha}_{\rm R}), \tag{4.7.3}$$

where $\mathcal{J}_{\rm R}$ is the red Lagrangian volumetric coefficient. A similar equation can be written for a differential blue volume.

4.7.2 Red and Blue Material Derivatives

The red material derivative of an appropriate field, ϕ, is defined with respect to the red-species velocity, $\mathbf{u}_{\rm R}$, as

$$\left(\frac{\mathrm{D}\phi}{\mathrm{D}t}\right)_{\rm R} \equiv \frac{\partial \phi}{\partial t} + \mathbf{u}_{\rm R} \cdot \nabla\phi. \tag{4.7.4}$$

The blue material derivative is defined in a similar fashion with respect to the blue-species velocity, $\mathbf{u}_{\rm B}$, as

$$\left(\frac{\mathrm{D}\phi}{\mathrm{D}t}\right)_{\rm B} \equiv \frac{\partial \phi}{\partial t} + \mathbf{u}_{\rm B} \cdot \nabla\phi. \tag{4.7.5}$$

Similar species material derivatives are defined in a multicomponent mixture.

Physically, the red or blue material derivative of ϕ is the rate of change of ϕ following a red or blue material point particle. It is not surprising that the red material derivative is typically written for a red property, while the blue material derivative is typically written for a blue property.

For example, if $\mathbf{X}_{\rm R}$ is the position of a red point particle and $\mathbf{X}_{\rm B}$ is the position of a blue point particle, then

$$\mathbf{u}_{\rm R} = \left(\frac{\mathrm{D}\mathbf{X}_{\rm R}}{\mathrm{D}t}\right)_{\rm R}, \qquad \mathbf{u}_{\rm B} = \left(\frac{\mathrm{D}\mathbf{X}_{\rm B}}{\mathrm{D}t}\right)_{\rm B}, \tag{4.7.6}$$

where the velocities on the left-hand sides are evaluated at $\mathbf{X}_{\rm R}$ or $\mathbf{X}_{\rm B}$.

The Eulerian derivatives of a shared property, ϕ, with respect to time, t, or position, $\mathbf{x}$, are common for both species. Consequently, because $\partial\phi/\partial t$ and $\nabla\phi$ are the same in the definition of the red and blue material derivatives,

$$\left(\frac{\mathrm{D}\phi}{\mathrm{D}t}\right)_{\mathrm{R}} - \left(\frac{\mathrm{D}\phi}{\mathrm{D}t}\right)_{\mathrm{B}} = (\mathbf{u}_{\mathrm{R}} - \mathbf{u}_{\mathrm{B}}) \cdot \nabla\phi. \tag{4.7.7}$$

In the context of mass transfer, the difference in the species velocities on the right-hand side is due to diffusion.

4.7.3 Mixture Material Derivative

The mixture material derivative is defined with respect to the mixture mass velocity, $\mathbf{u}$, as

$$\frac{\mathrm{D}\phi}{\mathrm{D}t} \equiv \frac{\partial\phi}{\partial t} + \mathbf{u} \cdot \nabla\phi. \tag{4.7.8}$$

For species X in a binary or multicomponent solution, we find that

$$\left(\frac{\mathrm{D}\phi}{\mathrm{D}t}\right)_{\mathrm{X}} = \frac{\mathrm{D}\phi}{\mathrm{D}t} + (\mathbf{u}_{\mathrm{X}} - \mathbf{u}) \cdot \nabla\phi. \tag{4.7.9}$$

The velocity difference on the right-hand side is the species X diffusion velocity.

Multiplying the definition of the red material derivative in (4.7.4) by ω_{R} and the definition of the blue material derivative in (4.7.5) by ω_{B}, adding the resulting expressions, and recalling that $\omega_{\mathrm{R}} + \omega_{\mathrm{B}} = 1$ and $\mathbf{u} = \omega_{\mathrm{R}}\mathbf{u}_{\mathrm{R}} + \omega_{\mathrm{B}}\mathbf{u}_{\mathrm{B}}$, we obtain a reassuring relation,

$$\frac{\mathrm{D}\phi}{\mathrm{D}t} = \omega_{\mathrm{R}} \left(\frac{\mathrm{D}\phi}{\mathrm{D}t}\right)_{\mathrm{R}} + \omega_{\mathrm{B}} \left(\frac{\mathrm{D}\phi}{\mathrm{D}t}\right)_{\mathrm{B}}. \tag{4.7.10}$$

The satisfaction of this equation hinges on the definition of the mixture mass velocity. This definition is not only appropriate for mass and momentum transport, but also consistent with the underlying kinematics.

In the case of a multicomponent mixture, we find that

$$\frac{\mathrm{D}\phi}{\mathrm{D}t} = \sum_{\mathrm{X}} \omega_{\mathrm{X}} \left(\frac{\mathrm{D}\phi}{\mathrm{D}t}\right)_{\mathrm{X}} \tag{4.7.11}$$

for any field function that is shared across species, ϕ, where the sum is over all mixture components. An example of a shared property could be the temperature.

4.7.4 Acceleration

The acceleration of a red point particle, a blue point particle, and a mixture point particle is given by the corresponding material derivative of the respective velocity,

$$\mathbf{a}_R = \left(\frac{D\mathbf{u}_R}{Dt}\right)_R, \qquad \mathbf{a}_B = \left(\frac{D\mathbf{u}_B}{Dt}\right)_B, \qquad \mathbf{a} = \frac{D\mathbf{u}}{Dt}. \tag{4.7.12}$$

Substituting the definition of the mixture mass velocity, we obtain

$$\mathbf{a} = \frac{\partial(\omega_R \mathbf{u}_R + \omega_B \mathbf{u}_B)}{\partial t} + \mathbf{u} \cdot \nabla \mathbf{u}. \tag{4.7.13}$$

Expanding the time derivative of products, we find that

$$\mathbf{a} = \frac{\partial \omega_R}{\partial t}\mathbf{u}_R + \frac{\partial \omega_B}{\partial t}\mathbf{u}_B + \omega_R \frac{\partial \mathbf{u}_R}{\partial t} + \omega_B \frac{\partial \mathbf{u}_B}{\partial t} + \mathbf{u} \cdot \nabla \mathbf{u}, \tag{4.7.14}$$

yielding

$$\mathbf{a} = \omega_R\, \mathbf{a}_R + \omega_B\, \mathbf{a}_B + \widehat{\mathbf{a}}, \tag{4.7.15}$$

where

$$\widehat{\mathbf{a}} = \frac{\partial \omega_R}{\partial t}(\mathbf{u}_R - \mathbf{u}_B) + \mathbf{u} \cdot \nabla \mathbf{u} - \omega_R\, \mathbf{u}_R \cdot \nabla \mathbf{u}_R - \omega_B\, \mathbf{u}_B \cdot \nabla \mathbf{u}_B \tag{4.7.16}$$

is a coupling acceleration term. We recall that $\mathbf{u} \cdot \nabla \mathbf{u}$, $\mathbf{u}_R \cdot \nabla \mathbf{u}_R$, and $\mathbf{u}_B \cdot \nabla \mathbf{u}_B$ are directional derivatives. We conclude that the acceleration of a mixture particle is not necessarily a weighted average of the accelerations of species point particles.

4.7.5 Reynolds Transport Equation

A Reynolds transport equation similar to that shown in (1.7.8) for a homogeneous fluid can be written for a red parcel, a blue parcel, or a mixture parcel, involving corresponding velocities. For parcel X, we obtain

$$\frac{d}{dt}\iiint_{X\ parcel} \phi\, dV = \iiint_{X\ parcel} \frac{\partial \phi}{\partial t}\, dV + \iint_{X\ parcel} \phi\, \mathbf{n} \cdot \mathbf{u}_X\, dS, \tag{4.7.17}$$

where ϕ is a suitable intensive property, $\mathbf{n}$ is the normal unit vector pointing *outward* from the parcel, and $\mathbf{n} \cdot \mathbf{u}$ is the normal velocity component. For a mixture parcel we obtain

$$\frac{d}{dt}\iiint_{\text{mixture parcel}} \phi\, dV = \iiint_{\text{mixture parcel}} \frac{\partial \phi}{\partial t}\, dV + \iint_{\text{mixture parcel}} \phi\, \mathbf{n} \cdot \mathbf{u}\, dS. \tag{4.7.18}$$

Note that different velocities are employed in the second integrals on the right-hand sides. In practice, Eq. (4.7.17) is written for a species X property, such as the species density, and Eq. (4.7.18) is written for a mixture property, such as the mixture density.

4.7.6 Parcel Superposition

Consider a mixture parcel and its superposed constituent species parcels at a certain time instant, as shown in Fig. 4.2. The integration domains on the right-hand sides of (4.7.17) and (4.7.18) are identical. Subtracting pairs of equations, we obtain

$$\frac{\mathrm{d}}{\mathrm{d}t}\iiint_{X\ parcel}\phi\,\mathrm{d}V = \frac{\mathrm{d}}{\mathrm{d}t}\iiint_{Y\ parcel}\phi\,\mathrm{d}V + \iint_{\text{parcel}}\phi\,\mathbf{n}\cdot(\mathbf{u}_{\mathrm{X}} - \mathbf{u}_{\mathrm{Y}})\,\mathrm{d}S \tag{4.7.19}$$

and

$$\frac{\mathrm{d}}{\mathrm{d}t}\iiint_{X\ parcel}\phi\,\mathrm{d}V = \frac{\mathrm{d}}{\mathrm{d}t}\iiint_{\text{mixture parcel}}\phi\,\mathrm{d}V + \iint_{\text{parcel}}\phi\,\mathbf{n}\cdot(\mathbf{u}_{\mathrm{X}} - \mathbf{u})\,\mathrm{d}S \tag{4.7.20}$$

for superposed parcels. The last integral on the right-hand side of (4.7.20) involves the X species lag velocity, $\Delta\mathbf{u}_{\mathrm{X}} \equiv \mathbf{u}_{\mathrm{X}} - \mathbf{u}$. Equations (4.7.19) and (4.7.20) will be recalled in Chaps. 5 and 7 for evaluating the rates of changes of parcel mass and momenta.

4.7.7 Exercise

4.7.1 Discuss the counterpart of (4.7.7) for a multicomponent mixture.

4.8 Fundamental Decomposition of Kinematics

The fundamental decomposition of kinematics for a homogeneous fluid was discussed in Sect. 1.3 with reference to the velocity-gradient tensor,

$$\mathbf{L} \equiv \nabla\mathbf{u}, \tag{4.8.1}$$

where $L_{ij} = \partial u_j/\partial x_i$. Velocity-gradient tensors and associated rates of expansion, vorticity, and rate-of-deformation tensors can be defined for the individual species.

4.8.1 Velocity-Gradient Tensor

Taking the gradient of the mixture velocity, $\mathbf{u} = \omega_{\mathrm{R}}\,\mathbf{u}_{\mathrm{R}} + \omega_{\mathrm{B}}\,\mathbf{u}_{\mathrm{B}}$, we find that the velocity gradient of the mixture is given by

$$\nabla\mathbf{u} = \omega_{\mathrm{R}}\nabla\mathbf{u}_{\mathrm{R}} + \omega_{\mathrm{B}}\nabla\mathbf{u}_{\mathrm{B}} + \widehat{\nabla\mathbf{u}}, \tag{4.8.2}$$

where

$$\widehat{\nabla\mathbf{u}} \equiv \nabla\omega_{\mathrm{R}}\otimes\mathbf{u}_{\mathrm{R}} + \nabla\omega_{\mathrm{B}}\otimes\mathbf{u}_{\mathrm{B}} \tag{4.8.3}$$

is a coupling velocity-gradient tensor and $\otimes$ denotes the tensor product. These expressions admit a direct generalization to an arbitrary number of species in a multicomponent mixture.

In the case of a binary mixture, because $\nabla \omega_B = -\nabla \omega_R$, we find that

$$\widehat{\nabla \mathbf{u}} = \nabla \omega_R \otimes (\mathbf{u}_R - \mathbf{u}_B). \tag{4.8.4}$$

In Fickian diffusion discussed in Sect. 6.1, that gradient $\nabla \omega_R$ is parallel to the difference in the species velocities, $\mathbf{u}_R - \mathbf{u}_B$, and the coupling tensor $\widehat{\nabla \mathbf{u}}$ is symmetric.

4.8.2 Rate of Expansion

The rate of expansion of the mixture, red, and blue species is the divergence of the corresponding velocity field,

$$\alpha \equiv \nabla \cdot \mathbf{u}, \qquad \alpha_R \equiv \nabla \cdot \mathbf{u}_R, \qquad \alpha_B \equiv \nabla \cdot \mathbf{u}_B. \tag{4.8.5}$$

Taking the divergence of the mixture mass velocity, $\mathbf{u} = \omega_R\, \mathbf{u}_R + \omega_B\, \mathbf{u}_B$, we find that

$$\alpha = \omega_R\, \alpha_R + \omega_B\, \alpha_B + \widehat{\alpha}, \tag{4.8.6}$$

where

$$\widehat{\alpha} \equiv \mathbf{u}_R \cdot \nabla \omega_R + \mathbf{u}_B \cdot \nabla \omega_B \tag{4.8.7}$$

is a coupling rate of expansion equal to the trace of $\widehat{\nabla \mathbf{u}}$. These expressions can be generalized directly to an arbitrary number of species in a multicomponent mixture.

In the case of a binary mixture, because $\nabla \omega_B = -\nabla \omega_R$, we find that the coupling rate of expansion is given by

$$\widehat{\alpha} \equiv (\mathbf{u}_R - \mathbf{u}_B) \cdot \nabla \omega_R, \tag{4.8.8}$$

which is zero when the gradient $\nabla \omega_R$ is normal to the difference in the species velocities, $\mathbf{u}_R - \mathbf{u}_B$. However, in Fickian diffusion, $\nabla \omega_R$ is parallel to $\mathbf{u}_R - \mathbf{u}_B$ and this term survives.

4.8.3 Vorticity

The vorticity of the mixture, red, and blue species is the curl of the corresponding velocity field,

$$\boldsymbol{\omega} = \nabla \times \mathbf{u}, \qquad \boldsymbol{\omega}_R = \nabla \times \mathbf{u}_R, \qquad \boldsymbol{\omega}_B = \nabla \times \mathbf{u}_B. \tag{4.8.9}$$

Taking the curl of the mixture velocity defined in (4.4.1), we find that

$$\boldsymbol{\omega} = \omega_R\, \boldsymbol{\omega}_R + \omega_B\, \boldsymbol{\omega}_B + \widehat{\boldsymbol{\omega}}, \tag{4.8.10}$$

where

$$\widehat{\boldsymbol{\omega}} \equiv \nabla \omega_R \times \mathbf{u}_R + \nabla \omega_B \times \mathbf{u}_B \tag{4.8.11}$$

is a coupling vorticity. These expressions can be generalized directly to an arbitrary number of species in a multicomponent mixture.

In the case of a binary mixture, because $\nabla \omega_{\mathrm{B}} = -\nabla \omega_{\mathrm{R}}$, we find that

$$\widehat{\boldsymbol{\omega}} = \nabla \omega_{\mathrm{R}} \times (\mathbf{u}_{\mathrm{R}} - \mathbf{u}_{\mathrm{B}}). \tag{4.8.12}$$

In Fickian diffusion, the gradient $\nabla \omega_{\mathrm{R}}$ is parallel to the difference in the species velocities, $\mathbf{u}_{\mathrm{R}} - \mathbf{u}_{\mathrm{B}}$, and $\widehat{\boldsymbol{\omega}}$ is identically zero. Expression (4.8.10) then shows that the vorticity of the mixture is a weighted average of the vorticity of the species. In fact, in Sect. 6.1 we will show that, in Fickian diffusion, $\boldsymbol{\omega}_{\mathrm{R}} = \boldsymbol{\omega}_{\mathrm{B}} = \boldsymbol{\omega}$.

4.8.4 Antisymmetric Part of the Velocity Gradient Tensor

The antisymmetric part of the velocity gradient, indicated by pointed brackets, is given by

$$< \nabla \mathbf{u} > = \omega_{\mathrm{R}} < \nabla \mathbf{u}_{\mathrm{R}} > + \omega_{\mathrm{B}} < \nabla \mathbf{u}_{\mathrm{B}} > + < \widehat{\nabla \mathbf{u}} >, \tag{4.8.13}$$

where

$$< \widehat{\nabla \mathbf{u}} > \equiv \frac{1}{2} (\nabla \omega_{\mathrm{R}} \otimes \mathbf{u}_{\mathrm{R}} - \mathbf{u}_{\mathrm{R}} \otimes \nabla \omega_{\mathrm{R}} + \nabla \omega_{\mathrm{B}} \otimes \mathbf{u}_{\mathrm{B}} - \mathbf{u}_{\mathrm{B}} \otimes \nabla \omega_{\mathrm{B}}) \tag{4.8.14}$$

is an antisymmetric coupling tensor. These expressions admit a direct generalization to an arbitrary number of species.

In the case of a binary mixture, because $\nabla \omega_{\mathrm{B}} = -\nabla \omega_{\mathrm{R}}$, we find that

$$< \widehat{\nabla \mathbf{u}} > = \frac{1}{2} \big(\nabla \omega_{\mathrm{R}} \otimes (\mathbf{u}_{\mathrm{R}} - \mathbf{u}_{\mathrm{B}}) - (\mathbf{u}_{\mathrm{R}} - \mathbf{u}_{\mathrm{B}}) \otimes \nabla \omega_{\mathrm{R}} \big). \tag{4.8.15}$$

In Fickian diffusion, $\nabla \omega_{\mathrm{R}}$ is parallel to $\mathbf{u}_{\mathrm{R}} - \mathbf{u}_{\mathrm{B}}$ and $< \widehat{\nabla \mathbf{u}} >$ is identically zero.

4.8.5 Symmetric Part of the Velocity Gradient Tensor

The symmetric part of the velocity gradient tensor, indicated by square brackets, is given by

$$[\nabla \mathbf{u}] = \omega_{\mathrm{R}} [\nabla \mathbf{u}_{\mathrm{R}}] + \omega_{\mathrm{B}} [\nabla \mathbf{u}_{\mathrm{B}}] + [\widehat{\nabla \mathbf{u}}], \tag{4.8.16}$$

where

$$[\widehat{\nabla \mathbf{u}}] = \frac{1}{2} (\nabla \omega_{\mathrm{R}} \otimes \mathbf{u}_{\mathrm{R}} + \mathbf{u}_{\mathrm{R}} \otimes \nabla \omega_{\mathrm{R}} + \nabla \omega_{\mathrm{B}} \otimes \mathbf{u}_{\mathrm{B}} + \mathbf{u}_{\mathrm{B}} \otimes \nabla \omega_{\mathrm{B}}) \tag{4.8.17}$$

is a symmetric coupling tensor. These expressions admit a direct generalization to an arbitrary number of species.

In the case of a binary mixture, because $\nabla \omega_{\mathrm{B}} = -\nabla \omega_{\mathrm{R}}$, we find that

$$[\widehat{\nabla \mathbf{u}}] = \frac{1}{2} \big(\nabla \omega_{\mathrm{R}} \otimes (\mathbf{u}_{\mathrm{R}} - \mathbf{u}_{\mathrm{B}}) + (\mathbf{u}_{\mathrm{R}} - \mathbf{u}_{\mathrm{B}}) \otimes \nabla \omega_{\mathrm{R}} \big). \tag{4.8.18}$$

In Fickian diffusion, $\nabla \omega_{\mathrm{R}}$ is parallel to $\mathbf{u}_{\mathrm{R}} - \mathbf{u}_{\mathrm{B}}$ and the two terms on the right-hand side make equal contributions.

4.8.6 Rate-of-Deformation Tensor

The mixture rate-of-deformation tensor is given by

$$\mathbf{E} \equiv [\nabla \mathbf{u}] - \frac{1}{3}\,\alpha\,\mathbf{I}, \tag{4.8.19}$$

and the red and blue species rate-of-deformation tensors are given by

$$\mathbf{E}_{\mathrm{R}} \equiv [\nabla \mathbf{u}_{\mathrm{R}}] - \frac{1}{3}\,\alpha_{\mathrm{R}}\,\mathbf{I} \qquad \mathbf{E}_{\mathrm{B}} \equiv [\nabla \mathbf{u}_{\mathrm{B}}] - \frac{1}{3}\,\alpha_{\mathrm{B}}\,\mathbf{I}. \tag{4.8.20}$$

Making substitutions, we obtain

$$\mathbf{E} = \omega_{\mathrm{R}}\mathbf{E}_{\mathrm{R}} + \omega_{\mathrm{B}}\mathbf{E}_{\mathrm{B}} + \widehat{\mathbf{E}}, \tag{4.8.21}$$

where

$$\widehat{\mathbf{E}} \equiv [\widehat{\nabla \mathbf{u}}] - \frac{1}{3}\,\widehat{\alpha}\,\mathbf{I} \tag{4.8.22}$$

is a coupling rate-of-deformation tensor. By construction, the traces of these rate-of-deformation tensors are zero.

In the case of a binary mixture, we find that

$$\widehat{\mathbf{E}} = \frac{1}{2}\left(\nabla\omega_{\mathrm{R}} \otimes (\mathbf{u}_{\mathrm{R}} - \mathbf{u}_{\mathrm{B}}) + (\mathbf{u}_{\mathrm{R}} - \mathbf{u}_{\mathrm{B}}) \otimes \nabla\omega_{\mathrm{R}}\right) - \frac{1}{3}\left(\mathbf{u}_{\mathrm{R}} \cdot \nabla\omega_{\mathrm{R}} + \mathbf{u}_{\mathrm{B}} \cdot \nabla\omega_{\mathrm{B}}\right)\mathbf{I}. \tag{4.8.23}$$

4.8.7 Exercise

4.8.1 Derive expression (4.8.23).

4.9 Essence of Diffusion

Assume that the molecular weights, and thus the molar masses of the red and blue molecules in a binary mixture are precisely or essentially the same. This will be true if one of the species is a tagged version of the other species, or if the red and blue molecules are isotopes. Since the two molecular species are virtually identical, it is tempting to suggest that the species velocities will also be the same at every location in a stationary or flowing mixture.

4.9.1 Mixture of Isotopes in a Box

Consider, however, a binary mixture of two gases in a box where the molecular fraction f_{R} is high on the left and low on the right, and correspondingly f_{B} is low on the left and high on the right, while the fluid density, ρ, is uniform throughout the box, as shown at the top box of Fig. 4.6.

We expect by intuition that diffusion mediated by random walks will cause the molecular fractions to become uniform throughout the box, while the overall density of the fluid remains uniform in a macroscopic state of hydrostatics, as shown at the bottom box of Fig. 4.6.

Fig. 4.6 Illustration of a binary mixture of gases in a box where the molecular fraction f_R is high on the left and low at the right, and correspondingly f_B is low on the left and high at the right, while the fluid density, ρ, is uniform throughout the box. Diffusion leads to homogenization, as shown at the bottom box

The red molecules can be regarded as cars and the blue molecules can be regarded as trucks in Pahrump, Nevada. Given enough time, the vehicles will be distributed in a homogeneous fashion due to random motions throughout Pahrump.

4.9.2 Random Crossings

Now consider a fixed vertical plane across the top box shown in Fig. 4.6. Because of random motions, red molecules will cross the plane from left to right and from right to left. However, since a higher number of red molecules reside on the left than on the right at an instant, the imbalance will lead to a net red species migration to the right until equilibration. The converse is true for the blue species. Consequently, the red species migrates to the right and the blue species migrates to the left.

4.9.3 Macroscopic State of Rest

Because a net mass transport has not taken place in our experiment, the mixture mass velocity remains zero throughout the box during the entire process,

$$\mathbf{u} = 0, \tag{4.9.1}$$

even though the species velocities $\mathbf{u}_R$ and $\mathbf{u}_B$ are both nonzero, related by

$$\omega_R \mathbf{u}_R = -\omega_B \mathbf{u}_B. \tag{4.9.2}$$

The opposite signs are expected on physical grounds since the species velocities represent motion into opposite directions. The overall process is described as *self-diffusion.*

4.9.4 Diffusion Alone Is Responsible for Different Species Velocities

The experiment discussed in this section indicates that the species velocities, $\mathbf{u}_R$ and $\mathbf{u}_B$, are the same only in a fluid with a uniform molecular fractions distributions, even when the two molecular species are identical isotopes. To be more precise, $\mathbf{u}_R$ and $\mathbf{u}_B$ are equal only at points where the gradients of the molecular fractions are zero. The assertion that $\mathbf{u}_R = \mathbf{u}_B$ for chemically identical species under any circumstances is incorrect.

Since $\mathbf{u} = \underline{0}$ in our experiment, the kinetic energy of the macroscopic motion is precisely zero. Consequently, computing the macroscopic kinetic energy as the sum of the species macroscopic kinetic energies based in $\mathbf{u}_R$ and $\mathbf{u}_B$ is not appropriate for a chemical solution.

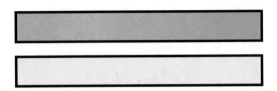

4.9.5 Diffusion in a Gas

Diffusion in a gas is typically mediated by random motions of the individual molecules, described and modeled as random walks, as discussed in Sect. 4.10. The higher the temperature, the more vigorous the random motion quantified by the root-square-mean (rms) value of the molecular velocities representing the kinetic energy of the fluid. Rigorous expressions for diffusion fluxes and associated diffusivities can be derived working in the framework of statistical mechanics and thermodynamics, as discussed in Chap. 6.

4.9.6 Diffusion in a Liquid or Solid

Diffusion in a liquid is mediated by similar, but not identical, physical mechanics involving random excursions. Lattice diffusion in the crystalic structure of a solid occurs by interstitial or substitutional advancement. In interstitial diffusion, a particle moves through the crystal free space. In substitutional diffusion, a particle moves by exchanging positions with other particles on a crystalic lattice.

4.9.7 Further Types of Diffusion

Ordinary diffusion associated with random walkers and random walks is discussed in Sects. 4.10–4.12. Further types of diffusion include forced diffusion associated with an external body force, thermal diffusion due to a temperature gradient, and pressure diffusion due to a pressure gradient.

4.9.8 Exercise

4.9.1 Discuss the self-diffusion experiment in an entomological context involving hopping red and blue grasshoppers.

4.10 Random Walkers

We have mentioned that the physical process of diffusion is attributed to random molecular excursions. To demonstrate this connection, we consider an idealized configuration where a particle, identified as a walker or jumper, hops from one site to another on a one-dimensional grid.

The collective motion of particles stepping to nearest neighbor sites amounts to ordinary diffusion discussed in this section, also called Fickian diffusion, whereas the collective motion of particles executing short or long flights to arbitrary sites according to a specified probability distribution amounts to fractional or some other type of nonlocal diffusion, as discussed in Chap. 6.

4.10.1 Walking Particles

To illustrate the physics of ordinary diffusion, we consider the evolution of a large collection of particles distributed over the x axis at positions x_i for $i = 0, \pm 1, \ldots$, separated by a specified interval, Δx. At a designated origin of time, each particle makes one step to the right by Δx with probability

q or to the left with probability s, where q and s are two free parameters regarded as Bernoulli probabilities ranging in the interval [0, 1], assigned such that $q + s \leq 1$. The procedure is then repeated for an indefinite number of cycles.

The probability that a particle does not make a step is $r = 1 - q - s$; when $q + s = 1$, all particles make a step since $r = 0$. When $q = 1$ and $s = 0$, all particles move to the right by Δx at each step. When $q = 0$ and $s = 1$, all particles move to the left by $-\Delta x$ at each step. In the last two cases, the initial particle distribution translates to the right or left unchanged.

4.10.2 Particle Population

At any time instant designated by the time level, n, the ith node hosts $m_i^{(n)}$ particles. To study the collective particle motion, we quantify the particle population in terms of a discrete number-density distribution defined as

$$p_i^{(n)} \equiv \frac{1}{N} m_i^{(n)}, \tag{4.10.1}$$

where $N = \sum m_i^{(n)}$ is the total number of particles. Since particles are neither generated nor disappear, $\sum p_i^{(n)} = 1$ at any time level, n.

4.10.3 Population Balance

After one step has been completed, a fraction r of particles at the ith node have remained, a fraction q of particles have arrived from the $i - 1$ node, and a fraction s of particles have arrived from the $i + 1$ node, so that

$$p_i^{(n+1)} = q\, p_{i-1}^{(n)} + r\, p_i^{(n)} + s\, p_{i+1}^{(n)}. \tag{4.10.2}$$

Rearranging, we obtain

$$p_i^{(n+1)} - p_i^{(n)} + \tfrac{1}{2}(q - s)\left(p_{i+1}^{(n)} - p_{i-1}^{(n)}\right) = \tfrac{1}{2}(q + s)\left(p_{i-1}^{(n)} - 2\, p_i^{(n)} + p_{i+1}^{(n)}\right). \tag{4.10.3}$$

Next, we introduce the duration of each step, Δt, and recast this equation into the form

$$\frac{p_i^{(n+1)} - p_i^{(n)}}{\Delta t} + V\, \frac{p_{i+1}^{(n)} - p_{i-1}^{(n)}}{2\Delta x} = \kappa\, \frac{p_{i-1}^{(n)} - 2\, p_i^{(n)} + p_{i+1}^{(n)}}{\Delta x^2}, \tag{4.10.4}$$

where V is a drift velocity and κ is a diffusivity with units of length squared divided by time, given by

$$V = (q - s)\, \frac{\Delta x}{\Delta t}, \qquad \kappa = \tfrac{1}{2}(q + s)\, \frac{\Delta x^2}{\Delta t}. \tag{4.10.5}$$

When $q = s$, the drift velocity is zero.

4.10.4 Convection–Diffusion Equation

The first fraction on the left-hand side of (4.10.4) is the first-order forward-difference approximation to the time derivative, $\partial p_i/\partial t$. The second fraction on the left-hand side is the second-order centered-difference approximation to the first spatial derivative, $\partial p/\partial x$, evaluated at the ith node. The fraction on the right-hand side is the second-order centered-difference approximation to the second spatial derivative, $\partial^2 p/\partial x^2$, evaluated at the ith node.

Consequently, the difference equation (4.10.4) can be regarded as the discrete representation of a convection–diffusion equation,

$$\frac{\partial p}{\partial t} + V\,\frac{\partial p}{\partial x} = \mathcal{D}\,\frac{\partial^2 p}{\partial x^2}, \tag{4.10.6}$$

where the drift velocity, V, is a convection velocity and $\mathcal{D}$ is an appropriate diffusivity.

In the context of finite-difference methods, Eq. (4.10.6) is the modified differential equation corresponding to the difference equation (4.10.4). Following a standard procedure that accounts for the effect of numerical diffusivity due to the backward discretization of the time derivative, $\partial p/\partial t$, we find that

$$\mathcal{D} = \kappa\,\Big(1 - \tfrac{1}{2}\,\frac{c^2}{\alpha}\Big), \tag{4.10.7}$$

where

$$c \equiv \frac{V\,\Delta t}{\Delta x}, \qquad \alpha \equiv \frac{\kappa\,\Delta t}{\Delta x^2} \tag{4.10.8}$$

are dimensionless parameters. Referring to (4.10.5), we find that $c = q - s$ and $\alpha = \frac{1}{2}\,(q+s)$. Making substitutions, we obtain

$$\mathcal{D} = \tfrac{1}{2}\,\big(q + s - (q - s)^2\big)\,\frac{\Delta x^2}{\Delta t}. \tag{4.10.9}$$

The expression enclosed by the outer parentheses on the right-hand side is nonnegative. When $q+s = 1$, we obtain

$$\mathcal{D} = 2\,q\,(1 - q)\,\frac{\Delta x^2}{\Delta t}. \tag{4.10.10}$$

Expression (4.10.9) will be reproduced in Sect. 4.11 by a different method.

4.10.5 Green's Function

Consider the solution of (4.10.6) subject to an impulse represented by a Dirac delta function centered at a point, x_0, denoted by $\delta(x - x_0)$, and another Dirac delta function in time centered at a time, t_0, denoted by $\delta(t - t_0)$,

$$\frac{\partial \mathcal{G}}{\partial t} + V\,\frac{\partial \mathcal{G}}{\partial x} = \mathcal{D}\,\frac{\partial^2 \mathcal{G}}{\partial x^2} + \delta(x - x_0)\,\delta(t - t_0). \tag{4.10.11}$$

The solution is the Green's function of the unsteady diffusion equation with units of inverse length, $\mathcal{G}(x - x_0, t - t_0)$. For convenience, we denote $\widehat{x} = x - x_0$ and $\hat{t} = t - t_0$.

Using the method of similarity solutions, or else by performing Fourier transforms, we find that the Green's function is given by

$$\mathcal{G}(\widehat{x}, \hat{t}) = \frac{1}{\sqrt{4\pi \mathcal{D}\hat{t}}} \exp\Big(-\frac{1}{4} \frac{(\widehat{x} - V\hat{t})^2}{\mathcal{D}\hat{t}} \Big) \tag{4.10.12}$$

for any $\widehat{x}$ and $\hat{t} \geq 0$. Note that time appears in three places on the right-hand side. The Green's function is an evolving Gaussian distribution centered at $\widehat{x} = V\hat{t}$ where the argument of the exponential term is zero.

The variance of the distribution around the peak is

$$\sigma^2(\hat{t}) \equiv \int_{-\infty}^{\infty} (\widehat{x} - V\hat{t})^2 \, \mathcal{G}(\widehat{x}, \hat{t}) \, \mathrm{d}\widehat{x} = 2\mathcal{D}\hat{t}, \tag{4.10.13}$$

where σ is the standard deviation with units of length. In terms of the standard deviation, the Green's function is given by

$$\mathcal{G}(\widehat{x}, \hat{t}) = \frac{1}{\sqrt{2\pi}\,\sigma(\hat{t})} \exp\Big(-\frac{1}{2} \frac{(\widehat{x} - V\hat{t})^2}{\sigma^2(\hat{t})} \Big). \tag{4.10.14}$$

Because the standard deviation has units of length, the argument of the exponential on the right-hand side is dimensionless, as required.

4.10.6 Numerical Simulation

Consider a collection of particles arranged in n_c columns along the x axis, where n_c is a specified integer. When n_c is odd, the particles are situations at notched positions

$$x_i = i \, \Delta x \quad \text{for} \quad i = -\tfrac{1}{2}\,(n_c - 1), \ldots, \tfrac{1}{2}\,(n_c - 1), \tag{4.10.15}$$

where Δx is a specified interval. When n_c is even, the particles are situations at notched positions

$$x_i = (i - \tfrac{1}{2})\,\Delta x \quad \text{for} \quad i = -\tfrac{1}{2}\,(n_c - 2), \ldots, \tfrac{1}{2}\,n_c. \tag{4.10.16}$$

When $n_c = 1$, we obtain one column corresponding to $i = 0$, located at the origin of the x axis, $x_0 = 0$. When $n_c = 2$, we obtain two columns corresponding to $i = 0, 1$, located at $x_0 = -\frac{1}{2}\,\Delta x$ and $x_1 = \frac{1}{2}\,\Delta x$.

The following Matlab code named *walkers1* performs a simulation, visualizes the evolving particle population, and simultaneously displays the evolving Green's function:

```
%==================
% Copyright by Constantine Pozrikidis, 2020
% All rights reserved
```

```
%
% simulation of random walkers to nearest neighbors
%=================

Dx = 0.574;     % spatial step (arbitrary)
Dt = 2.3;       % time step (arbitrary)

nsteps = 2*32; % number of time steps (arbitrary)

q = 0.40;
s = 0.20;

if(q+s>1)
 disp('q+s must be less than one')
 return
end

%---
% number of walkers (arbitrary)
% walkers will be assigned
% in 2*nc initial columns
%---

nc = 2;

nwc = 32*128;     % walkers per column (arbitrary)
nwalker = nc*nwc;

%---
% odd-even number of columns
%---

oddevenc = 1;
if(mod(nc,2)==0) oddevenc = 2; end

%---
% define notches
% and initial population
%---

notch = 50;  % arbitrary

ishift = notch+1;

if(oddevenc==1) xshift = 0.0; end
if(oddevenc==2) xshift = 0.5; end

for i=-notch:notch
 xnotch(i+ishift) = (i-xshift)*Dx;  % x position of notch
 pnotch(i+ishift) = 0;              % notch population
end

%---
% assign the initial x-position of the walkers
% in nc columns
% and compute the initial population
%---

if(oddevenc==1) klow =-(nc-1)/2; khigh = (nc-1)/2; end
if(oddevenc==2) klow =-nc/2+1;   khigh = nc/2;     end
```

```
Ic = 0; % walker counter

for i=1:nwc
  for k = klow:khigh
   Ic = Ic+1;
   xwalker(Ic) = k;
   pnotch(ishift+k) = pnotch(ishift+k)+1;
  end
end

%---
% plot the initial population
%---

figure(1)
hold on
xlabel('i','fontsize',15)
ylabel('population','fontsize',15)
plot(xnotch,pnotch/nwalker,'ko')

%---
% simulation
%---

time = 0.0;

%---
for n=1:nsteps
%---

 for i=1:nwalker

  j = xwalker(i);         % current x position of walker
  walk = 0;
  random = rand;

  if(random<q)                        % walk to the right
   walk =  1;
  elseif(random>q & random<(q+s))     % walk to the left
   walk = -1;
  end


 % update the x position of walker:

  xwalker(i) = j + walk;

  % update the notch population:

  pnotch(j+ishift)      = pnotch(j+ishift)      - 1;
  pnotch(j+ishift+walk) = pnotch(j+ishift+walk) + 1;

 end

 time = time + Dt;

 %---
 % Green's function
 %---
```

```
  V = (q-s)*Dx/Dt;
  D = 0.5*(q+s-(q-s)^2)*Dx^2/Dt;

  var = 2.0*D*time;
  sdv = sqrt(var);

  for i=-nbin:nbin
    xx = xnotch(i+ishift);
    GF = 1.0/(sqrt(2*pi)*sdv)*exp(-0.5*(xx-V*time)^2/var);
    Gnotch(i+ishift) = Dx*GF;  % scale with 1/Dx
  end

  clf
  hold on
  plot(xnotch,pnotch/nwalker,'k.')
  plot(xnotch,Gnotch,'r-')

end
```

The code makes use of the internal Matlab function *rand* to generate a random number, and thus decide whether a particle will move to the right, move to the left, or stay put at each step.

Snapshots of the evolving population distribution for $q = 0.4$, $s = 0.2$, and other simulation parameters defined in the code are shown in Fig. 4.7. The solid lines represent the distribution of the Green's function of the convection–diffusion equation scaled by $1/\Delta x$. The simulation reproduces with remarkable accuracy the translating and spreading Gaussian distribution.

4.10.7 Exercises

4.10.1 Derive the expression for the variance shown in (4.10.13).

4.10.2 Equation (4.10.2) can be restated as

$$p_i^{(n+1)} - p_i^{(n)} + (q-s)\left(p_i^{(n)} - p_{i-1}^{(n)}\right) = s\left(p_{i-1}^{(n)} - 2\,p_i^{(n)} + p_{i+1}^{(n)}\right), \tag{4.10.17}$$

which can be recast into the form

$$\frac{p_i^{(n+1)} - p_i^{(n)}}{\Delta t} + V\,\frac{p_i^{(n)} - p_{i-1}^{(n)}}{\Delta x} = (\kappa - \tfrac{1}{2}\,V\,\Delta x)\,\frac{p_{i-1}^{(n)} - 2\,p_i^{(n)} + p_{i+1}^{(n)}}{\Delta x^2}, \tag{4.10.18}$$

where V and κ are given in (4.10.6). The second fraction on the left-hand side is the first-order backward-difference approximation to the first spatial derivative, $\partial p/\partial x$. Write the counterpart of Eqs. (4.10.17) and (4.10.18) for the first-order forward-difference approximation to $\partial p/\partial x$.

4.11 Random Walks

A further illustration of the physical process of ordinary diffusion can be displayed by considering the trajectory of a randomly moving solitary particle over the x axis. The underlying nature of the motion is the same as that discussed in Sect. 4.10 for a particle population.

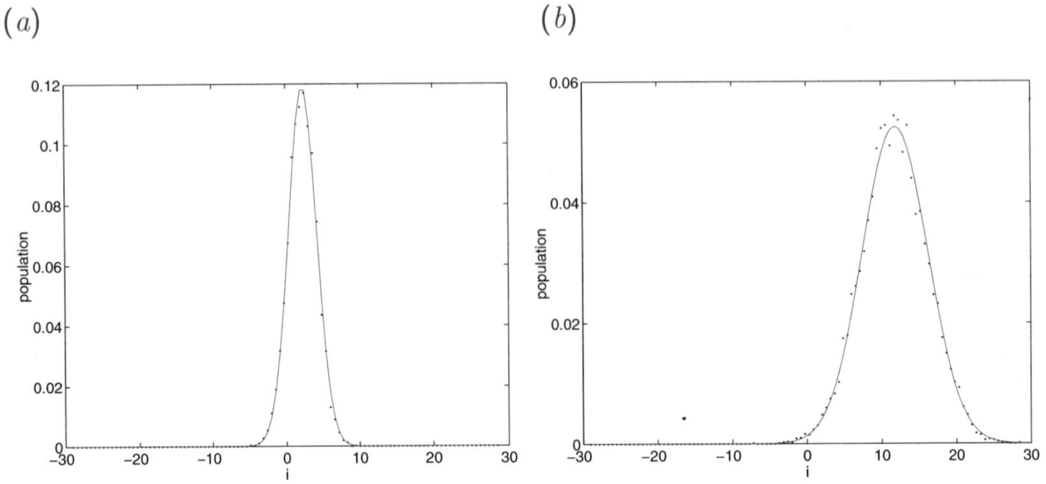

Fig. 4.7 Evolution of random walker population distribution for $q = 0.4$ and $s = 0.2$. The solid lines describe Gaussian distributions associated with the Green's function of the unsteady convection–diffusion equation, scaled by $1/\Delta x$

At a designated time, the solitary particle wakes up and starts making a step to the right by Δx with probability q or to the left by $-\Delta x$ with probability $1 - q$, where q is a Bernoulli probability ranging in the interval [0, 1]. The steps are then repeated for an indefinite number of cycles. The particle path is described as a random walk.

4.11.1 Particle Position After n Steps

After n steps have been made, the particle finds itself at position $x = k\Delta x$, where k is an integer given by

$$k = \sum_{i=1}^{n} \xi_i \tag{4.11.1}$$

and ξ_i is a collection of n independent random variables taking values ± 1 with probabilities $p(1) = q$ and $p(-1) = 1 - q$.

4.11.2 m Steps to the Right

If the sequence ξ_i involves m steps to the right and consequently $n - m$ steps to the left, then

$$k = m - (n - m) = 2m - n \tag{4.11.2}$$

for $m = 0, 1, \ldots, n$, corresponding to

$$k = -n, -(n-2), \ldots, (n-2), n. \tag{4.11.3}$$

Conversely,

$$m = \frac{1}{2}\,(k + n), \tag{4.11.4}$$

which is an integer in the range of k shown in (4.11.3).

4.11.3 dpd for m

The number of steps to the right after n steps, m, is a stochastic variable with a discrete probability distribution (dpd), denoted by φ_m, described by the binomial distribution, $\mathcal{B}_{m|n}(q)$,

$$\varphi_m = \mathcal{B}_{m|n}(q) \equiv \mathcal{C}^n_m\, q^m (1-q)^{n-m}, \tag{4.11.5}$$

where

$$\mathcal{C}^n_m \equiv \binom{n}{m} \equiv \frac{n\,!}{m\,!\,(n-m)\,!} \tag{4.11.6}$$

is the binomial coefficient and $n! = 1 \cdot 2 \cdots n$ is the factorial, subject to the convention that the factorial of zero is unity, $0\,! = 1$. Substituting the definition of the factorial, we find that

$$\mathcal{C}^n_m = \frac{(n-m+1)\cdots(n-1)\,n}{m\,!} = \prod_{a=1}^{m} \frac{n-a+1}{a}, \tag{4.11.7}$$

which shows that $\mathcal{C}^n_m$ is an mth-degree polynomial in n.

In combinatorics, the factorial of an integer, $n!$, expresses the number of ways by which n distinct objects can be arranged in an n-dimensional array, where each way represents a permutation. For large n, the factorial can be approximated with Stirling's formula

$$n\,! \simeq \sqrt{2\pi n}\,\left(\frac{n}{e}\right)^n. \tag{4.11.8}$$

The Stirling formula is remarkably accurate, introducing an error on the order of only 2% for n as low as 4.

Expression (4.11.5) provides us with φ_m as the product of (a) m single-event (Bernoulli) independent probabilities q, (b) $n - m$ complementary Bernoulli independent probabilities $1 - q$, and (c) the number of combinations by which right steps m are made in n steps, expressed by the binomial coefficient.

Using the binomial expansion,

$$(a+b)^n = \sum_{m=0}^{n} \mathcal{C}^n_m\, a^{n-m}\, b^m = \sum_{m=0}^{n} \binom{n}{m}\, a^{n-m}\, b^m, \tag{4.11.9}$$

setting $a = q$ and $b = 1 - q$, and noting that $(q + (1-q))^n = 1$, we confirm that all probabilities add to unity for any n,

$$\sum_{m=0}^{n} \varphi_m = \sum_{m=0}^{n} \mathcal{B}_{m|n}(q) = 1, \tag{4.11.10}$$

as required.

4.11.4 Expected Value and Variance for *m*

The expected value of m for a given number of steps, n, is given by

$$\overline{m} \equiv \sum_{m=0}^{n} m\, \mathcal{B}_{m|n}(q) = q\, n. \tag{4.11.11}$$

To derive this identity, we take the partial derivative of (4.11.9) with respect to b, and then set $a = q$ and $b = 1 - q$. The associated variance is

$$\sigma_m^2 \equiv \sum_{m=0}^{n} (m - \overline{m})^2\, \mathcal{B}_{m|n}(q), \tag{4.11.12}$$

where σ_m is the standard deviation. Expanding the square and using (4.11.10) and (4.11.11), we find that

$$\sigma_m^2 = \sum_{m=0}^{n} (m^2 - 2\, m\, \overline{m} + \overline{m}^2)\, \mathcal{B}_{m|n}(q) = \sum_{m=0}^{n} m^2\, \mathcal{B}_{m|n}(q) - q^2 n^2. \tag{4.11.13}$$

Now using the identity

$$\sum_{m=0}^{n} m^2\, \mathcal{B}_{m|n}(q) = q^2 n^2 + q\,(1 - q)\, n, \tag{4.11.14}$$

we obtain

$$\sigma_m^2 = q\,(1 - q)\, n. \tag{4.11.15}$$

As expected, the variance is zero when $q = 0$ or 1, and reaches a maximum at the mid-point, $q = \frac{1}{2}$. To derive (4.11.14), we take the second partial derivative of (4.11.9) with respect to b, simplify, and then set $a = q$ and $b = 1 - q$.

4.11.5 De Moivre–Laplace Theorem

The De Moivre–Laplace theorem states that, for large n, the binomial distribution can be approximated with a Gaussian distribution,

$$\mathcal{B}_{m|n}(q) \simeq \frac{1}{\sqrt{2\pi}\, \sigma_m} \exp\Big(-\tfrac{1}{2}\, \frac{(m - \overline{m})^2}{\sigma_m^2} \Big) \tag{4.11.16}$$

for m sufficiently close to $\overline{m} = qn$, where σ_m^2 is given in (4.11.15).

4.11.6 dpd for k

The particle position after n steps is determined by the integer k, which is related to m and n through (4.11.2). The integer k is a stochastic variable with a discrete probability distribution (dpd) denoted by ψ_k, given by

$$\psi_k = \varphi_{\frac{1}{2}(k+n)} = \mathcal{B}_{\frac{1}{2}(k+n)|n}(q) \tag{4.11.17}$$

in the range of k shown in (4.11.3), and $\psi_k = 0$ otherwise. The normalization condition

$$\sum_{k=-n}^{n} \psi_k = 1 \tag{4.11.18}$$

is satisfied.

Now referring to Eq. (4.11.2), we find that the expected value of k after n steps is

$$\overline{k} \equiv \sum_{k=-n}^{n} k\,\psi_k = 2\,\overline{m} - n = (2q-1)\,n, \tag{4.11.19}$$

and the associated variance is

$$\sigma_k^2 \equiv \sum_{k=0}^{n} (k-\overline{k})^2\,\psi_k = 4\sum_{m=0}^{n}(m-qn)^2\,\mathcal{B}_{m|n}(q) = 4\,\sigma_m^2. \tag{4.11.20}$$

Substituting the expression given in (4.11.15) for σ_m^2, we obtain

$$\sigma_k^2 = 4\,q\,(1-q)\,n. \tag{4.11.21}$$

The linear growth of the variance with respect to n is unmistakable evidence of diffusive motion.

We may introduce a time step, Δt, define the time elapsed after n steps, $t_n = n\Delta t$, and write

$$\overline{x} \equiv \overline{k}\,\Delta x = V t_n, \qquad \sigma_x^2 \equiv \sigma_k^2\,\Delta x^2 = 2\mathcal{D}\,t_n, \tag{4.11.22}$$

where

$$V = (2q-1)\,\frac{\Delta x}{\Delta t}, \qquad \mathcal{D} = 2\,q\,(1-q)\,\frac{\Delta x^2}{\Delta t} \tag{4.11.23}$$

are a drift velocity and a diffusivity.

For a large number of steps, n, we use the De Moivre–Laplace theorem to obtain

$$\psi_k \simeq \frac{1}{\sqrt{2\pi q\,(1-q)n}}\exp\Big(-\frac{1}{2}\,\frac{(\frac{1}{2}(k+n)-qn)^2}{q\,(1-q)n}\Big), \tag{4.11.24}$$

which can be simplified to

$$\psi_k \simeq \frac{1}{\sqrt{2\pi q(1-q)\,n}}\exp\Big(-\frac{1}{2}\,\frac{(k-(2q-1)n)^2}{4\,q\,(1-q)n}\Big). \tag{4.11.25}$$

This expression can be compared with the Green's function of the unsteady diffusion equation shown in (4.10.12), repeated below for convenience,

$$\mathcal{G}(\widehat{x}, \hat{t}) = \frac{1}{\sqrt{4\pi \mathcal{D}\hat{t}}} \exp\Big(-\frac{1}{4} \frac{(\widehat{x} - V\hat{t})^2}{\mathcal{D}\hat{t}} \Big). \tag{4.11.26}$$

Substituting $\hat{t} = n\Delta t$ and $\widehat{x} = k\Delta x$, we obtain the expressions given in (4.11.23) for V and $\mathcal{D}$. A difference by a factor of *two* is due to the restricted range of k.

4.11.7 Unbiased Walk

In the case of unbiased walk, $q = 1 - q = \frac{1}{2}$, we obtain

$$\varphi_m = \mathcal{B}_{m|n}(\tfrac{1}{2}) = \frac{1}{2^n} \binom{n}{m} \tag{4.11.27}$$

for $m = 1, \ldots, n$. The expected value is $\overline{m} = \frac{1}{2}\, n$, and the associated variance is $\sigma_m^2 = \frac{1}{4}\, n$.

Considering all possible particle paths, we find that if $n = 0$, then $\psi_0 = 1$; if $n = 1$, then $\psi_{-1} = \frac{1}{2}$ and $\psi_1 = \frac{1}{2}$; if $n = 2$, then $\psi_{-2} = \frac{1}{4}$, $\psi_0 = \frac{1}{2}$, and $\psi_2 = \frac{1}{4}$. Invoking (4.11.17), we find that

$$\psi_k = \frac{1}{2^n} \binom{n}{\frac{n+k}{2}} = \frac{1}{2^n} \frac{n\,!}{\big(\frac{n+k}{2}\big)!\big(\frac{n-k}{2}\big)!} \tag{4.11.28}$$

in the range of k shown in (4.11.3), and $\psi_k = 0$ otherwise. By symmetry, the expected value of k is zero, $\overline{k} = 0$, and the associated variance is $\sigma_k^2 = n$. The linear growth of the variance with respect to n is the hallmark of diffusive motion.

For a large number of steps, n, we use the De Moivre–Laplace theorem to obtain the Gaussian approximation

$$\psi_k \simeq \frac{2}{\sqrt{2\pi n}} \exp\Big(-\frac{1}{2} \frac{k^2}{n} \Big) \tag{4.11.29}$$

in the range of k shown in (4.11.3). The binomial probability distribution ψ_k is compared with the Gaussian distribution in Fig. 4.8 for $n = 16$. The agreement is excellent, even for this small number of random steps.

4.11.8 Walk with Pauses

In a generalization of the random walk problem discussed previously in this section, we assume that a particle wakes up and starts making a step to the right by Δx with probability q or to the left by $-\Delta x$ with probability s, where q and s are Bernoulli probabilities ranging in the interval $[\,0, 1\,]$, subject to the restriction that $q + s < 1$, as discussed in Sect. 4.10. The probability that the particle does not make a step is $r = 1 - q - s$. The steps are then repeated for an indefinite number of cycles.

If the particle path after n time units involves m steps to the right, ℓ pauses, and consequently $n - m - \ell$ steps to the left, then

$$k{=}m - (n - m - \ell){=}2m - n + \ell. \tag{4.11.30}$$

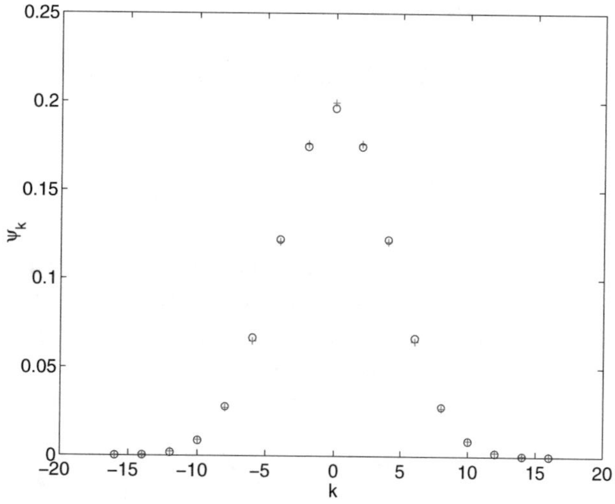

Fig. 4.8 Comparison of the probability distribution of one-dimensional random walker, ψ_k after $n = 16$ steps (∘), with the Gaussian distribution associated with the Green's function of the unsteady diffusion equation (+)

The expected value of m is qn, and the expected value of ℓ is rn. Consequently, the expected value of k is

$$\overline{k} = 2qn - n + rn = (q - s)\, n. \tag{4.11.31}$$

The associated variance is

$$\sigma_k^2 = 4\,\sigma_m^2 + \sigma_\ell^2 = 4qs + r\,(1 - r) = q + s - (q - s)^2. \tag{4.11.32}$$

This expression reproduces the formula for the diffusivity shown in (4.10.9).

4.11.9 Exercise

4.11.1 Confirm by numerical computation the identity (4.11.14).

4.12 Random Walkers with Continuous Steps

Further insights into the physics of diffusion can be obtained by considering an infinite number of particles distributed continuously along the entire x axis. After a small time interval Δt has elapsed, a particle located at x has jumped to the left or right by ξ with probability density function (pdf) $\phi(\xi)$, where $-\infty < \xi < \infty$. Note that the particle positions are not restricted to discrete sites.

By definition of a pdf, the probability that ξ lies inside an infinitesimal interval dξ is $\phi(\xi)\,\mathrm{d}\xi$, where $\phi(\xi)$ has units of inverse length. Normalization requires that

$$\int_{-\infty}^{\infty} \phi(\xi)\, \mathrm{d}\xi = 1. \tag{4.12.1}$$

In the case of symmetric walkers , $\phi(\xi) = \phi(-\xi)$. For the integral on the left-hand side to exist, $\phi(\xi)$ must decay faster than $1/|\xi|$, that is, it must behave like $1/|\xi|^\beta$, where $\beta > 1$.

4.12.1 Balance Equation

We denote the particle density distribution at position x and time t by $f(x,t)$, and write the balance equation

$$f(x, t+\Delta t) = f(x,t) + \int_{-\infty}^{\infty} \big(f(x-\xi, t) - f(x,t) \big)\, \phi(\xi)\, \mathrm{d}\xi. \tag{4.12.2}$$

The balance equation states that, after each time step of duration Δt, particles have arrived at position x from position $x - \xi$ with probability $\phi(\xi)$, and particles have departed from position x for greener pastures in proportion to $f(x)$. If $f(x)$ is initially flat, it will remain flat at all times, independent of the probability density function, $\phi(\xi)$.

Rearranging, (4.12.2), we obtain

$$\frac{f(x, t+\Delta t) - f(x,t)}{\Delta t} = \frac{1}{\Delta t} \int_{-\infty}^{\infty} \big(f(x-\xi, t) - f(x,t) \big)\, \phi(\xi)\, \mathrm{d}\xi. \tag{4.12.3}$$

The left-hand side is the first-order forward difference approximation to the time derivative, $\partial f/\partial t$.

4.12.2 Local Expansion

Now substituting into the integrand on the right-hand side of (4.12.3) the Maclaurin expansion

$$f(x-\xi, t) = f(x,t) - \frac{\partial f}{\partial x}\, \xi + \tfrac{1}{2}\, \frac{\partial^2 f}{\partial x^2}\, \xi^2 + \cdots, \tag{4.12.4}$$

we obtain

$$\frac{f(x, t+\Delta t) - f(x,t)}{\Delta t} = -V\, \frac{\partial f}{\partial x} + \kappa\, \frac{\partial^2 f}{\partial x^2} + \cdots, \tag{4.12.5}$$

where the dots represent cubic and higher-order terms associated with third- and higher-order derivatives, and

$$V = \frac{1}{\Delta t} \int_{-\infty}^{\infty} \xi\, \phi(\xi)\, \mathrm{d}\xi, \qquad \kappa = \tfrac{1}{2}\, \frac{1}{\Delta t} \int_{-\infty}^{\infty} \xi^2\, \phi(\xi)\, \mathrm{d}\xi \tag{4.12.6}$$

are a convection velocity and a diffusivity. If the probability density function, $\phi(\xi)$, is sufficient narrow, the terms represented by the dots can be neglected on the right-hand side.

For the integral defining the convection velocity V in (4.12.6) to exist, $\phi(\xi)$ must decay faster than $1/|\xi|^2$, that is, it must behave like $1/|\xi|^\beta$, where $\beta > 2$. For the integral defining the diffusivity to exist, $\phi(\xi)$ must decay faster than $1/|\xi|^3$.

Working as in Sect. 4.10 in context of finite-difference approximations, we find that solution of (4.12.5) is an approximation of the solution of the convection–diffusion equation (4.10.6), where $\mathcal{D} = \kappa - \frac{1}{2}\, V^2 \Delta t$.

4.12.3 Notched Positions

In the case of the random walkers discussed in Sect. 4.10, the probability density function is given by

$$\phi(\xi) = s\,\delta(\xi + \Delta x) + (1 - q - s)\,\delta(\xi) + q\,\delta(\xi - \Delta x), \tag{4.12.7}$$

where $\delta(\eta)$ is the Dirac delta function. Making substitutions, we derive expressions (4.10.5) for V and κ.

4.12.4 Mapped pdf

Assume that a random variable, ξ, defined in an interval, $[\xi_{min}, \xi_{max}]$, is related to another random variable, η, defined in a corresponding interval, $[\eta_{min}, \eta_{max}]$, by a monotonic function

$$\eta = \psi(\xi), \tag{4.12.8}$$

where $\eta_{min} = \psi(\xi_{min})$ and $\eta_{max} = \psi(\xi_{max})$. In agricultural applications, ξ can be the amount of rainfall and η can be the crop yield.

By way of this association, η is also a random variable whose pdf is denoted as $\hat{\phi}(\eta)$. We write

$$\phi(\xi)\,\mathrm{d}\xi = \pm\hat{\phi}(\eta)\,\mathrm{d}\eta \tag{4.12.9}$$

for corresponding differentials $\mathrm{d}\xi$ and $\mathrm{d}\eta$, and find that the two pdfs are related by the equation

$$\frac{\phi(\xi)}{\hat{\phi}(\eta)} = \pm\psi'(\xi) = \pm\frac{\mathrm{d}\eta}{\mathrm{d}\xi}, \tag{4.12.10}$$

where a prime denotes the first derivative with respect to ξ. The plus sign is chosen when $\psi' > 0$ and the minus sign is chosen when $\psi' < 0$ so that the ratio of the probabilities is nonnegative.

Substituting these expressions into Eq. (4.12.6), we find that

$$V = \pm\frac{1}{\Delta t}\int_{\eta_{min}}^{\eta_{max}} \xi\,\hat{\phi}(\eta)\,\mathrm{d}\eta, \qquad \kappa = \pm\frac{1}{2}\frac{1}{\Delta t}\int_{\eta_{min}}^{\eta_{max}} \xi^2\,\hat{\phi}(\eta)\,\mathrm{d}\eta, \tag{4.12.11}$$

where ξ is related to η by way of (4.12.8). For the integrals to exist, the pdf $\hat{\phi}(\eta)$ must decay sufficiently fast.

4.12.5 Intercepting Particles

As an application, we consider a dilute suspension of spherical particles with rough surfaces, deformable particles such as bubbles, drops, and cells, or charged particles undergoing simple shear flow along the x axis with velocity $u_x = ky$, as shown in Fig. 4.9, where k is the shear rate with units of inverse time. For simplicity, we assume that all particles lie in a monolayer located at $z = 0$.

The linear density of the interception frequency of a test particle A located at y_A with another particle B located y_B is

$$\zeta(\eta) = k\,c\,|2\eta|, \tag{4.12.12}$$

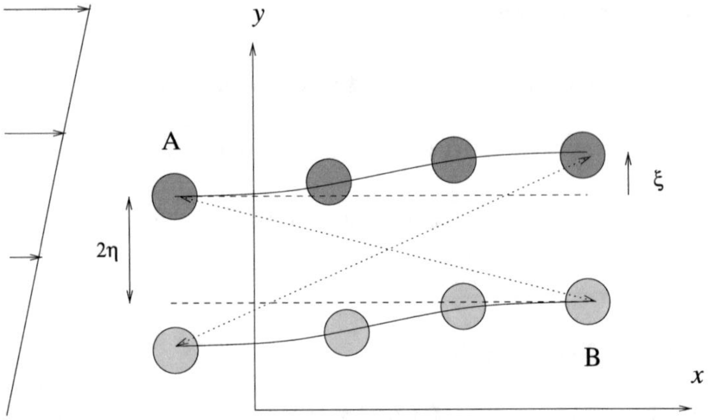

Fig. 4.9 Illustration of two particles, A and B, intercepting in simple shear flow. After each interception, particle A has been displaced by distance ξ that depends on the interception half-distance η

where c is the particle number density in the xy plane with units of number of particles over length squared, and

$$\eta = \frac{1}{2}\,(y_A - y_B) \tag{4.12.13}$$

is the particle half-separation before interception.

After an interception has been completed, particle A is displaced by a distance ξ that depends on the interception half-distance η, where $\xi > 0$ when $\eta > 0$ and $\xi < 0$ when $\eta < 0$, as shown in Fig. 4.9. If the particles are perfectly spherical, smooth, and non-deformable, we invoke reversibility in slow motion to argue that $\xi = 0$ for any η.

Repeating the preceding analysis, we set $\hat{\phi}(\eta) = \Delta t\, \zeta(\eta)$ and find that, by symmetry, $V = 0$. The random displacements amount to a diffusive process with diffusivity

$$\mathcal{D}=\frac{1}{2}\int_{-\infty}^{\infty} \xi^2\, \zeta(\eta)\, \mathrm{d}\eta = kc\int_{-\infty}^{\infty} \xi^2\, |\eta|\, \mathrm{d}\eta. \tag{4.12.14}$$

The displacement $\xi(\eta)$ is assumed to decay sufficiently fast for the diffusivity integral to be finite. To compute the diffusivity, the function $\xi(\eta)$ must be prescribed or reconstructed by numerical simulation of the particle interception.

4.12.6 Exercise

4.12.1 Compute V and $\mathcal{D}$ when $\phi(\xi)$ is the Gaussian distribution,

$$\phi(\xi) = \frac{1}{\sqrt{2\pi}\,\sigma}\, \exp\!\left(-\frac{1}{2}\,\frac{\xi^2}{\sigma^2}\right), \tag{4.12.15}$$

where σ is a specified standard deviation.

Species Balances

5

We have regarded an inhomogeneous mixture parcel consisting of different species as an instantaneous superposition of single-species parcels that interact by chemical reactions and intermolecular force fields as they move through one other in a flow. Species parcel properties and rates of evolution may be defined and then combined to yield expressions for effective mixture properties as functions of constitution. The diffusive mass flux arises naturally in these equations in terms of the differences between the species and mixture velocities, constituting diffusion velocities.

5.1 Species Mass Balances

The mass of a red parcel consisting of a fixed collection of red material point particles is given by

$$m_R \equiv \iiint_{\text{red parcel}} \rho_R \, dV, \tag{5.1.1}$$

where ρ_R is the red density and the integration is performed over the instantaneous red parcel volume. We recall that a red point particle at the surface of a red parcel remains that surface of the parcel at all times in the presence or absence of diffusion.

A chemical reaction between red and blue molecules may cause the mass of a red parcel to increase or decrease in time, so that

$$\frac{d m_R}{dt} = \iiint_{\text{red parcel}} \dot{r}_R \, dV, \tag{5.1.2}$$

where $\dot{r}_R$ is the *volumetric rate of red mass generation* with units of mass over volume and time.

Combining the last two equations, we obtain

$$\frac{d}{dt} \iiint_{\text{red parcel}} \rho_R \, dV = \iiint_{\text{red parcel}} \dot{r}_R \, dV. \tag{5.1.3}$$

C. Pozrikidis, *Transport Processes Primer*,
https://doi.org/10.1007/978-1-4939-9909-5_5

A similar equation can be written for a blue parcel,

$$\frac{\mathrm{d}}{\mathrm{d}t}\iiint_{\text{blue parcel}} \rho_{\mathrm{B}}\,\mathrm{d}V = \iiint_{\text{blue parcel}} \dot{r}_{\mathrm{B}}\,\mathrm{d}V, \tag{5.1.4}$$

where $\dot{r}_{\mathrm{B}}$ is the volumetric rate of blue mass generation. To ensure mass conservation, we require that

$$\dot{r}_{\mathrm{R}} + \dot{r}_{\mathrm{B}} = 0 \tag{5.1.5}$$

in a binary mixture. No other physical mechanism, including diffusion, contributes to the rate of change of the mass of a red or blue parcel.

For an arbitrary number of species in a multicomponent mixture, the mass conservation constraint takes the form

$$\sum_{\mathrm{X}} \dot{r}_{\mathrm{X}} = 0, \tag{5.1.6}$$

where the sum extends over all species.

5.1.1 Red Material Derivative

The time derivative on the left-hand side of (5.1.3) can be transferred into the integral as a *red* material derivative, yielding

$$\iiint_{\text{red parcel}} \left(\frac{\mathrm{D}(\rho_{\mathrm{R}}\,\delta V_{\mathrm{R}})}{\mathrm{D}t}\right)_{\mathrm{R}} = \iiint_{\text{red parcel}} \dot{r}_{\mathrm{R}}\,\delta V_{\mathrm{R}}, \tag{5.1.7}$$

where δV_{R} is a small red material volume. Discarding the integral sign on both sides, and rearranging, we obtain a differential red mass evolution equation,

$$\frac{1}{\delta V_{\mathrm{R}}}\left(\frac{\mathrm{D}(\rho_{\mathrm{R}}\,\delta V_{\mathrm{R}})}{\mathrm{D}t}\right)_{\mathrm{R}} = \dot{r}_{\mathrm{R}}. \tag{5.1.8}$$

A similar equation can be written for the blue species or any species in a multicomponent solution.

5.1.2 Reynolds Transport Equation

Applying the Reynolds transport equation shown in (4.7.17) for the density and velocity of a red parcel, that is, setting $\phi = \rho_{\mathrm{R}}$, we obtain

$$\frac{\mathrm{d}}{\mathrm{d}t}\iiint_{\text{red parcel}} \rho_{\mathrm{R}}\,\mathrm{d}V = \iiint_{\text{red parcel}} \frac{\partial \rho_{\mathrm{R}}}{\partial t}\,\mathrm{d}V + \iint_{\text{red parcel}} \rho_{\mathrm{R}}\,\mathbf{n}\cdot\mathbf{u}_{\mathrm{R}}\,\mathrm{d}S, \tag{5.1.9}$$

where the normal unit vector, $\mathbf{n}$, points outward from the parcel. A similar equation can be written for a blue parcel. If a red parcel and a blue parcel are superposed at a particular instant in time, and only then, the integration domains and normal unit vector, $\mathbf{n}$, are shared.

5.1.3 Eulerian Mass Balance

Setting the right-hand side of (5.1.9) equal to the right-hand side of (5.1.3), we obtain

$$\iiint_{\text{red parcel}} \frac{\partial \rho_R}{\partial t}\, dV + \iint_{\text{red parcel}} \rho_R\, \mathbf{n} \cdot \mathbf{u}_R\, dS = \iiint_{\text{red parcel}} \dot{r}_R\, dV. \tag{5.1.10}$$

A similar equation can be written for a blue parcel or any species parcel in a multicomponent solution. We will see that diffusion enters these balance equations by way of the species velocities in the second term on the left-hand side.

5.1.4 Control Volumes

Equation (5.1.10) can be transformed into a red mass balance mass over a control volume, $\mathcal{V}_c$, defined by the instantaneous red parcel shape,

$$\iiint_{\mathcal{V}_c} \frac{\partial \rho_R}{\partial t}\, dV = \iint_{\mathcal{S}_c} \rho_R\, \mathbf{n}^{\text{in}} \cdot \mathbf{u}_R\, dS + \iiint_{\mathcal{V}_c} \dot{r}_R\, dV, \tag{5.1.11}$$

where $\mathcal{S}_c$ is the boundary of the control volume and $\mathbf{n}^{\text{in}}$ is the inward unit normal vector, $\mathbf{n}^{\text{in}} = -\mathbf{n}$. The left-hand side is the rate of species mass accumulation inside a stationary or frozen control volume.

5.1.5 Parcel Superposition

If a red parcel and a blue parcel are superposed at some time instant, and only then, equations (5.1.3) and (5.1.4) can be added to yield a mass conservation condition,

$$\frac{d}{dt} \iiint_{\text{red parcel}} \rho_R\, dV + \frac{d}{dt} \iiint_{\text{blue parcel}} \rho_B\, dV = 0. \tag{5.1.12}$$

Addition is possible only because superposition implies that the volume of integration is the same at some instant for both integrals on the right-hand sides of (5.1.3) and (5.1.4). The sum on the left-hand side of (5.1.12) can be extended to an arbitrary number of species.

In fact, equation (5.1.12) implies that

$$\frac{d}{dt} \iiint_{\text{mixture parcel}} \rho\, dV = 0, \tag{5.1.13}$$

where the mixture parcel moves with the mixture mass velocity, $\mathbf{u}$. To demonstrate this, we apply equation (4.7.20) for $\phi = \rho_X$, and sum over all species to obtain

$$\sum_X \frac{d}{dt} \iiint_{\text{X parcel}} \rho_X\, dV = \frac{d}{dt} \iiint_{\text{mixture parcel}} \rho\, dV + \iint_{\text{parcel}} \mathbf{n} \cdot \Big(\sum_X (\rho_X \mathbf{u}_X) - \rho \mathbf{u}\Big)\, dS. \tag{5.1.14}$$

Equation (5.1.13) arises by noting that the left-hand side is zero by mass conservation, while the integrand of the second integral on the right-hand side is identically zero by the definition of the mixture mass velocity.

5.1.6 Y Mass in an X Parcel

We may apply equation (4.7.19) for $\phi = \rho_Y$ and obtain

$$\frac{d}{dt} \iiint_{X\ \text{parcel}} \rho_Y \, dV = \frac{d}{dt} \iiint_{Y\ \text{parcel}} \rho_Y \, dV + \iint_{\text{parcel}} \rho_Y \, \mathbf{n} \cdot (\mathbf{u}_X - \mathbf{u}_Y) \, dS, \tag{5.1.15}$$

yielding

$$\frac{d}{dt} \iiint_{X\ \text{parcel}} \rho_Y \, dV = \iiint_{X\ \text{parcel}} \dot{r}_Y \, dV + \iint_{X\ \text{parcel}} \rho_Y \, \mathbf{n} \cdot (\mathbf{u}_X - \mathbf{u}_Y) \, dS, \tag{5.1.16}$$

which provides us with the rate of change of the Y mass contained in an X parcel.

5.1.7 Exercise

5.1.1 Add equation (5.1.11) over all species and discuss the resulting equation.

5.2 Species Density Evolution Equations

Evolution equations for the densities of the red and blue species can be derived in terms of the red and blue velocities working as in Chap. 1 for homogeneous fluids. The presence of another species is significant only insofar as to allow for chemical conversions.

To derive evolution differential equations for the red density and mass fraction, we use the divergence theorem to convert the surface integral in (5.1.10) into a volume integral over a red parcel surface. Discarding the volume integral signs, we derive the evolution equation

$$\frac{\partial \rho_R}{\partial t} + \nabla \cdot (\rho_R \mathbf{u}_R) = \dot{r}_R. \tag{5.2.1}$$

In terms of the red material derivative, we obtain

$$\left(\frac{D\rho_R}{Dt}\right)_R + \rho_R \, \nabla \cdot \mathbf{u}_R = \dot{r}_R. \tag{5.2.2}$$

Similar equations can be derived for the blue species. These equations differ from corresponding equations for homogeneous fluids only by the presence of the mass generation terms on the right-hand sides. The red material derivative of the species densities in (5.2.2) is not necessarily zero, even when the mixture is precisely, nearly, or approximately incompressible.

5.2.1 Multicomponent Solution

The density evolution equation for species X in a multicomponent solution takes one of the equivalent forms

$$\frac{\partial \rho_X}{\partial t} + \nabla \cdot (\rho_X \mathbf{u}_X) = \dot{r}_X, \qquad \left(\frac{D\rho_X}{Dt}\right)_X + \rho_X \nabla \cdot \mathbf{u}_X = \dot{r}_X, \tag{5.2.3}$$

where $\dot{r}_X$ is the volumetric rate of species X mass production. We recall that $\sum \dot{r}_X = 0$ to satisfy mass conservation, where the sum is over all species.

5.2.2 From Species to Mixture

Adding the species density evolution equation (5.2.1) to its counterpart for the blue species, and recalling that $\rho = \rho_R + \rho_B$ and $\dot{r}_R + \dot{r}_B = 0$, we obtain the familiar continuity equation for the fluid density of the mixture,

$$\frac{\partial \rho}{\partial t} + \nabla \cdot (\rho\, \mathbf{u}) = 0, \tag{5.2.4}$$

where ρ is the mixture density and $\mathbf{u}$ is the mixture mass velocity. In terms of the mixture material derivative,

$$\frac{D\rho}{Dt} + \rho\, \nabla \cdot \mathbf{u} = 0. \tag{5.2.5}$$

Integrating this equation over a mixture parcel, we obtain the parcel mass conservation law expressed by (5.1.13).

If a mixture is incompressible, $D\rho/Dt = 0$, and therefore $\nabla \cdot \mathbf{u} = 0$, as dictated by the continuity equation, which requires that

$$\nabla \cdot (\omega_R \mathbf{u}_R) = -\nabla \cdot (\omega_B \mathbf{u}_B), \tag{5.2.6}$$

where neither side is necessarily zero.

5.2.3 Evolution of Mass Fractions

Substituting into the evolution equation (5.2.1) the definition of the red species density, $\rho_R = \omega_R \rho$, where ω_R is the red mass fraction, and expanding the derivatives on the left-hand side, we obtain

$$\omega_R \frac{\partial \rho}{\partial t} + \rho \frac{\partial \omega_R}{\partial t} + \omega_R\, \mathbf{u}_R \cdot \nabla \rho + \rho\, \nabla \cdot (\omega_R \mathbf{u}_R) = \dot{r}_R. \tag{5.2.7}$$

Rearranging, we obtain the evolution equation

$$\omega_R \left(\frac{D\rho}{Dt}\right)_R + \rho \left(\frac{\partial \omega_R}{\partial t} + \nabla \cdot (\omega_R \mathbf{u}_R)\right) = \dot{r}_R. \tag{5.2.8}$$

A similar evolution equation can be derived for the blue mass fraction,

$$\omega_B \left(\frac{D\rho}{Dt}\right)_B + \rho \left(\frac{\partial \omega_B}{\partial t} + \nabla \cdot (\omega_B \mathbf{u}_B)\right) = \dot{r}_B. \tag{5.2.9}$$

Adding the red and blue equations, setting

$$\frac{D\rho}{Dt} = \omega_R \left(\frac{D\rho}{Dt}\right)_R + \omega_B \left(\frac{D\rho}{Dt}\right)_B \tag{5.2.10}$$

according to (4.7.10), and recalling the definition of the mixture velocity, $\mathbf{u} = \omega_R \mathbf{u}_R + \omega_B \mathbf{u}_B$, we recover the Lagrangian form of the mixture continuity equation shown in (5.2.5).

When the mixture is incompressible or the mixture density can be assumed to be uniform over the fluid, the material derivatives on the left-hand sides of (5.2.8) and (5.2.9) are both zero.

5.2.4 Diffusion in a Box

As an application, we consider the one-dimensional diffusion experiment shown in Fig. 4.4 where the horizontal box extends from $x = a$ to b, and follow the motion of a red material parcel extending from $x = a$ to $X_R(t)$, where $X_R(t)$ is the position of a red point particle. The red mass contained in this parcel is

$$m_R = \int_a^{X_R(t)} \rho_R(x)\, dx. \tag{5.2.11}$$

Using the Leibnitz rule, we find that the rate of change of the parcel mass is

$$\frac{d}{dt} \int_a^{X_R(t)} \rho_R(x)\, dx = \frac{dX_R}{dt}\, \rho_R(X_R) + \int_a^{X_R(t)} \frac{\partial \rho_R}{\partial t}(x)\, dx. \tag{5.2.12}$$

Setting $dX_R/dt = u_R$ and using the evolution equation (5.2.1), we obtain

$$\frac{d}{dt} \int_a^{X_R(t)} \rho_R(x)\, dx = (\rho_R u_R)(X_R) + \int_a^{X_R(t)} \left(- \frac{\partial(\rho_R u_R)}{\partial x} + \dot{r}_R \right) dx. \tag{5.2.13}$$

Performing the integration, and assuming that $u_R = 0$ at the left end of the box located at $x = a$, we find that

$$\frac{d}{dt} \int_a^{X_R(t)} \rho_R(x)\, dx = \int_a^{X_R(t)} \dot{r}_R\, dx, \tag{5.2.14}$$

which confirms that the evolving red parcel mass is conserved in the absence of chemical conversion.

5.2.5 Exercise

5.2.1 Demonstrate that adding (5.2.8) and (5.2.9) yields the mixture continuity equation.

5.3 Rankin–Hugoniot Species Density Condition

The rate of change of the red density recorded by an observer who moves with arbitrary velocity $\mathbf{v}_R$ in three-dimensional space is given by the traveler's derivative

$$\frac{d\rho_R}{dt} = \frac{\partial \rho_R}{\partial t} + \mathbf{v}_R \cdot \nabla \rho_R, \tag{5.3.1}$$

which is analogous to the material derivative. Using the red-density evolution equation (5.2.1), we obtain

$$\frac{d\rho_R}{dt} = \mathbf{v}_R \cdot \nabla \rho_R - \nabla \cdot (\rho_R \mathbf{u}_R) + \dot{r}_R. \tag{5.3.2}$$

Based on this equation, we find that the rate of change of the red density jump across a moving front of red-density discontinuity moving with velocity $\mathbf{v}_R$ is

$$\frac{d\Delta\rho_R}{dt} = \mathbf{v}_R \cdot \nabla \Delta\rho_R - \nabla \cdot \big(\Delta(\rho_R \mathbf{u}_R)\big) + \Delta\dot{r}_R, \tag{5.3.3}$$

where Δ denotes the discontinuity or jump from one side to another.

All terms in equation (5.3.3) are finite, except for two terms involving spatial derivatives normal to the evolving front inherent in the gradient, $\nabla\Delta\rho_R$, and divergence, $\nabla \cdot \big(\Delta(\rho_R \mathbf{u}_R)\big)$. To ensure regularity, we concentrate on these bothersome terms and require that, to leading order,

$$0 = (\mathbf{n} \cdot \mathbf{v}_R)\,(\mathbf{n} \cdot \nabla\Delta\rho_R) - \mathbf{n} \cdot \nabla\big(\Delta(\rho_R \mathbf{n} \cdot \mathbf{u}_R)\big), \tag{5.3.4}$$

where $\mathbf{n}$ is the unit vector normal to the front and $\mathbf{n} \cdot \mathbf{v}_R$ is the normal velocity of the front. Discarding the normal derivative operator and rearranging, we obtain the Rankin–Hugoniot density condition

$$\mathbf{n} \cdot \mathbf{v}_R = \frac{\Delta(\rho_R \mathbf{n} \cdot \mathbf{u}_R)}{\Delta\rho_R}. \tag{5.3.5}$$

If the species velocity is continuous across the front, then $\mathbf{n} \cdot \mathbf{v}_R = \mathbf{n} \cdot \mathbf{u}$.

A similar equation can be written for the blue species,

$$\mathbf{n} \cdot \mathbf{v}_B = \frac{\Delta(\rho_B \mathbf{n} \cdot \mathbf{u}_B)}{\Delta\rho_B}. \tag{5.3.6}$$

Note that the red and blue fronts and corresponding front velocities are not necessarily the same.

The Rankin–Hugoniot species density condition essentially relate the species density discontinuity to the normal-momentum discontinuity at a moving front.

5.3.1 One-Dimensional Flow

In the idealized case of one-dimensional flow along the x axis, we obtain the expression

$$v_{R_x} = \frac{\Delta(\rho_R\, u_{R_x})}{\Delta\rho_R}, \tag{5.3.7}$$

where u_{R_x} is the red species velocity. If the velocity is continuous across the front, then $v_{R_x} = u_{R_x}$.

Equation (5.3.7) can be derived by performing a red mass balance over a fixed interval $[a, b]$ on the x axis that includes the discontinuity, $x_{R_d}(t)$, that is, $a < x_{R_d}(t) < b$ within a time interval of interest. The red mass contained in the interval is

$$m_R = \int_a^b \rho_R \, dx. \tag{5.3.8}$$

The mass balance requires that

$$\frac{dm_R}{dt} = (\rho_R u_{R_x})_{x=a} - (\rho_R u_{R_x})_{x=b} + \int_a^b \dot{r}_R \, dx. \tag{5.3.9}$$

Writing

$$m_R = \int_a^{x_{R_d}(t)} \rho_R \, dx + \int_{x_{R_d}(t)}^b \rho_R \, dx, \tag{5.3.10}$$

and differentiating by way of the Leibnitz rule, we find that

$$\frac{dm_R}{dt} = -\Delta\rho_R \frac{dx_{R_d}}{dt} + \int_a^b \frac{\partial \rho_R}{\partial t} \, dx, \tag{5.3.11}$$

where

$$\Delta\rho_R = \rho^+(x_{R_d}) - \rho^-(x_{R_d}) \tag{5.3.12}$$

and the superscripts $\pm$ indicate evaluation at the left or right edge of the discontinuity.

Now using the one-dimensional version of the red species continuity equation,

$$\frac{\partial \rho_R}{\partial t} + \frac{\partial(\rho_R u_{R_x})}{\partial x} = \dot{r}_R, \tag{5.3.13}$$

we express (5.3.11) as

$$\frac{dm_R}{dt} = -\Delta\rho_R \frac{dx_{R_d}}{dt} - \int_a^b \frac{\partial(\rho_R u_{R_x})}{\partial x} \, dx + \int_a^b \dot{r}_R \, dx. \tag{5.3.14}$$

Performing the integration in the second term on the right-hand side, we find that

$$\frac{dm_R}{dt} = -\Delta\rho_R \frac{dx_{R_d}}{dt} + (\rho_R u_{R_x})_{x=a} - (\rho_R u_{R_x})_{x=b} + \Delta(\rho_R u_{R_x}) + \int_a^b \dot{r}_R \, dx, \tag{5.3.15}$$

where

$$\Delta(\rho_R u_{R_x}) \equiv (\rho u_{R_x})^+ - (\rho u_{R_x})^- \tag{5.3.16}$$

and all terms are evaluated at the point of discontinuity, x_{R_d}. Substituting expression (5.3.15) into (5.3.9), simplifying, rearranging, and setting $dx_{R_d}/dt = v_{R_x}$, we recover equation (5.3.7).

5.3.2 Exercise

5.3.1 Simplify and discuss (5.3.5) for one-dimensional flow on the x axis.

5.4 Species Molar Balances

The volumetric rates of mass generation, $\dot{r}_\mathrm{R}$ and $\dot{r}_\mathrm{B}$, are related to the volumetric rates of mole generation, $\dot{r}^*_\mathrm{R}$ and $\dot{r}^*_\mathrm{B}$, by the equations

$$\dot{r}_\mathrm{R} = M_\mathrm{R}\,\dot{r}^*_\mathrm{R}, \qquad \dot{r}_\mathrm{B} = M_\mathrm{B}\,\dot{r}^*_\mathrm{B}, \tag{5.4.1}$$

where M_R and M_B are the red and blue molar masses. The simultaneous usage of a dot (to indicate time rate) and an asterisk (to indicate molar rate) is cumbersome yet necessary. The mass conservation equation (5.1.5) requires that

$$M_\mathrm{R}\,\dot{r}^*_\mathrm{R} + M_\mathrm{B}\,\dot{r}^*_\mathrm{B} = 0. \tag{5.4.2}$$

In practice, the rates $\dot{r}^*_\mathrm{R}$ and $\dot{r}^*_\mathrm{B}$ are determined by the chemical reaction stoichiometry and kinetics.

For an arbitrary number of species in a multicomponent mixture, the constraint (5.4.2) takes the form

$$\sum_\mathrm{X} M_\mathrm{X}\dot{r}^*_\mathrm{X} = 0, \tag{5.4.3}$$

where M_X is the molar mass of the X species, ensuring that the total rate of mass generation is zero.

5.4.1 Number of Red and Blue Molecules

In terms of the red molar concentration, c_R, the number of red molecules contained in a red parcel is given by

$$n_\text{red parcel} \equiv \iiint_\text{red parcel} c_\mathrm{R}\,\mathrm{d}V, \tag{5.4.4}$$

where the integration is performed over the instantaneous red parcel volume.

A chemical reaction between red and blue molecules may cause the number of red molecules in a parcel to increase or decrease in time, so that

$$\frac{\mathrm{d}\,n_\text{red parcel}}{\mathrm{d}t} = \iiint_\text{red parcel} \dot{r}^*_\mathrm{R}\,\mathrm{d}V. \tag{5.4.5}$$

Combining the last two equations, we obtain

$$\frac{\mathrm{d}}{\mathrm{d}t}\iiint_\text{red parcel} c_\mathrm{R}\,\mathrm{d}V = \iiint_\text{red parcel} \dot{r}^*_\mathrm{R}\,\mathrm{d}V. \tag{5.4.6}$$

A similar equation can be written for a blue parcel,

$$\frac{\mathrm{d}}{\mathrm{d}t}\iiint_{\text{blue parcel}} c_{\mathrm{B}}\,\mathrm{d}V=\iiint_{\text{blue parcel}}\dot{r}^*_{\mathrm{B}}\,\mathrm{d}V. \tag{5.4.7}$$

No other physical mechanism, including diffusion, contributes to the rate of change of the number of molecules in a red or blue parcel.

The time derivative on the right-hand side of (5.4.6) can be transferred into the integral as a *red* material derivative, yielding

$$\iiint_{\text{red parcel}}\Big(\frac{\mathrm{D}(c_{\mathrm{R}}\,\delta V_{\mathrm{R}})}{\mathrm{D}t}\Big)_{\mathrm{R}}=\iiint_{\text{red parcel}}\dot{r}^*_{\mathrm{R}}\,\delta V_{\mathrm{R}}, \tag{5.4.8}$$

where δV_{R} is a small red material volume. Discarding the integral sign, we obtain the evolution equation

$$\frac{1}{\delta V_{\mathrm{R}}}\Big(\frac{\mathrm{D}(c_{\mathrm{R}}\,\delta V_{\mathrm{R}})}{\mathrm{D}t}\Big)_{\mathrm{R}}=\dot{r}^*_{\mathrm{R}}. \tag{5.4.9}$$

A similar equation can be written for the blue species or any species in a multicomponent solution.

5.4.2 Exercise

5.4.1 Derive a relation between $\dot{r}^*_{\mathrm{R}}$ and $\dot{r}^*_{\mathrm{B}}$ when two blue molecules combine to yield one red molecule.

5.5 Species Concentration Evolution Equations

Working as in Sect. 5.2, we derive red and blue molar concentration evolution equations in a binary solution. For the red species, we derive the evolution equation

$$\frac{\partial c_{\mathrm{R}}}{\partial t}+\nabla\cdot(c_{\mathrm{R}}\mathbf{u}_{\mathrm{R}})=\dot{r}^*_{\mathrm{R}}. \tag{5.5.1}$$

In terms of the red material derivative,

$$\Big(\frac{\mathrm{D}c_{\mathrm{R}}}{\mathrm{D}t}\Big)_{\mathrm{R}}+c_{\mathrm{R}}\,\nabla\cdot\mathbf{u}_{\mathrm{R}}=\dot{r}^*_{\mathrm{R}}. \tag{5.5.2}$$

We recall that the red material derivative is defined with respect to the red velocity, $\mathbf{u}_{\mathrm{R}}$.

For the blue species, we derive the evolution equation

$$\frac{\partial c_{\mathrm{B}}}{\partial t}+\nabla\cdot(c_{\mathrm{B}}\mathbf{u}_{\mathrm{B}})=\dot{r}^*_{\mathrm{B}}. \tag{5.5.3}$$

In terms of the blue material derivative,

$$\Big(\frac{\mathrm{D}c_{\mathrm{B}}}{\mathrm{D}t}\Big)_{\mathrm{B}}+c_{\mathrm{B}}\,\nabla\cdot\mathbf{u}_{\mathrm{B}}=\dot{r}^*_{\mathrm{B}}. \tag{5.5.4}$$

We recall that the blue material derivative is defined with respect to the blue velocity, $\mathbf{u}_{\mathrm{B}}$.

5.5.1 Mixture Molar Concentration

Now we recall that the mixture molar concentration is given by $c = c_R + c_B$, and set $c_R = c f_R$ and $c_B = c f_B$. Adding equations (5.5.1) and (5.5.3), we obtain

$$\frac{\partial c}{\partial t} + \nabla \cdot (c\,\mathbf{u}^*) = \dot{r}_R^* + \dot{r}_B^*, \tag{5.5.5}$$

where

$$\mathbf{u}^* \equiv f_R\,\mathbf{u}_R + f_B\,\mathbf{u}_B \tag{5.5.6}$$

is the mixture molar velocity, which is generally different than the mixture mass velocity,

$$\mathbf{u} \equiv \omega_R\,\mathbf{u}_R + \omega_B\,\mathbf{u}_B. \tag{5.5.7}$$

The mixture molar velocity is equal to the mixture mass velocity only in the case of equal molar masses.

The sum on the right-hand side of (5.5.5) is not necessarily zero. Unlike mass, moles are not necessarily conserved in the presence of a chemical reaction. For example, if two blue molecules combine into a red molecule, a counting argument suggests that $\dot{r}_B^* = -2\,\dot{r}_R^*$.

5.5.2 Uniform Mixture Concentration

At steady state and for uniform mixture molar concentration, c, equation (5.5.5) yields

$$\nabla \cdot \mathbf{u}^* = \frac{1}{c}\,(\dot{r}_R^* + \dot{r}_B^*). \tag{5.5.8}$$

Uniform mixture concentration occurs when the mixture is an ideal gas under uniform pressure and temperature.

5.5.3 Evolution of Parcel Number of Molecules

Equation (5.5.5) can be restated as

$$\frac{\partial c}{\partial t} + \nabla \cdot (c\,\mathbf{u}) = \nabla \cdot \big(c\,(\mathbf{u} - \mathbf{u}^*)\big) + \dot{r}_R^* + \dot{r}_B^*. \tag{5.5.9}$$

Integrating this equation over a mixture parcel and using the divergence theorem, we obtain a parcel evolution equation,

$$\frac{\mathrm{d}}{\mathrm{d}t} \iiint_{\text{mixture parcel}} c\,\mathrm{d}V = \iint_{\text{mixture parcel}} c\,\mathbf{n} \cdot (\mathbf{u} - \mathbf{u}^*)\,\mathrm{d}S + \iiint_{\text{mixture parcel}} (\dot{r}_R^* + \dot{r}_B^*)\,\mathrm{d}V, \tag{5.5.10}$$

where $\mathbf{n}$ is the unit vector normal to the parcel surface. The left-hand side is generally nonzero even in the absence of a chemical reaction.

5.5.4 Predator–Prey Equations

Suppose that c_R is the concentration of a prey, and c_B is the concentration of a predator. In the Lotka–Volterra model, the rates of prey and predator generation are given by

$$r_R^* = \alpha\, c_R - \beta\, c_R c_B, \quad r_B^* = \delta\, c_R\, c_B - \gamma\, c_B, \tag{5.5.11}$$

where α, β, γ, and δ are specified positive parameters. The one-dimensional species concentration evolution equations over the x axis read

$$\frac{\partial c_R}{\partial t} + \frac{\partial (c_R u_R)}{\partial x} = \alpha\, c_R - \beta\, c_R c_B \tag{5.5.12}$$

and

$$\frac{\partial c_B}{\partial t} + \frac{\partial (c_R u_B)}{\partial x} = \delta\, c_R\, c_B - \gamma\, c_B. \tag{5.5.13}$$

For closure, the velocities u_R and u_B must be specified.

We may assume that prey move toward dense vegetation and minimum population of predators, and predators move toward maximum population of prey. The behavior of the system can be studied by numerical solution of the governing differential equations.

5.5.5 Exercise

5.5.1 Demonstrate that adding (5.2.8) and (5.2.9) yields the mixture continuity equation.

5.6 Species Parcel Momentum

The momentum of a red parcel is defined in terms of the red species density, ρ_R, and velocity, $\mathbf{u}_R$, as

$$\mathbf{M}_R \equiv \iiint_{\text{red parcel}} \rho_R \mathbf{u}_R \, dV. \tag{5.6.1}$$

The rate of change of the parcel momentum is

$$\frac{d\mathbf{M}_R}{dt} = \iiint_{\text{red parcel}} \Big(\frac{D(\rho_R\, \mathbf{u}_R\, \delta V_R)}{Dt} \Big)_R, \tag{5.6.2}$$

where $(D/Dt)_R$ is the red material derivative and δV_R is a small red material volume.

5.6.1 Species Particle Acceleration

Expanding the red material derivative of the product in (5.6.2), and recalling that

$$\Big(\frac{D\,(\rho_R \delta V_R)}{Dt} \Big)_R = \dot{r}_R\, \delta V_R, \tag{5.6.3}$$

we obtain

$$\frac{\mathrm{d}\,\mathbf{M}_\mathrm{R}}{\mathrm{d}t} = \iiint_\mathrm{parcel} \Big(\rho_\mathrm{R} \left(\frac{\mathrm{D}\mathbf{u}_\mathrm{R}}{\mathrm{D}t}\right)_\mathrm{R} + \dot{r}_\mathrm{R}\mathbf{u}_\mathrm{R} \Big)\, \mathrm{d}V. \tag{5.6.4}$$

The integrand involves the red point-particle acceleration,

$$\mathbf{a}_\mathrm{R} = \left(\frac{\mathrm{D}\mathbf{u}_\mathrm{R}}{\mathrm{D}t}\right)_\mathrm{R}, \tag{5.6.5}$$

and also accounts for the effect of chemical conversion due to a chemical reaction. Similar equations can be written for a blue parcel or any monochromatic species parcel in a multicomponent mixture.

5.6.2 Conservative Form

Now expanding the red material derivative of the product in (5.6.2) in a different way, and recalling that the rate of change of a red material volume is given by

$$\left(\frac{\mathrm{D}\,\delta V_\mathrm{R}}{\mathrm{D}t}\right)_\mathrm{R} = (\boldsymbol{\nabla}\cdot\mathbf{u}_\mathrm{R})\,\delta V_\mathrm{R}, \tag{5.6.6}$$

we obtain

$$\frac{\mathrm{d}\,\mathbf{M}_\mathrm{R}}{\mathrm{d}t} = \iiint_\text{red parcel} \Big(\left(\frac{\mathrm{D}(\rho_\mathrm{R}\,\mathbf{u}_\mathrm{R})}{\mathrm{D}t}\right)_\mathrm{R} + \rho_\mathrm{R}\,\mathbf{u}_\mathrm{R}\,\boldsymbol{\nabla}\cdot\mathbf{u}_\mathrm{R} \Big)\, \mathrm{d}V. \tag{5.6.7}$$

Expressing the material derivative inside the integral in terms of Eulerian derivatives,

$$\left(\frac{\mathrm{D}(\rho_\mathrm{R}\,\mathbf{u}_\mathrm{R})}{\mathrm{D}t}\right)_\mathrm{R} = \frac{\partial(\rho_\mathrm{R}\,\mathbf{u}_\mathrm{R})}{\partial t} + \mathbf{u}_\mathrm{R}\cdot\boldsymbol{\nabla}(\rho_\mathrm{R}\,\mathbf{u}_\mathrm{R}), \tag{5.6.8}$$

we obtain

$$\frac{\mathrm{d}\,\mathbf{M}_\mathrm{R}}{\mathrm{d}t} = \iiint_\text{red parcel} \Big(\frac{\partial(\rho_\mathrm{R}\,\mathbf{u}_\mathrm{R})}{\partial t} + \mathbf{u}_\mathrm{R}\cdot\boldsymbol{\nabla}(\rho_\mathrm{R}\,\mathbf{u}_\mathrm{R}) + \rho_\mathrm{R}\,\mathbf{u}_\mathrm{R}\,\boldsymbol{\nabla}\cdot\mathbf{u}_\mathrm{R} \Big)\, \mathrm{d}V. \tag{5.6.9}$$

The second and third terms inside the integral can be combined to yield

$$\frac{\mathrm{d}\,\mathbf{M}_\mathrm{R}}{\mathrm{d}t} = \iiint_\text{red parcel} \Big(\frac{\partial(\rho_\mathrm{R}\,\mathbf{u}_\mathrm{R})}{\partial t} + \boldsymbol{\nabla}\cdot(\rho_\mathrm{R}\,\mathbf{u}_\mathrm{R}\otimes\mathbf{u}_\mathrm{R}) \Big)\, \mathrm{d}V, \tag{5.6.10}$$

where $\rho_\mathrm{R}\,\mathbf{u}_\mathrm{R}\otimes\mathbf{u}_\mathrm{R}$ is the red momentum tensor with components

$$[\,\rho_\mathrm{R}\,\mathbf{u}_\mathrm{R}\otimes\mathbf{u}_\mathrm{R}\,]_{ij} = \rho_\mathrm{R}\,u_{\mathrm{R}_i}u_{\mathrm{R}_j}. \tag{5.6.11}$$

The trace of the red momentum tensor is the square of the magnitude of the red velocity multiplied by the red density.

Using the divergence theorem to convert the term involving the divergence inside the integral into a surface integral in (5.6.9), we obtain

$$\frac{\mathrm{d}\,\mathbf{M}_\mathrm{R}}{\mathrm{d}t} = \iiint_{\text{red parcel}} \frac{\partial(\rho_\mathrm{R}\,\mathbf{u}_\mathrm{R})}{\partial t}\,\mathrm{d}V + \iint_{\text{red parcel}} (\rho_\mathrm{R}\,\mathbf{u}_\mathrm{R})\,(\mathbf{n}\cdot\mathbf{u}_\mathrm{R})\,\mathrm{d}S, \tag{5.6.12}$$

where $\mathbf{n}$ is the unit normal vector pointing outward from the parcel. Equation (5.6.12) could have been derived by applying the Reynolds transport equation expressed by (4.7.17) for the momentum of a red parcel, that is, by setting $\phi = \rho_\mathrm{R}\mathbf{u}_\mathrm{R}$.

5.6.3 Reconciliation

To reconcile (5.6.10) with (5.6.4), we recall the red species continuity equation

$$\frac{\partial\rho_\mathrm{R}}{\partial t} + \nabla\cdot(\rho_\mathrm{R}\,\mathbf{u}_\mathrm{R}) = \dot{r}_\mathrm{R}, \tag{5.6.13}$$

and obtain from (5.6.10)

$$\frac{\mathrm{d}\,\mathbf{M}_\mathrm{R}}{\mathrm{d}t} = \iiint_{\text{red parcel}} \Big(\rho_\mathrm{R}\,\Big(\frac{\partial\mathbf{u}_\mathrm{R}}{\partial t} + \mathbf{u}_\mathrm{R}\cdot\nabla\mathbf{u}_\mathrm{R}\Big) + \dot{r}_\mathrm{R}\mathbf{u}_\mathrm{R}\Big)\,\mathrm{d}V, \tag{5.6.14}$$

which is the Eulerian form of (5.6.4). Comparing (5.6.10) with (5.6.4), we find that

$$\rho_\mathrm{R}\left(\frac{\mathrm{D}\mathbf{u}_\mathrm{R}}{\mathrm{D}t}\right)_\mathrm{R} + \dot{r}_\mathrm{R}\mathbf{u}_\mathrm{R} = \frac{\partial(\rho_\mathrm{R}\,\mathbf{u}_\mathrm{R})}{\partial t} + \nabla\cdot(\rho_\mathrm{R}\,\mathbf{u}_\mathrm{R}\otimes\mathbf{u}_\mathrm{R}). \tag{5.6.15}$$

The second term on the left-hand side does not appear in the absence of a chemical reaction.

Similar equations can be written for a blue parcel. The species equations can be added only if the parcels occupy the same volume in space at a particular instant.

5.6.4 Exercise

5.6.1 Explain why the nominal mixture point-particle acceleration, $\mathbf{a} \equiv \mathrm{D}\mathbf{u}/\mathrm{D}t$, is not necessarily equal to the weighted sum, $\omega_\mathrm{R}\mathbf{a}_\mathrm{R} + \omega_\mathrm{B}\mathbf{a}_\mathrm{B}$.

5.7 Species Parcel Angular Momentum

The angular momentum of a red parcel is given by

$$\mathbf{A}_\mathrm{R} \equiv \iiint_{\text{red parcel}} \rho_\mathrm{R}\,(\mathbf{X}_\mathrm{R} - \mathbf{x}_\mathrm{R})\times\mathbf{u}_\mathrm{R}\,\mathrm{d}V, \tag{5.7.1}$$

where $\mathbf{X}_\mathrm{R}$ is the position of a red point particle inside the parcel and $\mathbf{x}_\mathrm{R}$ is an arbitrarily specified point that can be set at the origin of the working coordinate system.

The rate of change of a parcel's angular momentum is given by an ordinary time derivative that can be transferred inside the integral as a red material derivative,

$$\frac{\mathrm{d}\,\mathbf{A}_\mathrm{R}}{\mathrm{d}t} = \iiint_{\text{red parcel}} \left(\frac{\mathrm{D}(\rho_\mathrm{R}\,(\mathbf{X}_\mathrm{R} - \mathbf{x}_\mathrm{R}) \times \mathbf{u}_\mathrm{R}\,\delta V_\mathrm{R})}{\mathrm{D}t}\right)_\mathrm{R}, \tag{5.7.2}$$

where δV_R is a small red material volume.

5.7.1 Evolution in Terms of Acceleration

Expanding the red material derivative of the product in (5.7.2), and recalling equation (5.6.3), we obtain

$$\frac{\mathrm{d}\,\mathbf{A}_\mathrm{R}}{\mathrm{d}t} = \iiint_{\text{red parcel}} \Big(\rho_\mathrm{R}\left(\frac{\mathrm{D}((\mathbf{X}_\mathrm{R} - \mathbf{x}_\mathrm{R}) \times \mathbf{u}_\mathrm{R})}{\mathrm{D}t}\right)_\mathrm{R} + \dot{r}_\mathrm{R}\,(\mathbf{X}_\mathrm{R} - \mathbf{x}_\mathrm{R}) \times \mathbf{u}_\mathrm{R}\Big)\,\mathrm{d}V. \tag{5.7.3}$$

Expanding further the material derivative inside the integral, we obtain

$$\frac{\mathrm{d}\,\mathbf{A}_\mathrm{R}}{\mathrm{d}t} = \iiint_{\text{red parcel}} \Big(\rho_\mathrm{R}\left(\frac{\mathrm{D}\,\mathbf{X}_\mathrm{R}}{\mathrm{D}t}\right)_\mathrm{R} \times \mathbf{u}_\mathrm{R} + \rho_\mathrm{R}\,(\mathbf{X}_\mathrm{R} - \mathbf{x}_\mathrm{R}) \times \left(\frac{\mathrm{D}\mathbf{u}_\mathrm{R}}{\mathrm{D}t}\right)_\mathrm{R}$$
$$+\,\dot{r}_\mathrm{R}\,(\mathbf{X}_\mathrm{R} - \mathbf{x}_\mathrm{R}) \times \mathbf{u}_\mathrm{R}\Big)\,\mathrm{d}V. \tag{5.7.4}$$

We note that the red material derivative of the position of a red point particle is the red velocity, $(\mathrm{D}\mathbf{X}_\mathrm{R}/\mathrm{D}t)_\mathrm{R} = \mathbf{u}_\mathrm{R}$, and the cross product of $\mathbf{u}_\mathrm{R}$ with itself is zero, and obtain

$$\frac{\mathrm{d}\,\mathbf{A}_\mathrm{R}}{\mathrm{d}t} = \iiint_{\text{red parcel}} (\mathbf{X}_\mathrm{R} - \mathbf{x}_\mathrm{R}) \times \Big(\rho_\mathrm{R}\left(\frac{\mathrm{D}\mathbf{u}_\mathrm{R}}{\mathrm{D}t}\right)_\mathrm{R} + \dot{r}_\mathrm{R}\mathbf{u}_\mathrm{R}\Big)\,\mathrm{d}V. \tag{5.7.5}$$

At this final stage, the red point-particle position, $\mathbf{X}_\mathrm{R}$, can be replaced by the position, $\mathbf{x}$, inside the red parcel volume. The integrand involves the red point-particle acceleration,

$$\mathbf{a}_\mathrm{R} = \left(\frac{\mathrm{D}\mathbf{u}_\mathrm{R}}{\mathrm{D}t}\right)_\mathrm{R}, \tag{5.7.6}$$

and also accounts for the effect of chemical conversion due to a chemical reaction.

Similar equations can be written for a blue parcel or any monochromatic species parcel in a multicomponent mixture.

5.7.2 Evolution in Terms of Eulerian Derivatives

Now expanding the red material derivative in a different way inside the integral, in (5.7.2), and using the fundamental law of kinematics expressed by (5.6.6), we obtain

$$\frac{\mathrm{d}\,\mathbf{A}_\mathrm{R}}{\mathrm{d}t} = \iiint_{\text{red parcel}} \Big(\left(\frac{\mathrm{D}(\rho_\mathrm{R}\,(\mathbf{X}_\mathrm{R} - \mathbf{x}_\mathrm{R}) \times \mathbf{u}_\mathrm{R})}{\mathrm{D}t}\right)_\mathrm{R} + \rho_\mathrm{R}\,(\mathbf{X}_\mathrm{R} - \mathbf{x}_\mathrm{R}) \times \mathbf{u}_\mathrm{R}\,(\nabla\cdot\mathbf{u}_\mathrm{R})\Big)\,\mathrm{d}V. \tag{5.7.7}$$

Expressing the red material derivative inside the integrand in terms of Eulerian derivatives and combining two terms, we obtain a conservative form inside the integral,

$$\frac{\mathrm{d}\,\mathbf{A}_\mathrm{R}}{\mathrm{d}t} = \iiint_{\text{red parcel}} \Big(\frac{\partial(\rho_\mathrm{R}\,(\mathbf{X}_\mathrm{R}-\mathbf{x}_\mathrm{R})\times\mathbf{u}_\mathrm{R})}{\partial t} + \nabla\cdot\big(\rho_\mathrm{R}\,\mathbf{u}_\mathrm{R}\otimes\big((\mathbf{X}_\mathrm{R}-\mathbf{x}_\mathrm{R})\times\mathbf{u}_\mathrm{R}\big)\big)\Big)\,\mathrm{d}V, \tag{5.7.8}$$

where $\rho_\mathrm{R}\,\mathbf{u}_\mathrm{R}\otimes\mathbf{u}_\mathrm{R}$ is the red momentum tensor. Using the divergence theorem to convert the term involving the divergence inside the integral into a surface integral, and replacing $\mathbf{X}_\mathrm{R}$ with the position vector, $\mathbf{x}$, we obtain

$$\frac{\mathrm{d}\,\mathbf{A}_\mathrm{R}}{\mathrm{d}t} = \iiint_{\text{red parcel}} \frac{\partial(\rho_\mathrm{R}(\mathbf{x}-\mathbf{x}_\mathrm{R})\times\mathbf{u}_\mathrm{R})}{\partial t}\,\mathrm{d}V + \iint_{\text{red parcel}} (\rho_\mathrm{R}\,(\mathbf{x}-\mathbf{x}_\mathrm{R})\times\mathbf{u}_\mathrm{R})\,(\mathbf{n}\cdot\mathbf{u}_\mathrm{R})\,\mathrm{d}S, \tag{5.7.9}$$

where $\mathbf{n}$ is the unit normal vector pointing outward from the parcel.

5.7.3 Reconciliation

To reconcile (5.7.5) with (5.7.8), we expand further the first term inside the integral on the right-hand side of (5.7.8), and obtain

$$\begin{aligned}\frac{\mathrm{d}\,\mathbf{A}_\mathrm{R}}{\mathrm{d}t} = \iiint_{\text{red parcel}} \Big(\Big(\frac{\mathrm{D}\,\mathbf{X}_\mathrm{R}}{\mathrm{D}t}\Big)_\mathrm{R} \times (\rho_\mathrm{R}\,\mathbf{u}_\mathrm{R}) + (\mathbf{X}_\mathrm{R}-\mathbf{x}_\mathrm{R})\times\Big(\frac{\mathrm{D}(\rho_\mathrm{R}\,\mathbf{u}_\mathrm{R})}{\mathrm{D}t}\Big)_\mathrm{R} \\ + \rho_\mathrm{R}\,(\mathbf{X}_\mathrm{R}-\mathbf{x}_\mathrm{R})\times\mathbf{u}_\mathrm{R}\,(\nabla\cdot\mathbf{u}_\mathrm{R})\Big)\,\mathrm{d}V.\end{aligned} \tag{5.7.10}$$

The red material derivative of the position of a red point particle in the first term inside the integral is the red velocity, $(\mathrm{D}\mathbf{X}_\mathrm{R}/\mathrm{D}t)_\mathrm{R} = \mathbf{u}_\mathrm{R}$, and the cross product of $\mathbf{u}_\mathrm{R}$ with $\rho_\mathrm{R}\mathbf{u}_\mathrm{R}$ is zero. We conclude that

$$\frac{\mathrm{d}\,\mathbf{A}_\mathrm{R}}{\mathrm{d}t} = \iiint_{\text{red parcel}} (\mathbf{X}_\mathrm{R}-\mathbf{x}_\mathrm{R})\times\Big(\Big(\frac{\mathrm{D}(\rho_\mathrm{R}\,\mathbf{u}_\mathrm{R})}{\mathrm{D}t}\Big)_\mathrm{R} + \rho_\mathrm{R}\,\mathbf{u}_\mathrm{R}\,(\nabla\cdot\mathbf{u}_\mathrm{R})\Big)\,\mathrm{d}V. \tag{5.7.11}$$

Expressing the material derivative inside the integrand in terms of Eulerian derivatives and rearranging, we obtain

$$\frac{\mathrm{d}\,\mathbf{A}_\mathrm{R}}{\mathrm{d}t} = \iiint_{\text{red parcel}} (\mathbf{X}_\mathrm{R}-\mathbf{x}_\mathrm{R})\times\Big(\frac{\partial(\rho_\mathrm{R}\,\mathbf{u}_\mathrm{R})}{\partial t} + \nabla\cdot(\rho_\mathrm{R}\,\mathbf{u}_\mathrm{R}\otimes\mathbf{u}_\mathrm{R})\Big)\,\mathrm{d}V. \tag{5.7.12}$$

Now using the red species continuity equation (5.6.13), we obtain

$$\frac{\mathrm{d}\,\mathbf{A}_\mathrm{R}}{\mathrm{d}t} = \iiint_{\text{red parcel}} (\mathbf{X}_\mathrm{R}-\mathbf{x}_\mathrm{R})\times\Big(\rho_\mathrm{R}(\frac{\partial\mathbf{u}_\mathrm{R}}{\partial t} + \mathbf{u}_\mathrm{R}\cdot\nabla\mathbf{u}_\mathrm{R}) + \dot{r}_\mathrm{R}\mathbf{u}_\mathrm{R}\Big)\,\mathrm{d}V. \tag{5.7.13}$$

which is the Eulerian form of the integrand in (5.7.5).

Comparing (5.7.5) with (5.7.8), we obtain

$$\begin{aligned}(\mathbf{X}_\mathrm{R}-\mathbf{x}_\mathrm{R})\times\Big(\rho_\mathrm{R}\Big(\frac{\mathrm{D}\mathbf{u}_\mathrm{R}}{\mathrm{D}t}\Big)_\mathrm{R} + \dot{r}_\mathrm{R}\mathbf{u}_\mathrm{R}\Big) = \frac{\partial(\rho_\mathrm{R}\,(\mathbf{X}_\mathrm{R}-\mathbf{x}_\mathrm{R})\times\mathbf{u}_\mathrm{R})}{\partial t} \\ + \nabla\cdot\big(\rho_\mathrm{R}\,\mathbf{u}_\mathrm{R}\otimes\big((\mathbf{X}_\mathrm{R}-\mathbf{x}_\mathrm{R})\times\mathbf{u}_\mathrm{R}\big).\end{aligned} \tag{5.7.14}$$

Note that the red rate of mass generation, $\dot{r}_\mathrm{R}$ appears on the left but not on the right-hand side.

5.7.4 Exercise

5.7.1 Simplify (5.7.5) for a stationary red species.

5.8 Diffusive Mass Flux

We have referred to diffusion as a physical process causing differences in the red and blue species velocities in a binary mixture under nonuniform species distributions. Consideration of the motion of monochromatic parcels through each other in an evolving mixture has led us to the notion of mixture lag velocities relative to the species velocities. In the case of chemical mixtures, the lag velocities are diffusion velocities attributed to random molecular excursions.

To quantify the kinematic action of diffusion, we revisit the red and blue parcel mass evolution equations, recall the definition of the mixture mass velocity as a weighted average of the species velocities, $\mathbf{u} = \omega_R \mathbf{u}_R + \omega_B \mathbf{u}_B$, and formulate the difference between the species and mixture velocities.

5.8.1 Species Parcel Mass Conservation

The red parcel mass conservation equation (5.1.10) is repeated below for convenience,

$$\iiint_{\text{red parcel}} \frac{\partial \rho_R}{\partial t}\, dV + \iint_{\text{red parcel}} \rho_R\, \mathbf{n} \cdot \mathbf{u}_R\, dS = \iiint_{\text{red parcel}} \dot{r}_R\, dV. \tag{5.8.1}$$

Rearranging the second term on the left-hand side, we obtain

$$\iiint_{\text{red parcel}} \frac{\partial \rho_R}{\partial t}\, dV + \iint_{\text{red parcel}} \rho_R\, \mathbf{n} \cdot \mathbf{u}\, dS = - \iint_{\text{red parcel}} \mathbf{n} \cdot \boldsymbol{j}_R\, dS + \iiint_{\text{red parcel}} \dot{r}_R\, dV, \tag{5.8.2}$$

where

$$\boldsymbol{j}_R \equiv \rho_R\, (\mathbf{u}_R - \mathbf{u}) \tag{5.8.3}$$

is the *diffusive mass flux* for the red species. Note that the mixture velocity, $\mathbf{u}$, appears inside the second integral on the left-hand side of (5.8.2).

5.8.2 Binary Mixture

In the case of a binary mixture, we substitute into (5.8.3) the definition of the mixture mass velocity, $\mathbf{u} = \omega_R \mathbf{u}_R + \omega_B \mathbf{u}_B$, and simplify to obtain

$$\boldsymbol{j}_R = \rho\, \omega_R\, \omega_B\, (\mathbf{u}_R - \mathbf{u}_B), \tag{5.8.4}$$

where $\rho = \rho_R + \rho_B$ is the mixture density.

5.8.3 Blue Parcels

Similar equations can be written for a blue parcel,

$$\iiint_{\text{blue parcel}} \frac{\partial \rho_B}{\partial t} \, dV + \iint_{\text{blue parcel}} \rho_B \, \mathbf{n} \cdot \mathbf{u} \, dS = - \iint_{\text{blue parcel}} \mathbf{n} \cdot \boldsymbol{j}_B \, dS + \iiint_{\text{blue parcel}} \dot{r}_B \, dV, \tag{5.8.5}$$

where

$$\boldsymbol{j}_B \equiv \rho_B \, (\mathbf{u}_B - \mathbf{u}) \tag{5.8.6}$$

is the *diffusive mass flux* of the blue species. We note again that the mixture velocity, $\mathbf{u}$, appears on the left-hand side of (5.8.5).

5.8.4 Binary Mixture

In the case of a binary mixture, the blue mass flux is given by

$$\boldsymbol{j}_B \equiv \rho \, \omega_R \, \omega_B \, (\mathbf{u}_B - \mathbf{u}_R), \tag{5.8.7}$$

which is equal in magnitude and opposite in direction to that for the red species defined in (5.8.4).

5.8.5 The Sum of the Species Diffusive Mass Fluxes Is Zero

By definition, the sum of the species diffusive mass fluxes is zero,

$$\boldsymbol{j}_R + \boldsymbol{j}_B = \rho_R \, (\mathbf{u}_R - \mathbf{u}) + \rho_B \, (\mathbf{u}_B - \mathbf{u}) = \mathbf{0}. \tag{5.8.8}$$

To ensure mass conservation, empirical physical laws providing us with expressions for the diffusive mass fluxes, such as those discussed in Chap. 6, must satisfy with this constraint.

5.8.6 Control Volumes

Equations (5.8.2) and (5.8.5) can be transformed into balances of red or blue mass over a control volume, $\mathcal{V}_c$, defined by the instantaneous shape of a red or blue parcel,

$$\iiint_{\mathcal{V}_c} \frac{\partial \rho_X}{\partial t} \, dV = \iint_{\mathcal{S}_c} \rho_X \, \mathbf{n}^{\text{in}} \cdot \mathbf{u} \, dS + \iint_{\mathcal{S}_c} \mathbf{n}^{\text{in}} \cdot \boldsymbol{j}_X \, dS + \iiint_{\mathcal{V}_c} \dot{r}_X \, dV, \tag{5.8.9}$$

where X can be R, B, or any other color, $\mathcal{S}_c$ is the boundary of the control volume, and $\mathbf{n}^{\text{in}}$ is the inward unit normal vector.

The left-hand side of (5.8.9) is the rate of species mass accumulation inside a stationary or frozen control volume. The first term on the right-hand side is the rate of convective transport into the control volume under the influence of the mixture velocity, $\mathbf{u}$. The second term on the right-hand side is the rate of inward diffusive transport. The last term expresses the effect of a chemical reaction. It is important to observe that when a diffusive mass flux is employed, convection is defined with respect to the mixture velocity, $\mathbf{u}$.

Equation (5.8.9) could have been written at the outset as a stand-alone transport equation. However, the physical interpretation of the underlying balance and the meaning of the diffusive mass flux would have to be explained carefully and confirmed in hindsight.

5.8.7 Mass Balances Over a Mixture Parcel

Using the Reynolds transport equation in the reverse direction, we find that equations (5.8.2) and (5.8.5) express species mass balances over a nominal red–blue mixture parcel that evolves with the mixture velocity, $\mathbf{u}$,

$$\frac{\mathrm{d}}{\mathrm{d}t}\iiint_{\text{mixture parcel}} \rho_{\mathrm{R}}\,\mathrm{d}V = -\iint_{\text{mixture parcel}} \mathbf{n}\cdot\boldsymbol{j}_{\mathrm{R}}\,\mathrm{d}S + \iiint_{\text{mixture parcel}} \dot{r}_{\mathrm{R}}\,\mathrm{d}V \tag{5.8.10}$$

and

$$\frac{\mathrm{d}}{\mathrm{d}t}\iiint_{\text{mixture parcel}} \rho_{\mathrm{B}}\,\mathrm{d}V = -\iint_{\text{mixture parcel}} \mathbf{n}\cdot\boldsymbol{j}_{\mathrm{B}}\,\mathrm{d}S + \iiint_{\text{mixture parcel}} \dot{r}_{\mathrm{B}}\,\mathrm{d}V, \tag{5.8.11}$$

where the integration domains are the same in both equations. Similar equations can be written for any species in a multicomponent mixture.

In fact, equations (5.8.10) and (5.8.11) arise directly from (4.7.20) by setting $\phi = \rho_{\mathrm{R}}$ or ρ_{B}, invoking the definition of the diffusive mass flux, and rearranging. These balances confirm that the species diffusive mass fluxes pertain to the boundary of a mixture parcel moving with the mixture mass velocity.

Equations (5.8.10) and (5.8.11) are consistent with the generic natural law expressed by (1.6.5) written for a mixture parcel,

$$\frac{\mathrm{d}}{\mathrm{d}t}\iiint_{\text{mixture parcel}} \phi\,\mathrm{d}V = -\iint_{\text{mixture parcel}} \mathbf{n}\cdot\boldsymbol{\xi}\,\mathrm{d}S + \iiint_{\text{mixture parcel}} \dot{\pi}\,\mathrm{d}V + \iint_{\text{mixture parcel}} \dot{\nu}\,\mathrm{d}S, \tag{5.8.12}$$

where $\phi = \rho_{\mathrm{X}}$, $\boldsymbol{\xi} = \boldsymbol{j}_{\mathrm{X}}$, $\dot{\pi} = \dot{r}_{\mathrm{X}}$, and $\dot{\nu} = 0$.

Equations (5.8.10) and (5.8.11) could have been written at the outset, and the underlying differential equations of mass conservation could have been derived working in the reverse direction. Understanding and accepting these equations are keys to internalizing the process of species transport in a binary or multicomponent mixture.

Adding equations (5.8.10) and (5.8.11), and recalling that $\boldsymbol{j}_{\mathrm{R}} + \boldsymbol{j}_{\mathrm{B}} = \mathbf{0}$ and $\dot{r}_R + \dot{r}_{\mathrm{B}} = 0$, we obtain equation (5.1.13),

$$\frac{\mathrm{d}}{\mathrm{d}t}\iiint_{\text{mixture parcel}} \rho\,\mathrm{d}V = 0, \tag{5.8.13}$$

where $\rho = \rho_{\mathrm{R}} + \rho_B$ is the mixture density. This equation is precisely the same as that for a homogeneous fluid.

5.8.8 Stationary Mixture

When the mixture is locally or globally stationary, $\mathbf{u} = \mathbf{0}$, the diffusive mass fluxes are given by

$$j_R = \rho_R \mathbf{u}_R = \rho\, \omega_R \mathbf{u}_R, \qquad j_B = \rho_B \mathbf{u}_B = \rho\, \omega_B \mathbf{u}_B, \tag{5.8.14}$$

where $\boldsymbol{j}_R = -\boldsymbol{j}_B$ at any position and time.

5.8.9 Stationary Species

When the blue species is stationary, $\mathbf{u}_B = 0$, the mixture velocity is given by $\mathbf{u} = \omega_R \mathbf{u}_R$; consequently,

$$j_R = \rho_R\, (\mathbf{u}_R - \mathbf{u}) = \rho_R\, \omega_B\, \mathbf{u}_R = \rho\, \omega_R\, \omega_B \mathbf{u}_R \tag{5.8.15}$$

and

$$j_B = -\rho_B \mathbf{u} = -\rho_B\, \omega_R \mathbf{u}_R = -\rho\, \omega_R\, \omega_B \mathbf{u}_R, \tag{5.8.16}$$

where $\boldsymbol{j}_R = -\boldsymbol{j}_B$ at any position and time. An example of a stationary species is discussed in Sect. 6.4 with reference to a coyote.

5.8.10 Validation of Diffusion Flux Laws

The species diffusive fluxes are described by empirical and other assumed laws in terms of the species mass fractions and other driving gradients, as discussed in Chap. 6. Suppose that expressions have been proposed for the diffusive mass fluxes, denoted by $\boldsymbol{j}_R^{prop}$ and $\boldsymbol{j}_B^{prop}$. To be validated, these expressions must be tested against the condition $\boldsymbol{j}_R + \boldsymbol{j}_B = \mathbf{0}$. If this condition is satisfied, then $\boldsymbol{j}_R^{prop}$ and $\boldsymbol{j}_B^{prop}$ could be accepted as $\boldsymbol{j}_R$ and $\boldsymbol{j}_B$.

If the mandatory condition $\boldsymbol{j}_R + \boldsymbol{j}_B = \mathbf{0}$ is not satisfied, each proposed mass flux can be resolved into two components,

$$j_R^{prop} = j_R - \rho_R\, \mathbf{w}, \qquad j_B^{prop} = j_B - \rho_B\, \mathbf{w}, \tag{5.8.17}$$

where

$$\mathbf{w} = -\frac{1}{\rho}\, (j_R^{prop} + j_B^{prop}) \tag{5.8.18}$$

and

$$j_R = \omega_B\, j_R^{prop} - \omega_R\, j_B^{prop}, \qquad j_B = -j_R. \tag{5.8.19}$$

The vector field $\mathbf{w}$ may then be regarded as a convection velocity to be added to the nominal mixture velocity, $\mathbf{u}$.

5.8.11 Multicomponent Mixture

The diffusive mass flux of species X in a multicomponent mixture is defined as

$$\boldsymbol{j}_X \equiv \rho_X(\mathbf{u}_X - \mathbf{u}), \tag{5.8.20}$$

where $\mathbf{u} = \sum \omega_X \mathbf{u}_X$ is the mixture velocity. By definition of the mixture mass or momentum velocity,

$$\sum_X \boldsymbol{j}_X = \mathbf{0}, \tag{5.8.21}$$

where the sum is over all components participating in the mixture.

5.8.12 Exercise

5.8.1 Confirm that expressions (5.8.18) and (5.8.19) satisfy the decomposition (5.8.17).

5.9 Diffusion Velocities

In the present physical context, the mixture lag velocities in a binary mixture, defined as $\mathbf{u}_R - \mathbf{u}$ and $\mathbf{u}_B - \mathbf{u}$, are red and blue diffusion velocities given by

$$\mathbf{u}_R^{\text{diff}} \equiv \mathbf{u}_R - \mathbf{u} = \frac{\boldsymbol{j}_R}{\rho_R}, \qquad \mathbf{u}_B^{\text{diff}} \equiv \mathbf{u}_B - \mathbf{u} = \frac{\boldsymbol{j}_B}{\rho_B}, \tag{5.9.1}$$

where $\boldsymbol{j}_R$ and $\boldsymbol{j}_B$ are diffusive mass fluxes. These expressions suggest the disturbing possibility that a diffusion velocity may become infinite at a point where the species density is zero. To prevent the diffusion velocity from diverging, the associated diffusive mass flux must also be zero.

Since the sum of the diffusive mass fluxes should be zero, $\boldsymbol{j}_R + \boldsymbol{j}_B = \mathbf{0}$, it must be that

$$\omega_R\, \mathbf{u}_R^{\text{diff}} + \omega_B\, \mathbf{u}_B^{\text{diff}} = \mathbf{0}. \tag{5.9.2}$$

The sum on the left-hand side can be extended to any number of species.

$$\sum \omega_X\, \mathbf{u}_X^{\text{diff}} = \mathbf{0}. \tag{5.9.3}$$

In the case of a binary mixture, we substitute into (5.9.1) the definition $\mathbf{u} = \omega_R \mathbf{u}_R + \omega_B u_B$ and find that

$$\mathbf{u}_R^{\text{diff}} = \omega_B\,(\mathbf{u}_R - \mathbf{u}_B), \qquad \mathbf{u}_B^{\text{diff}} = \omega_R\,(\mathbf{u}_B - \mathbf{u}_R), \tag{5.9.4}$$

which shows that the diffusion velocities are apportioned according to the mass fractions.

5.9.1 Species Velocities in Terms of Diffusion Fluxes

Rearranging expressions (5.9.1), we obtain expressions for the species velocities in terms of the mixture velocity and diffusion velocities,

$$\mathbf{u}_R = \mathbf{u} + \mathbf{u}_R^{\text{diff}} = \mathbf{u} + \frac{\boldsymbol{j}_R}{\rho_R}, \qquad \mathbf{u}_B = \mathbf{u} + \mathbf{u}_B^{\text{diff}} = \mathbf{u} + \frac{\boldsymbol{j}_B}{\rho_B}. \tag{5.9.5}$$

Subtracting these equations, we find that

$$\mathbf{u}_R - \mathbf{u}_B = \mathbf{u}_R^{\text{diff}} - \mathbf{u}_B^{\text{diff}}. \tag{5.9.6}$$

Knowledge of the mixture velocity, $\mathbf{u}$, species fluxes, $\boldsymbol{j}_R$ and $\boldsymbol{j}_B$, and species mass densities, ρ_R and ρ_B, suffices for evaluating the species velocities, $\mathbf{u}_R$ and $\mathbf{u}_B$ at a point.

5.9.2 Stationary Species

Equations (5.9.5) suggest that one of the species can be nominally stationary, $\mathbf{u}_R = \mathbf{0}$ or $\mathbf{u}_B = \mathbf{0}$, yet the corresponding diffusive flux may be nonzero, $\boldsymbol{j}_R \neq \mathbf{0}$ or $\boldsymbol{j}_B \neq \mathbf{0}$, provided that $\mathbf{u} \neq \mathbf{0}$. The stationary state of a species does not refer to the diffusion velocity but rather to the species velocity.

5.9.3 Kinetic Energy

The square of the magnitude of the red velocity is given by

$$|\mathbf{u}_R|^2 = |\mathbf{u}|^2 + |\mathbf{u}_R^{\text{diff}}|^2 + 2\,\mathbf{u} \cdot \mathbf{u}_R^{\text{diff}}, \tag{5.9.7}$$

and the square of the magnitude of the blue velocity is given by a similar expression. Using (5.9.2), we find that

$$\rho_R |\mathbf{u}_R|^2 + \rho_B |\mathbf{u}_B|^2 = \rho\, |\mathbf{u}|^2 + \rho_R |\mathbf{u}_R^{\text{diff}}|^2 + \rho_B |\mathbf{u}_B^{\text{diff}}|^2. \tag{5.9.8}$$

In terms of diffusion fluxes,

$$\rho_R |\mathbf{u}_R|^2 + \rho_B |\mathbf{u}_B|^2 = \rho\, |\mathbf{u}|^2 + \frac{1}{\rho_R} |\boldsymbol{j}_R|^2 + \frac{1}{\rho_B} |\boldsymbol{j}_B|^2. \tag{5.9.9}$$

The sums can be extended to any number of species in a multicomponent solution,

$$\sum_X \rho_X |\mathbf{u}_X|^2 = \rho\, |\mathbf{u}|^2 + \sum_X \frac{1}{\rho_X} |\boldsymbol{j}_X|^2. \tag{5.9.10}$$

The sum of the specific kinetic energies of the species, expressed by the left-hand side, is no greater than the specific kinetic energy of the mixture expressed by the first term on the right-hand side.

5.9.4 Point-Particle Motion

Red point particles move with velocity $\mathbf{u}_R$ and blue point particles move with velocity $\mathbf{u}_B$. Only in the absence of diffusion red and blue particles move with the same velocity.

The position of a red point particle, $\mathbf{X}_R(t)$, evolves according to the advection equation

$$\frac{d\mathbf{X}_R}{dt} = \mathbf{u}_R(\mathbf{X}_R, t) = \mathbf{u}(\mathbf{X}_R, t) + \mathbf{u}_R^{\text{diff}}(\mathbf{X}_R, t), \tag{5.9.11}$$

the position of a blue point particle, $\mathbf{X}_B(t)$, evolves according to the analogous advection equation

$$\frac{d\mathbf{X}_B}{dt} = \mathbf{u}_B(\mathbf{X}_B, t) = \mathbf{u}(\mathbf{X}_B, t) + \mathbf{u}_B^{\text{diff}}(\mathbf{X}_B, t), \tag{5.9.12}$$

and the position of a mixture point particle evolves according to the advection equation

$$\frac{d\mathbf{X}}{dt} = \mathbf{u}(\mathbf{X}, t), \tag{5.9.13}$$

where $\mathbf{u}$ is the mixture mass velocity. We recall that a mixture point particle does not consist of the same collection of molecules. When the mixture is macroscopically stationary, the position of a mixture point particle remains constant in time. The ordinary differential equations governing point-particle motion can be integrated in time using standard numerical methods.

5.9.5 Chaotic Advection

The motion of red, blue, or mixture point particles under the influence of the velocity field stated in (5.9.11), (5.9.12), and (5.9.13) may exhibit features of chaotic advection due to the mixture velocity, $\mathbf{u}$, or the diffusion velocities. It is known that chaotic advection may occur in unsteady two-dimensional flow or in steady or unsteady three-dimensional flow. The effect of diffusion on chaotic advection is not clear.

5.9.6 Exercise

5.9.1 Derive an expression for the initial rate of separation of a red and a blue point particles that happen to be at the same location.

5.10 Fundamental Decomposition of Kinematics

The fundamental decomposition of kinematics for a mixture was discussed in Sect. 4.8 with reference to the mixture velocity gradient tensor. The expressions derived in that section can be restated in terms of the diffusion mass fluxes by substituting expressions (5.9.5) for $\mathbf{u}_R$ and $\mathbf{u}_B$ in the terms of the mixture velocity, $\mathbf{u}$, and species diffusion fluxes.

5.10.1 Velocity-Gradient Tensor

Recalling that the mass fractions ω_R and ω_B add to unity, we find that the coupling velocity-gradient tensor defined in (4.8.3) takes the form

$$\widehat{\nabla \mathbf{u}} = \frac{1}{\rho_R} \nabla \omega_R \otimes \boldsymbol{j}_R + \frac{1}{\rho_B} \nabla \omega_B \otimes \boldsymbol{j}_B. \tag{5.10.1}$$

The symmetric part is given by

$$[\widehat{\nabla \mathbf{u}}] = \frac{1}{2\rho_R} (\nabla \omega_R \otimes \boldsymbol{j}_R + \boldsymbol{j}_R \otimes \nabla \omega_R) + \frac{1}{2\rho_B} (\nabla \omega_B \otimes \boldsymbol{j}_B + \boldsymbol{j}_B \otimes \nabla \omega_B), \tag{5.10.2}$$

and the skew-symmetric part is given by

$$< \widehat{\nabla \mathbf{u}} >= \frac{1}{2\rho_R} (\nabla \omega_R \otimes \boldsymbol{j}_R - \boldsymbol{j}_R \otimes \nabla \omega_R) + \frac{1}{2\rho_B} (\nabla \omega_B \otimes \boldsymbol{j}_B - \boldsymbol{j}_B \otimes \nabla \omega_B). \tag{5.10.3}$$

Conversely, we find that

$$\nabla \mathbf{u}_R = \nabla \mathbf{u} + \nabla (\frac{1}{\rho_R} \boldsymbol{j}_R) = \nabla \mathbf{u} + \frac{1}{\rho_R} \nabla \boldsymbol{j}_R - \frac{1}{\rho_R^2} \boldsymbol{j}_R \otimes \nabla \rho_R, \tag{5.10.4}$$

and a similar expression for the blue species or any species in a multicomponent solution.

5.10.2 Rate of Expansion

In the case of a binary mixture, we obtain from (4.8.6) and (4.8.7)

$$\alpha = \omega_R \, \alpha_R + \omega_B \, \alpha_B + \frac{1}{\rho_R} \boldsymbol{j}_R \cdot \nabla \omega_R + \frac{1}{\rho_B} \boldsymbol{j}_B \cdot \nabla \omega_B. \tag{5.10.5}$$

Using the red species continuity equation (5.2.2) to write

$$\rho \, \omega_R \, \alpha_R = -\left(\frac{D\rho_R}{Dt}\right)_R + \dot{r}_R \tag{5.10.6}$$

and a similar equation for the blue species, and substituting these expressions into the first two terms on the right-hand side of (5.10.5), we obtain

$$\alpha\rho = -\left(\frac{D\rho_R}{Dt}\right)_R - \left(\frac{D\rho_B}{Dt}\right)_B + \boldsymbol{j}_R \cdot \nabla \ln \omega_R + \boldsymbol{j}_B \cdot \nabla \ln \omega_B, \tag{5.10.7}$$

which relates the rate of expansion of the mixture, α, to the rates of change of the densities of red and blue point particles represented by the material derivatives.

Conversely, we find that

$$\alpha_R = \alpha + \nabla \cdot (\frac{1}{\rho_R} \boldsymbol{j}_R) = \alpha + \frac{1}{\rho_R} \nabla \cdot \boldsymbol{j}_R - \frac{1}{\rho_R^2} \boldsymbol{j}_R \cdot \nabla \rho_R, \tag{5.10.8}$$

and a similar equation for the blue species.

5.10.3 Incompressible Mixtures

In the case of an incompressible mixture, $\mathrm{D}\rho/\mathrm{D}t = 0$ and therefore $\nabla \cdot \mathbf{u} = 0$, as required by the continuity equation. The species constraint (5.2.6), repeated below for convenience,

$$\nabla \cdot (\omega_{\mathrm{R}}\mathbf{u}_{\mathrm{R}}) = -\nabla \cdot (\omega_{\mathrm{B}}\mathbf{u}_{\mathrm{B}}), \tag{5.10.9}$$

takes the form

$$\nabla \cdot (\frac{\boldsymbol{j}_{\mathrm{R}}}{\rho}) = -\nabla \cdot (\frac{\boldsymbol{j}_{\mathrm{B}}}{\rho}), \tag{5.10.10}$$

which is an identity in light of the requirement that $\boldsymbol{j}_{\mathrm{R}} = -\boldsymbol{j}_{\mathrm{B}}$.

5.10.4 Exercise

5.10.1 Confirm that the sum of the last two terms on the right-hand side of (5.10.5) is the trace of $\widehat{\nabla \mathbf{u}}$.

5.11 Density Evolution Equations Redux

Evolution equations for the red and blue densities, ρ_{R} and ρ_{B}, and associated mass fractions, ω_{R} and ω_{B}, can be derived in terms of the mixture velocity, $\mathbf{u}$, and mass diffusive fluxes, $\boldsymbol{j}_{\mathrm{R}}$ and $\boldsymbol{j}_{\mathrm{B}}$, or corresponding diffusion velocities.

5.11.1 Eulerian Form

Using the divergence theorem to convert the surface integrals in (5.8.2) and (5.8.5) into volume integrals over a red parcel surface, and then discarding the volume integral signs, we obtain the red-species evolution equation

$$\frac{\partial \rho_{\mathrm{R}}}{\partial t} + \nabla \cdot (\rho_{\mathrm{R}}\mathbf{u}) = -\nabla \cdot \boldsymbol{j}_{\mathrm{R}} + \dot{r}_{\mathrm{R}}. \tag{5.11.1}$$

For the blue species, we obtain the corresponding evolution equation

$$\frac{\partial \rho_{\mathrm{B}}}{\partial t} + \nabla \cdot (\rho_{\mathrm{B}}\mathbf{u}) = -\nabla \cdot \boldsymbol{j}_{\mathrm{B}} + \dot{r}_{\mathrm{B}}. \tag{5.11.2}$$

Adding these equations and recalling that $\boldsymbol{j}_{\mathrm{R}} + \boldsymbol{j}_{\mathrm{B}} = \mathbf{0}$ and $\dot{r}_{\mathrm{R}} + \dot{r}_{\mathrm{B}} = 0$, we recover the familiar continuity equation involving the mixture density, ρ, and mass velocity, $\mathbf{u}$, in the Eulerian conservative form $\partial\rho/\partial t + \nabla \cdot (\rho\mathbf{u}) = 0$.

Alternatively, equation (5.11.1) can be derived by rearranging the second term on the left-hand side of equation (5.2.1) to obtain

$$\frac{\partial \rho_{\mathrm{R}}}{\partial t} + \nabla \cdot \big(\rho_{\mathrm{R}}(\mathbf{u} + (\mathbf{u}_{\mathrm{R}} - \mathbf{u})\big) = \dot{r}_{\mathrm{R}}, \tag{5.11.3}$$

recalling that $\boldsymbol{j}_{\mathrm{R}} \equiv \rho_{\mathrm{R}}(\mathbf{u}_{\mathrm{R}} - \mathbf{u})$, and transferring a partial term to the right-hand side.

A similar derivation can be written for the blue species or any species X in a multicomponent solution,

$$\frac{\partial \rho_X}{\partial t} + \nabla \cdot (\rho_X \mathbf{u}) = -\nabla \cdot \boldsymbol{j}_X + \dot{r}_X, \tag{5.11.4}$$

where $\boldsymbol{j}_X = \rho_X (\mathbf{u}_X - \mathbf{u})$ is the X mass flux.

5.11.2 Lagrangian Form

In terms of the mixture material derivative, D/Dt, the red-species equation (5.11.1) takes the form

$$\frac{D\rho_R}{Dt} + \rho_R \nabla \cdot \mathbf{u} = -\nabla \cdot \boldsymbol{j}_R + \dot{r}_R, \tag{5.11.5}$$

where $\nabla \cdot \mathbf{u}$ is the mixture rate of expansion. For the blue species, or any species X in a multicomponent solution,

$$\frac{D\rho_X}{Dt} + \rho_X \nabla \cdot \mathbf{u} = -\nabla \cdot \boldsymbol{j}_X + \dot{r}_X. \tag{5.11.6}$$

Adding all the species equations, we obtain the familiar continuity equation $D\rho/Dt + \rho \nabla \cdot \mathbf{u} = 0$.

5.11.3 Evolution of Mass Fractions

An evolution equation for the red mass fraction, ω_R, can be obtained by substituting $\rho_R = \rho\, \omega_R$ into the right-hand side of (5.11.5), and then expanding the material derivative to obtain

$$\rho \frac{D\omega_R}{Dt} + \omega_R \frac{D\rho}{Dt} + \rho_R \nabla \cdot \mathbf{u} = -\nabla \cdot \boldsymbol{j}_R + \dot{r}_R. \tag{5.11.7}$$

Using the continuity equation for the mixture to set

$$\frac{D\rho}{Dt} = -\rho \nabla \cdot \mathbf{u} \tag{5.11.8}$$

in the second term on the left-hand side, and simplifying, we obtain

$$\rho \frac{D\omega_R}{Dt} = -\nabla \cdot \boldsymbol{j}_R + \dot{r}_R. \tag{5.11.9}$$

A similar equation can be derived for the blue species,

$$\rho \frac{D\omega_B}{Dt} = -\nabla \cdot \boldsymbol{j}_B + \dot{r}_B. \tag{5.11.10}$$

We see that diffusive mass fluxes and species conversion rates contribute to the rate of change of the red or blue mass fraction following the motion of a nominal mixture point particle that moves with the mixture mass velocity, $\mathbf{u}$. Adding the last two equations, we obtain zeros on both sides. Similar equations can be written for each species in a multicomponent mixture consisting of an arbitrary number of species.

5.11.4 Alternative Derivation

Equation (5.11.9) can also be derived from the red density evolution equation (5.2.8), repeated below for convenience,

$$\omega_R \left(\frac{D\rho}{Dt}\right)_R + \rho\left(\frac{\partial \omega_R}{\partial t} + \nabla \cdot (\omega_R \mathbf{u}_R)\right) = \dot{r}_R, \tag{5.11.11}$$

where $(D/Dt)_R$ is the red material derivative. Substituting into the material derivative and into the last term on the left-hand side $\mathbf{u}_R = \mathbf{u} + \frac{1}{\rho_R}\, \boldsymbol{j}_R$, we obtain

$$\omega_R \frac{D\rho}{Dt} + \frac{1}{\rho}\, \boldsymbol{j}_R \cdot \nabla\rho + \rho\left(\frac{\partial \omega_R}{\partial t} + \nabla \cdot (\omega_R \mathbf{u}) + \nabla \cdot \left(\frac{\boldsymbol{j}_R}{\rho}\right)\right) = \dot{r}_R. \tag{5.11.12}$$

Now using the continuity equation (5.11.8) to evaluate $D\rho/Dt$, and simplifying, we obtain (5.11.9).

5.11.5 Stationary Mixture

In the absence of global motion, $\mathbf{u} = \mathbf{0}$, the red mass fraction is governed by the equation

$$\rho\, \frac{\partial \omega_R}{\partial t} = -\nabla \cdot \boldsymbol{j}_R + \dot{r}_R, \tag{5.11.13}$$

and the blue mass fraction is governed by a similar equation.

5.11.6 Exercise

5.11.1 Explain why adding (5.11.9) and (5.11.10) results in zeros on both sides.

5.12 Red and Blue Concentrations

The discussion of species concentration evolution equations in Sect. 5.5 can be reframed in terms of the mixture molar velocity defined as

$$\mathbf{u}^* = f_R\, \mathbf{u}_R + f_B\, \mathbf{u}_B, \tag{5.12.1}$$

and associated diffusive molar fluxes introduced in this section, $\boldsymbol{j}_R^*$ and $\boldsymbol{j}_B^*$, where f_R and f_B are molar fractions and an asterisk denotes molar-based definitions. We recall that the mixture molar velocity, $\mathbf{u}^*$, is generally not the same as the mixture mass velocity, $\mathbf{u}$, which is defined with respect to the mass fractions, ω_R and ω_B.

5.12.1 Evolution of Species Molar Concentrations

For the red species, we find that

$$\frac{\partial c_R}{\partial t} + \nabla \cdot (c_R \mathbf{u}^*) = -\nabla \cdot j_R^* + \dot{r}_R^*, \tag{5.12.2}$$

where c_R is the red species molar concentration, $\dot{r}_R^* = \dot{r}_R / M_R$ is the volumetric rate of red mole generation or consumption, M_R is the red molar mass, and

$$j_R^* \equiv c_R (\mathbf{u}_R - \mathbf{u}^*) \tag{5.12.3}$$

is the red diffusive molar flux.

For a binary mixture, we write $c_R = c\, f_R$ and substitute the definition of $\mathbf{u}^*$ to find that

$$j_R^* = c\, f_R\, f_B\, (\mathbf{u}_R - \mathbf{u}_B), \tag{5.12.4}$$

where c is the mixture molar concentration. Comparing this equation with the corresponding equation (5.8.4) for the diffusive mass flux, repeated below for convenience,

$$j_R = \rho\, \omega_R\, \omega_B\, (\mathbf{u}_R - \mathbf{u}_B), \tag{5.12.5}$$

we find that

$$j_R = \frac{\rho}{c} \frac{\omega_R \omega_B}{f_R f_B}\, j_R^*, \tag{5.12.6}$$

where ρ is the mixture density. Now setting $\rho = cM$, $\omega_R = f_R M_R / M$, and $\omega_B = f_B M_B / M$, we find that

$$j_R = \frac{M_R M_B}{M}\, j_R^*, \tag{5.12.7}$$

where M is the generally position-dependent mixture molar mass. In a binary mixture, the mass and molar fluxes point in the same direction.

Similar equations can be written for a blue parcel,

$$\frac{\partial c_B}{\partial t} + \nabla \cdot (c_B \mathbf{u}^*) = -\nabla \cdot j_B^* + \dot{r}_B^*, \tag{5.12.8}$$

where $\dot{r}_B^* = \dot{r}_B / M_B$ is the volumetric rate of blue mole generation or consumption, and

$$j_B^* \equiv c_B (\mathbf{u}_B - \mathbf{u}^*) \tag{5.12.9}$$

is the blue diffusive molar flux.

For a binary mixture, we substitute the definition of $\mathbf{u}^*$ and find that

$$j_B^* = c\, f_R\, f_B\, (\mathbf{u}_B - \mathbf{u}_R). \tag{5.12.10}$$

The blue diffusive mass flux is given by

$$j_B = \frac{M_R M_B}{M} j_B^*, \tag{5.12.11}$$

where M is the generally position-dependent mixture molar mass.

Referring to the definition of the diffusive molar fluxes, we derive the constraint

$$j_R^* + j_B^* = \mathbf{0}. \tag{5.12.12}$$

Proposed laws that provide us with expressions for the diffusive molar flux must conform with this condition in order to ensure mass conservation.

The sum of all diffusive molar fluxes must also be zero in a multicomponent solution,

$$\sum_X j_X^* = \mathbf{0}. \tag{5.12.13}$$

5.12.2 Evolution of Mixture Concentration

Adding the evolution equations for the two species of a binary mixture, we obtain

$$\frac{\partial c}{\partial t} + \nabla \cdot (c\mathbf{u}^*) = \dot{r}_R^* + \dot{r}_B^*. \tag{5.12.14}$$

The sum on the right-hand side is zero only in the case of equal molar masses. The total number of moles accounting for the mixture concentration, c, may increase or decrease due to a chemical reaction.

5.12.3 Evolution of Molar Fractions

To develop an evolution equation for the red molar fraction, f_R, we substitute into (5.12.2) the definition $c_R = cf_R$, and obtain

$$\frac{\partial (cf_R)}{\partial t} + \nabla \cdot (cf_R\mathbf{u}^*) = -\nabla \cdot j_R^* + \dot{r}_R^*. \tag{5.12.15}$$

Expanding the derivatives on the left-hand side, using (5.12.14), and rearranging, we obtain

$$c\left(\frac{\partial f_R}{\partial t} + \mathbf{u}^* \cdot \nabla f_R\right) = -\nabla \cdot j_R^* + \dot{r}_R^* - f_R\,(\dot{r}_R^* + \dot{r}_B^*). \tag{5.12.16}$$

Simplifying, we obtain

$$c\left(\frac{\partial f_R}{\partial t} + \mathbf{u}^* \cdot \nabla f_R\right) = -\nabla \cdot j_R^* + f_B\,\dot{r}_R^* - f_R\dot{r}_B^*. \tag{5.12.17}$$

A similar equation can be written for the blue species. Adding the red and blue equations, we obtain zeros on both sides.

In the case of a multicomponent mixture, we derive the evolution equation

$$c\left(\frac{\partial f_R}{\partial t} + \mathbf{u}^* \cdot \nabla f_R\right) = -\nabla \cdot \boldsymbol{j}_R^* + \dot{r}_R^* - f_R \sum_X \dot{r}_X^*, \tag{5.12.18}$$

where the sum is over all species. Adding all species equations, we obtain zeros on both sides.

5.12.4 Diffusive Molar Flux Validation

Suppose that two expressions for the diffusive molar fluxes have been suggested, denoted by $\boldsymbol{j}_R^{*\text{prop}}$ and $\boldsymbol{j}_B^{*\text{prop}}$. To be validated, these expressions must be tested against the requirement $\boldsymbol{j}_R^* + \boldsymbol{j}_B^* = \mathbf{0}$. If this condition is satisfied, then $\boldsymbol{j}_R^{*\text{prop}}$ and $\boldsymbol{j}_B^{*\text{prop}}$ could be accepted as $\boldsymbol{j}_R^*$ and $\boldsymbol{j}_B^*$.

If the mandatory condition $\boldsymbol{j}_R^* + \boldsymbol{j}_B^* = \mathbf{0}$ is not satisfied, the proposed fluxes could be resolved as

$$\boldsymbol{j}_R^{*\text{prop}} = \boldsymbol{j}_R^* - c_R\,\mathbf{w}, \qquad \boldsymbol{j}_B^{*\text{prop}} = \boldsymbol{j}_B^* - c_B\,\mathbf{w}, \tag{5.12.19}$$

where

$$\mathbf{w} = -\frac{1}{c}\,(\boldsymbol{j}_R^{*\text{prop}} + \boldsymbol{j}_B^{*\text{prop}}) \tag{5.12.20}$$

and

$$\boldsymbol{j}_R^* = f_B\,\boldsymbol{j}_R^{*\text{prop}} - f_R\,\boldsymbol{j}_B^{*\text{prop}}, \qquad \boldsymbol{j}_B^* = -\boldsymbol{j}_R^*. \tag{5.12.21}$$

The velocity $\mathbf{w}$ is accepted as a convection velocity to be added to the mixture velocity, $\mathbf{u}^*$.

5.12.5 Exercise

5.12.1 Discuss the notion of molar diffusion species velocities and their relation to the mass diffusion species velocities.

Diffusion Laws

6

Empirical, semi-empirical, and theoretical expressions for the species diffusive mass fluxes in terms of the species mass or molecular fractions can be derived by observation, intuition, or rigorously in the context of statistical mechanics and thermodynamics. Fick's law and a fractional diffusion law discussed in this chapter are favored in a broad range of science and engineering applications.

Fick's law is a local law, relating the diffusive mass flux at a point to the gradient of the species mass or molecular fraction at that point. The fractional diffusion law is a nonlocal law, involving the entire mass or molecular fraction field in an effectively unbounded domain. Fick's law is associated with random walks to the nearest neighbor sites, as discussed in Sect. 4.10, whereas the fractional diffusion law is associated with random flights to all available sites, subject to a power-law probability distribution. Other probability distributions with heavy tails can be employed. Fick's, fractional, and other generalized laws involving the heat flux are reviewed in this chapter.

6.1 Fick's Law

Fick's law of diffusion is the counterpart of Fourier's law of heat conduction discussed in Sect. 3.10. However, the underlying physical contexts and enabling physical mechanisms are not the same.

6.1.1 Fick's Law in Terms of Mass Fractions

According to the standard version of Fick's law, the diffusive mass fluxes in a binary mixture of red and blue species are given by

$$j_R \equiv \rho_R\,(\mathbf{u}_R - \mathbf{u}) = -\rho\,\mathcal{D}_{RB}\,\nabla\omega_R \tag{6.1.1}$$

and

$$j_B \equiv \rho_B\,(\mathbf{u}_B - \mathbf{u}) = -\rho\,\mathcal{D}_{BR}\,\nabla\omega_B, \tag{6.1.2}$$

where ρ is the mixture density, ω_R and ω_B are the species mass fractions, and $\mathcal{D}_{RB}$ and $\mathcal{D}_{BR}$ are binary interspecies diffusivities with units of length squared divided by time.

C. Pozrikidis, *Transport Processes Primer*,
https://doi.org/10.1007/978-1-4939-9909-5_6

Since $\omega_R + \omega_B = 1$, the mandatory constraint $\boldsymbol{j}_R + \boldsymbol{j}_B = \boldsymbol{0}$ is satisfied, provided that the diffusivities are symmetric,

$$\mathcal{D}_{RB} = \mathcal{D}_{BR}. \tag{6.1.3}$$

The physical origin of this symmetry condition can be traced to the micromechanics of random walks. In applications, the diffusivity is allowed to change in space and time or even be generalized into a diffusivity tensor.

6.1.2 Fick's Law in Terms of Molar Fractions

Using relation (4.2.11) between the mass fractions and molecular fractions, ω_R and f_R, repeated below for convenience,

$$\nabla \omega_R = \frac{M_R M_B}{M^2} \nabla f_R, \tag{6.1.4}$$

and its counterpart for the blue species, we obtain

$$\boldsymbol{j}_R = -\rho\, \mathcal{D}_{RB} \frac{M_R M_B}{M^2} \nabla f_R, \qquad \boldsymbol{j}_B = -\rho\, \mathcal{D}_{BR} \frac{M_R M_B}{M^2} \nabla f_B, \tag{6.1.5}$$

where M_R and M_B are species molar masses and $M \equiv f_R M_R + f_B M_B$ is the mixture molar mass. We recall that M is generally a function of position by way of its dependence on the molecular fractions.

Recalling that the mixture molar concentration is given by $c = \rho/M$, we obtain the alternative expressions

$$\boldsymbol{j}_R = -\frac{c^2}{\rho} \mathcal{D}_{RB} M_R M_B \nabla f_R, \qquad \boldsymbol{j}_B = -\frac{c^2}{\rho} \mathcal{D}_{BR} M_R M_B \nabla f_B. \tag{6.1.6}$$

Since $f_R + f_B = 1$ by construction, the sum of the red and blue fluxes is zero.

6.1.3 Fick's Law in Terms of Molar-Mass Concentrations

Substituting into (6.1.1) the relation $\omega_R = \phi_R M_R$, where ϕ_R is the molar-mass concentration, we obtain

$$\boldsymbol{j}_R = -\rho\, \mathcal{D}_{RB} M_R \nabla \phi_R, \qquad \boldsymbol{j}_B = -\rho\, \mathcal{D}_{BR} M_B \nabla \phi_B. \tag{6.1.7}$$

Since $\phi_R M_R + \phi_B M_B = 1$ by construction, the two fluxes add to zero, as required.

Equation (6.1.7) could be obtained by substituting into (6.1.5) relations (4.2.8) and (4.2.9) between the gradient of the molar-mass concentrations and the molecular fractions, repeated below for convenience,

$$\nabla \phi_R = \frac{M_B}{M^2} \nabla f_R, \qquad \nabla \phi_B = \frac{M_R}{M^2} \nabla f_B. \tag{6.1.8}$$

6.1.4 Diffusive Flux Orientation

It is reassuring that the diffusive mass flux of the red species is parallel to the gradient of the mass fraction ω_R, to the gradient of the red molecular fraction f_R, and also to the gradient of the molar-mass concentration ϕ_R. A similar observation can be made about the blue species.

6.1.5 Diffusion Velocities

We recall that $\rho_R = \omega_R \rho$ and $\rho_B = \omega_B \rho$, and find that the red and blue species diffusion velocities arising from Fick's law are given by

$$\mathbf{u}_R^{\text{diff}} \equiv \frac{1}{\rho_R} \, \boldsymbol{j}_R = -\mathcal{D}_{RB} \, \nabla \ln \omega_R \tag{6.1.9}$$

and

$$\mathbf{u}_B^{\text{diff}} \equiv \frac{1}{\rho_B} \, \boldsymbol{j}_B = -\mathcal{D}_{BR} \, \nabla \ln \omega_B . \tag{6.1.10}$$

We note that the gradient of the logarithm of the mass fractions appears naturally in these expressions.

An example of a red species diffusion velocity distribution for a prescribed red mass fraction distribution described by the inverse tangent function,

$$\omega_R = \tfrac{1}{2} - \frac{1}{\pi} \arctan \frac{x}{a}, \tag{6.1.11}$$

is shown with the solid line (Fig. 6.1), where a is a specified length. The associated mass flux, j_R, is described by the dot-dashed line, and the corresponding diffusion velocity, u_R^{diff}, is described by the dashed line.

Fig. 6.1 Distribution of the red species mass fraction described by (6.1.11) (solid line), associated red species mass flux scaled by $\mathcal{D}_{RB}/(a\rho)$ (dot-dashed line), and corresponding red species diffusion velocity scaled by $\mathcal{D}_{RB}/a$ (dashed line)

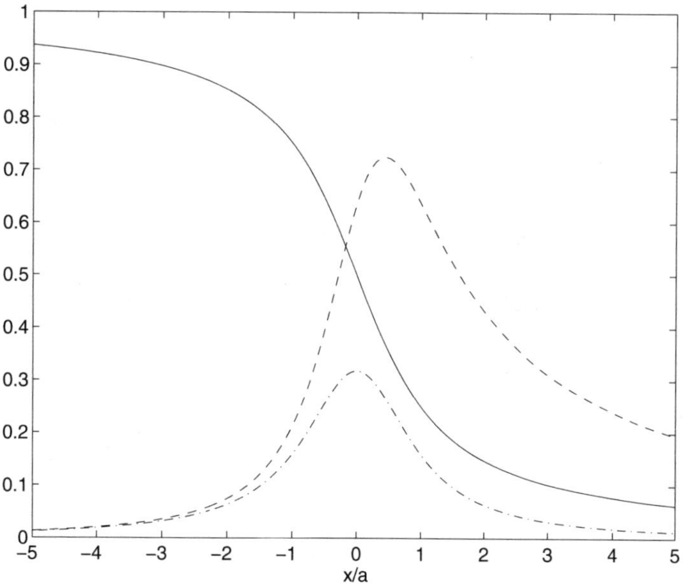

6.1.6 Difference in Species Velocities

The difference in the species velocities is equal to the difference in the diffusion velocities,

$$\mathbf{u}_R - \mathbf{u}_B = \mathbf{u}_R^{\text{diff}} - \mathbf{u}_B^{\text{diff}} = -\mathcal{D}_{RB}\,\nabla \ln \frac{\omega_R}{\omega_B} = -\mathcal{D}_{RB}\,\frac{1}{\omega_R \omega_B}\,\nabla \omega_R. \tag{6.1.12}$$

Substituting $\omega_R = f_R M_R/M$ and $\omega_B = f_B M_B/M$, we obtain

$$\mathbf{u}_R - \mathbf{u}_B = \mathbf{u}_R^{\text{diff}} - \mathbf{u}_B^{\text{diff}} = -\mathcal{D}_{RB}\,\nabla \ln \frac{f_R}{f_B} = -\mathcal{D}_{RB}\,\frac{1}{f_R f_B}\,\nabla f_R. \tag{6.1.13}$$

Similar expressions can be derived in terms of the molar-mass concentrations, ϕ_R and ϕ_B.

6.1.7 Fick's Law for the Molar Flux

Equation (5.12.4) provides us with an expression for the red molar flux, repeated below for convenience,

$$\boldsymbol{j}_R^* \equiv c_R\,(\mathbf{u}_R - \mathbf{u}^*) = c\,f_R\,f_B\,(\mathbf{u}_R - \mathbf{u}_B), \tag{6.1.14}$$

where c is the mixture molar concentration.

Substituting the last expression in (6.1.13) for the species velocity difference, $\mathbf{u}_R - \mathbf{u}_B$, and repeating for the blue species, we obtain Fick's law for the molar flux,

$$\boldsymbol{j}_R^* = -c\,\mathcal{D}_{RB}\,\nabla f_R, \qquad \boldsymbol{j}_B^* = -c\,\mathcal{D}_{BR}\,\nabla f_B. \tag{6.1.15}$$

The constraint $\boldsymbol{j}_R^* + \boldsymbol{j}_B^* = \mathbf{0}$ is satisfied, provided that $\mathcal{D}_{RB} = \mathcal{D}_{BR}$.

Comparing expressions (6.1.15) and (6.1.5), and recalling that $\rho = cM$, we obtain

$$\boldsymbol{j}_R = \frac{M_R M_B}{M}\,\boldsymbol{j}_R^*, \qquad \boldsymbol{j}_B = \frac{M_R M_B}{M}\,\boldsymbol{j}_B^*, \tag{6.1.16}$$

as shown previously in (5.12.7) and (5.12.11), where M is the generally position-dependent mixture molar mass.

6.1.8 Multicomponent Mixture

Fick's law (6.1.5) for species X in a multicomponent mixture takes the form

$$\boldsymbol{j}_X = \rho\,\frac{M_X}{M^2}\sum_Y{}'\,\mathcal{D}_{XY}\,M_Y\,\nabla f_Y, \tag{6.1.17}$$

where $\mathcal{D}_{XY}$ are pairwise diffusivities and the prime after the sum indicates exclusion of the term $Y = X$.

6.1.9 Exercise

6.1.1 Derive an expression for the difference $\mathbf{u}_R - \mathbf{u}_B$ in terms of the molar-mass concentrations, ϕ_R and ϕ_B.

6.2 Fick's Diffusion Kinematics

The red and blue species diffusion velocities arising from Fick's law were given in (6.1.9) and (6.1.10) as

$$\mathbf{u}_R^{\text{diff}} = -\mathcal{D}_{RB}\, \nabla \ln \omega_R, \qquad \mathbf{u}_B^{\text{diff}} = -\mathcal{D}_{BR}\, \nabla \ln \omega_B. \tag{6.2.1}$$

We will assume that the diffusivity, $\mathcal{D}_{RB}$, is uniform over the entire binary fluid.

6.2.1 Velocity-Gradient Tensor

Taking the gradient of the definitions $\mathbf{u}_R = \mathbf{u} + \mathbf{u}_R^{\text{diff}}$ and $\mathbf{u}_B = \mathbf{u} + \mathbf{u}_B^{\text{diff}}$, we find that the species velocity-gradient tensors are given by

$$\nabla \mathbf{u}_R = \nabla \mathbf{u} - \mathcal{D}_{RB}\, \nabla \otimes \nabla \ln \omega_R \tag{6.2.2}$$

and

$$\nabla \mathbf{u}_B = \nabla \mathbf{u} - \mathcal{D}_{BR}\, \nabla \otimes \nabla \ln \omega_B, \tag{6.2.3}$$

where $\nabla \otimes \nabla$ is the symmetric operator of second derivatives. The mixture velocity gradient tensor was given in (4.8.2) as

$$\nabla \mathbf{u} = \omega_R \nabla \mathbf{u}_R + \omega_B \nabla \mathbf{u}_B + \widehat{\nabla \mathbf{u}}, \tag{6.2.4}$$

where $\widehat{\nabla \mathbf{u}}$ is a coupling tensor given in (4.8.3). Substituting Fick's law, we find that

$$\widehat{\nabla \mathbf{u}} = -\mathcal{D}_{RB}\left(\nabla \omega_R \otimes \nabla \ln \omega_R + \nabla \omega_B \otimes \nabla \ln \omega_B \right), \tag{6.2.5}$$

which can be restated as

$$\widehat{\nabla \mathbf{u}} = -\frac{1}{\rho\, \mathcal{D}_{RB}} \Big(\frac{1}{\rho_R} \boldsymbol{j}_R \otimes \boldsymbol{j}_R + \frac{1}{\rho_B} \boldsymbol{j}_B \otimes \boldsymbol{j}_B \Big). \tag{6.2.6}$$

Since $\boldsymbol{j}_B = -\boldsymbol{j}_R$, the two terms can be consolidated on the right-hand side.

Substituting into (6.2.4) expressions (6.2.2), (6.2.3), and (6.2.5), we derive an identity,

$$\omega_R\, \nabla \otimes \nabla \ln \omega_R + \omega_B\, \nabla \otimes \nabla \ln \omega_B = -\nabla \omega_R \otimes \nabla \ln \omega_R - \nabla \omega_B \otimes \nabla \ln \omega_B. \tag{6.2.7}$$

6.2.2 Rate of Expansion

Recalling that the Laplacian is the divergence of the gradient, $\nabla^2 = \boldsymbol{\nabla} \cdot \boldsymbol{\nabla}$, we find that the mixture and species rates of expansion, $\alpha = \boldsymbol{\nabla} \cdot \mathbf{u}$, $\alpha_R \equiv \boldsymbol{\nabla} \cdot \mathbf{u}_R$, and $\alpha_B \equiv \boldsymbol{\nabla} \cdot \mathbf{u}_B$, are related by

$$\alpha_R = \alpha - \mathcal{D}_{RB}\, \nabla^2 \ln \omega_R, \qquad \alpha_B = \alpha - \mathcal{D}_{RB}\, \nabla^2 \ln \omega_B. \tag{6.2.8}$$

The rate of expansion of the mixture is the trace of the velocity-gradient tensor. Using the preceding expressions for the velocity gradient, we find that

$$\alpha = \text{trace}(\boldsymbol{\nabla}\mathbf{u}) = \omega_R \alpha_R + \omega_B \alpha_B - \mathcal{D}_{RB} \frac{1}{\omega_R \omega_B} |\boldsymbol{\nabla}\omega_R|^2. \tag{6.2.9}$$

Substituting expressions (6.2.8) for the species rates of expansion, we obtain the identity

$$\omega_R \omega_B \left(\omega_R \nabla^2 \ln \omega_R + \omega_B \nabla^2 \ln \omega_B\right) = -|\boldsymbol{\nabla}\omega_R|^2 = -|\boldsymbol{\nabla}\omega_B|^2, \tag{6.2.10}$$

which also arises from the trace of both sides of the more general identity (6.2.7).

6.2.3 Vorticity

Since the Fickian coupling velocity-gradient tensor $\widehat{\boldsymbol{\nabla}\mathbf{u}}$ is symmetric, the coupling vorticity tensor is identically zero, and we may write $\boldsymbol{\omega} = \omega_R \boldsymbol{\omega}_R + \omega_B \boldsymbol{\omega}_B$, where $\boldsymbol{\omega} \equiv \boldsymbol{\nabla} \times \mathbf{u}$ is the vorticity vector is defined as the curl of the velocity. In fact, using expressions (6.2.1), and recalling that the curl of the gradient is identically zero, we find that

$$\boldsymbol{\omega}_R = \boldsymbol{\omega}_B = \boldsymbol{\omega}, \tag{6.2.11}$$

that is, the species and mixture vorticities are all the same.

6.2.4 Exercise

6.2.1 Confirm identity (6.2.10).

6.3 Species Fraction Evolution Equations

Self-contained evolution equations for the species mass and molar fractions can be derived by substituting Fick's law into the species density evolution equations derived in Chap. 5.

6.3.1 Mass Fraction Evolution Equation

Substituting Fick's law into the red mass fraction evolution equation (5.11.9), repeated below for convenience,

$$\rho \frac{D\omega_R}{Dt} = -\boldsymbol{\nabla} \cdot \boldsymbol{j}_R + \dot{r}_R, \tag{6.3.1}$$

we obtain

$$\rho \frac{\mathrm{D}\omega_{\mathrm{R}}}{\mathrm{D}t} = \nabla \cdot (\mathcal{D}_{\mathrm{RB}}\, \rho\, \nabla \omega_{\mathrm{R}}) + \dot{r}_{\mathrm{R}}, \tag{6.3.2}$$

where the material derivative, $\mathrm{D}/\mathrm{D}t$, is defined with respect to the mixture mass velocity, $\mathbf{u}$.

For uniform mixture density, ρ, and uniform diffusivity, $\mathcal{D}_{\mathrm{RB}}$, we obtain the convection–diffusion equation

$$\rho \frac{\mathrm{D}\omega_{\mathrm{R}}}{\mathrm{D}t} = \rho\, \mathcal{D}_{\mathrm{RB}}\, \nabla^2 \omega_{\mathrm{R}} + \dot{r}_{\mathrm{R}}. \tag{6.3.3}$$

To derive the right-hand side, we have noted that $\nabla^2 = \nabla \cdot \nabla$, that is, the Laplacian is the divergence of the gradient.

6.3.2 Stationary Mixture

In the absence of global motion, $\mathbf{u} = \mathbf{0}$, the red species mass fraction evolution equation simplifies to

$$\rho \frac{\partial\, \omega_{\mathrm{R}}}{\partial t} = -\nabla \cdot \boldsymbol{j}_{\mathrm{R}} + \dot{r}_{\mathrm{R}}. \tag{6.3.4}$$

For uniform density and diffusivity, we substitute Fick's law and obtain

$$\rho \frac{\partial\, \omega_{\mathrm{R}}}{\partial t} = \rho\, \mathcal{D}_{\mathrm{RB}}\, \nabla^2 \omega_{\mathrm{R}} + \dot{r}_{\mathrm{R}}, \tag{6.3.5}$$

which is a standard unsteady diffusion or heat conduction equation in the presence of a source.

6.3.3 Molar Fraction Evolution Equation

Substituting Fick's law (6.1.15) into the red molar fraction evolution equation (5.12.17), repeated below for convenience,

$$c\left(\frac{\partial f_{\mathrm{R}}}{\partial t} + \mathbf{u}^* \cdot \nabla f_{\mathrm{R}}\right) = -\nabla \cdot \boldsymbol{j}_{\mathrm{R}}^* + f_{\mathrm{B}}\, \dot{r}_{\mathrm{R}}^* - f_{\mathrm{R}} \dot{r}_{\mathrm{B}}^*, \tag{6.3.6}$$

and assuming uniform mixture concentration, c, and diffusivity, $\mathcal{D}_{\mathrm{RB}}$, we obtain the governing equation

$$c\left(\frac{\partial f_{\mathrm{R}}}{\partial t} + \mathbf{u}^* \cdot \nabla f_{\mathrm{R}}\right) = c\, \mathcal{D}_{\mathrm{RB}}\, \nabla^2 f_{\mathrm{R}} + f_{\mathrm{B}}\, \dot{r}_{\mathrm{R}}^* - f_{\mathrm{R}} \dot{r}_{\mathrm{B}}^*. \tag{6.3.7}$$

The expression inside the parentheses on the left-hand side is *not* the material derivative of f_{R}. A similar equation can be written for the blue species. Simplified equations can be derived for a stationary mixture.

6.3.4 Exercise

6.3.1 Derive an evolution equation for the molar-mass fraction, ϕ_R.

6.4 Coyote

The bottom of an empty whiskey barrel in a ghost town of the Nevada dessert is coated with a thin layer of whiskey. The ghost town used to be a boomtown at a time of promised opportunity followed by despair. Now the boomtown is occupied by coyotes, mice, and rats. The gas mixture in the barrel contains whiskey (R) and air (B), as shown in Fig. 6.2. A coyote perched at the top of the open barrel finds the opportunity to sniff the smell.

For simplicity, we denote the x velocity component by u and the x component of the mass flux by j, where a species subscript can be added, as required, and the x axis points upward from the bottom to the top of the barrel. All other velocity components and fluxes are zero.

6.4.1 Species and Mixture Velocities

Common sense suggests that air in the barrel is stationary, $u_B = 0$, and therefore the mixture mass and molar velocities are given by

$$u = \omega_R u_R, \qquad u^* = f_R u_R. \tag{6.4.1}$$

We recall that the mass and molar fractions are related by $\omega_R = M_R f_R / M$, where M is the mixture molar mass.

The red mass flux is given by

$$j_R \equiv \rho_R (u_R - u) = \rho_R (u_R - \omega_R u_R) = \rho_R (1 - \omega_R)\, u_R = \rho_R\, \omega_B\, u_R, \tag{6.4.2}$$

yielding

$$j_R = \rho\, \omega_R \omega_B\, u_R = \rho\, \omega_B\, u. \tag{6.4.3}$$

The blue mass flux is given by

$$j_B = -j_R. \tag{6.4.4}$$

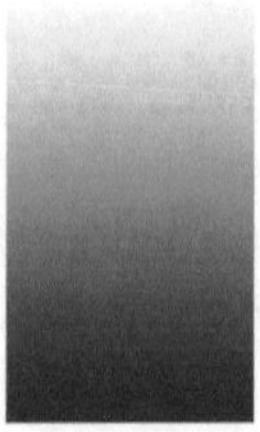

Fig. 6.2 Distribution of whiskey vapor (red species) in an open barrel. A coyote is enjoying the whiskey aroma at the top. The air (blue species) is stationary inside the barrel, while the vapor moves upward due to diffusion

The red molar flux is given by

$$j_R^* \equiv c_R(u_R - u^*) = c_R(u_R - f_R u_R) = c_R\,(1 - f_R)\,u_R = c_R\,f_B\,u_R, \tag{6.4.5}$$

yielding

$$j_R^* = c\,f_R f_B\,u_R = c\,f_B\,u^*. \tag{6.4.6}$$

The blue molar flux is given by

$$j_B^* = -j_R^*. \tag{6.4.7}$$

6.4.2 Uniform c

It is reasonable to assume that the mixture concentration, c, is uniform inside the barrel, determined by the temperature, T, and pressure, p, according to an equation of state. An implied assumption is that both T and p are uniform inside the barrel. For example, if the mixture behaves like an ideal gas, $c = p/RT$.

Mass conservation requires that the mixture molar velocity, u^*, should also be constant, and therefore the species molar fluxes should be proportional to the blue molar fraction at steady state.

6.4.3 Evolution Equations

In the absence of a chemical reaction, the general red molar concentration evolution equation (5.5.1) reads

$$\frac{\partial c_R}{\partial t} + \frac{\partial (c_R u_R)}{\partial x} = 0 \tag{6.4.8}$$

for $x_{\text{bottom}} \le x \le x_{\text{top}}$. It is convenient to set the origin at the surface of the liquid whiskey at the bottom so that $x_{\text{bottom}} = 0$ and $x_{\text{top}} = h$.

At steady state, we use the last expression in (6.4.5) to obtain

$$\frac{d}{dx}\left(\frac{1}{f_B}\,j_R^*\right) = 0. \tag{6.4.9}$$

Integrating this elementary ordinary differential equation, we find that

$$j_R^* = \gamma\,c\,\frac{\mathcal{D}}{h}\,f_B, \tag{6.4.10}$$

where $\mathcal{D}$ is the binary diffusivity and γ is a dimensionless constant to be determined from the boundary conditions for f_R at the barrel bottom and top. Note that the ratio $\mathcal{D}/h$ has units of velocity.

6.4.4 Fick's Law

Now substituting into the left-hand side of (6.4.10) Fick's law for the red molar flux,

$$j_R^* = -c\,\mathcal{D}\,\frac{d\,f_R}{dx}, \tag{6.4.11}$$

we obtain a governing ordinary differential equation,

$$-\,h\,\frac{d\,f_R}{dx} = \gamma\, f_B. \tag{6.4.12}$$

Setting $f_R = 1 - f_B$ and rearranging, we obtain

$$h\,\frac{d\,\ln f_B}{dx} = \gamma. \tag{6.4.13}$$

Integrating in x from the bottom up where $x = 0$ to an arbitrary elevation, x, we obtain

$$\ln\frac{f_B(x)}{f_B(0)} = \gamma\,\frac{x}{h}. \tag{6.4.14}$$

Rearranging, we obtain

$$f_B(x) = f_B(0)\,\exp\left(\gamma\,\frac{x}{h}\right), \tag{6.4.15}$$

which reveals an exponential distribution.

Now applying (6.4.15) at the top where $x = h$, we obtain an expression for the constant γ in terms of boundary values at the top and bottom,

$$\gamma = \ln\frac{f_B(h)}{f_B(0)}. \tag{6.4.16}$$

Substituting this expression into (6.4.15), we obtain

$$f_B(x) = f_B(0)\left(\frac{f_B(h)}{f_B(0)}\right)^{x/h}. \tag{6.4.17}$$

The associated distribution of the red species arises by setting $f_B = 1 - f_R$.

If p_w is the whiskey vapor pressure, then $f_R(0) = p_w/p$, where p is the mixture pressure; consequently, $f_B(0) = (p - p_w)/p$.

6.4.5 Summary

We have found that the red molar flux and velocity are given by

$$j_R^* = \gamma\,c\,\frac{\mathcal{D}}{h}\,f_B, \qquad u_R = \gamma\,\frac{\mathcal{D}}{h}\,\frac{1}{f_R}, \tag{6.4.18}$$

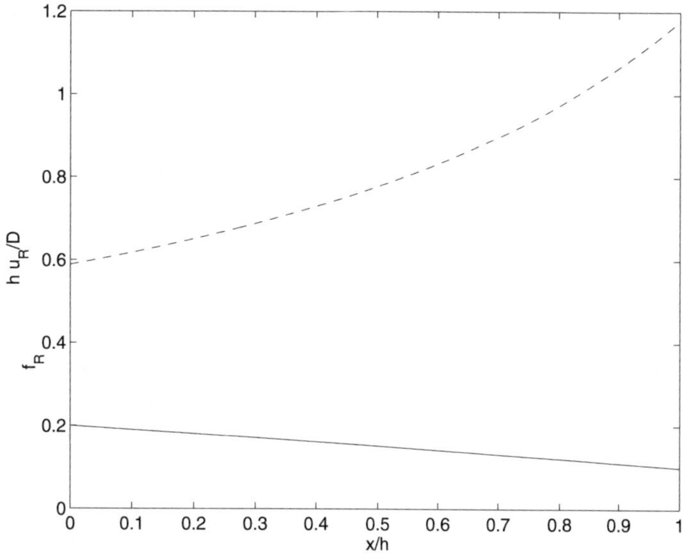

Fig. 6.3 Distribution of the whiskey mass fraction $\omega_{\rm R}$ (solid line) and associated whiskey vapor velocity, $u_{\rm R}$, scaled with $\mathcal{D}_{\rm RB}/h$ (broken line) for $fa_{\rm R_{bottom}} = 0.2$ and $f_{\rm R_{top}} = 0.1$, where h is the barrel height

the blue molar flux and velocity are given by

$$j^*_{\rm B} = -\gamma\, c\, \frac{\mathcal{D}}{h}\, f_{\rm B}, \qquad u_{\rm B} = 0, \tag{6.4.19}$$

and the mixture molar velocity is constant, given by

$$u^* = \gamma\, \frac{\mathcal{D}}{h}. \tag{6.4.20}$$

A singularity in the red velocity, $u_{\rm R}$, occurs at a point where the red density is zero.

It is worth pointing out that fluid parcels of pure blue air are still inside the barrel, even though the corresponding diffusion flux defined with respect to the mixture velocity is nonzero. The air inside a kitchen with an open door is still, in spite of the escaping aroma of a simmering soup.

A typical whiskey mass fraction distribution and associated velocity are plotted with the solid and broken lines in Fig. 6.3. We note that the velocity of the whiskey vapor is nonzero at the bottom of the barrel, $x = 0$, where $\omega_{\rm R} > 0$, and increases monotonically toward the coyote.

6.4.6 Motion of Point Particles

Blue point particles are still, while red point particles move upward with velocity $u_{\rm R}$. The position of a red point particle, $X_{\rm R}$, is governed by the differential equation

$$\frac{{\rm d}X_{\rm R}}{{\rm d}t} = u_{\rm R}(X_{\rm R}) = \gamma\, \frac{\mathcal{D}}{h}\, \frac{1}{1 - f_{\rm B}(0)\, \exp\left(\gamma\, \frac{X_{\rm R}}{h}\right)}. \tag{6.4.21}$$

Separating the variables and integrating from the bottom up to a position corresponding to an arbitrary time, t, we obtain

$$\gamma\, \frac{\mathcal{D}}{h}\, t = \int_0^{X_{\rm R}(t)} \left(1 - f_{\rm B}(0)\, \exp\left(\gamma\, \frac{x}{h}\right)\right) {\rm d}x. \tag{6.4.22}$$

Performing the integration and rearranging, we derive an implicit equation for $X_R(t)$,

$$\gamma^2 \frac{\mathcal{D}}{h^2} t = \gamma \hat{X}_R - f_B(0) \left(\exp\left(\gamma \hat{X}_R\right) - 1 \right), \tag{6.4.23}$$

where $\hat{X}_R \equiv X_R/h$. The time it takes for a red point particle to travel from bottom to top arises by setting $\hat{X}_R = 1$.

6.4.7 Depletion of Whiskey Layer

As red point particles move upward with velocity u_R, the whiskey layer at the bottom of the barrel is depleted. If η is the thickness of the liquid layer, then

$$\rho_w \frac{d\eta}{dt} = -\rho_R u_R, \tag{6.4.24}$$

where ρ_w is the density of the liquid layer. Setting $\rho_R = c M_R f_R$, substituting the derived expression for the red velocity, and noting that the barrel height is $L = \eta + h$, we find that

$$\frac{d\eta}{dt} = -\gamma\, c \frac{M_R}{\rho_w} \frac{\mathcal{D}}{L - \eta}. \tag{6.4.25}$$

Integrating, we obtain

$$(L - \eta)^2 = (L - \eta_0)^2 + 2\gamma\, c \frac{M_R}{\rho_w} \mathcal{D} t, \tag{6.4.26}$$

where η_0 is the initial liquor thickness. The time for the liquid layer to be depleted arises by setting $\eta = 0$.

6.4.8 Exercise

6.4.1 Discuss the relation between coyotes and gray wolves.

6.5 Red and Blue Gas

Fickian diffusion laws can be developed in terms of a chemical potential that is related to the molar fractions by theoretical or experimentally deduced equations. In the present context, the chemical potential for diffusion can be regarded as the counterpart of the potential due to a force field.

6.5.1 Fick's Law

Consider a mixture of a red and blue gas with strong or weak intermolecular force fields. In its general form, Fick's law for the red diffusive molar flux is expressed by the equation

$$j_R^* = -\mathcal{D}_{RB}^* \frac{1}{RT} c_R \nabla \mu_R^*, \tag{6.5.1}$$

where $\mathcal{D}^*_{\rm RB}$ is a binary diffusivity, R is the ideal-gas constant, and T is the absolute temperature. The specific volume chemical potential, $\mu^*_{\rm R}$, with units of energy per mole, is the chemical energy contained in one mole of red molecules in the mixture.

We acknowledge that $\mu^*_{\rm R}$ depends on $c_{\rm R}$ for spatially uniform mixture molar concentration, c, and write

$$\nabla \mu^*_{\rm R} = \left(\frac{\partial \mu^*_{\rm R}}{\partial c_{\rm R}}\right)_c \nabla c_{\rm R} \tag{6.5.2}$$

at constant temperature and pressure. Substituting $c_{\rm R} = c f_{\rm R}$, we obtain

$$\boldsymbol{j}^*_{\rm R} = -\mathcal{D}^*_{\rm RB} \, \frac{c}{RT} \left(\frac{\partial \mu^*_{\rm R}}{\partial \ln f_{\rm R}}\right)_c \nabla f_{\rm R}. \tag{6.5.3}$$

This expression does *not* account for variations in c in the available domain of transport.

6.5.2 Chemical Activity

The red chemical potential, $\mu^*_{\rm R}$, is related to the molar fraction, $f_{\rm R}$, by the species activity, $a_{\rm R}$, as

$$\mu^*_{\rm R} = \mu^*_{\rm R_0} + RT \ln a_{\rm R}, \tag{6.5.4}$$

where $\mu^*_{\rm R_0}$ is the chemical potential at zero temperature or unit (reference) activity. The species activity of an ideal solution is equal to the molar fraction, $\gamma_{\rm R} = f_{\rm R}$.

The activity is related to the molar fraction, $f_{\rm R}$, by an activity coefficient, $\gamma_{\rm R}(f_{\rm R})$, as $a_{\rm R} = \gamma_{\rm R} f_{\rm R}$, so that

$$\mu^*_{\rm R} = \mu^*_{\rm R_0} + RT \ln(\gamma_{\rm R} f_{\rm R}). \tag{6.5.5}$$

The activity coefficient of an ideal solution is unity, $\gamma_{\rm R} = 1$. Differentiating, we find that

$$\frac{\partial \mu^*_{\rm R}}{\partial \ln f_{\rm R}} = RT \, \frac{\partial \ln a_{\rm R}}{\partial \ln f_{\rm R}} = RT \left(1 + \frac{\partial \ln \gamma_{\rm R}}{\partial \ln f_{\rm R}}\right), \tag{6.5.6}$$

and conclude that

$$\boldsymbol{j}^*_{\rm R} = -c \, \mathcal{D}^{\rm c}_{\rm RB} \, \nabla f_{\rm R}, \tag{6.5.7}$$

where

$$\mathcal{D}^{\rm c}_{\rm RB} \equiv \mathcal{D}_{\rm RB} \, \frac{\partial \ln a_{\rm R}}{\partial \ln f_{\rm R}} = \mathcal{D}^*_{\rm RB} \left(1 + \frac{\partial \ln \gamma_{\rm R}}{\partial \ln f_{\rm R}}\right) \tag{6.5.8}$$

is the chemical diffusivity.

6.5.3 Self-diffusion

Assume that the molecular weights of the red and blue molecules are the same, $M_R = M_B$, which means that the red and blue molecules are isotopes, and the concentration c is uniform throughout the fluid. In that case, Eq. (6.5.7) describes *tracer diffusion* or *self-diffusion* in the absence of chemical potential gradient with tracer diffusivity or self-diffusivity equal to $\mathcal{D}^c_{RB}$.

6.5.4 Exercise

6.5.1 Discuss the physical interpretation of the activity coefficient.

6.6 Generalized Local Diffusive Flux Laws

In a generalized formulation, the diffusive mass fluxes in a binary mixture are related to a mixture exchange chemical potential,

$$\mu \equiv \frac{\mu_R}{M_R} - \frac{\mu_B}{M_B}, \tag{6.6.1}$$

and to the temperature field, T, by the equation

$$j_R = -\alpha \, \nabla\mu - \beta \, \nabla T, \tag{6.6.2}$$

where α and β are pressure and temperature dependent coefficients, μ_R and μ_B are red and blue chemical potentials, and M_R and M_B are the red and blue molar masses (e.g., Landau, L. D. & Lifsitz, E. M. (1987) *Fluid Mechanics*, Second Edition, Pergamon Press, p. 230.)

The exchange potential, μ, also discussed in Sect. 8.8 with regard to Gibbs' fundamental equation of thermodynamics, quantifies the differential change of the specific internal energy of the mixture when the blue species is exchanged for the red species by a corresponding differential change in the mass fractions.

We may regard μ a function of T, p, and ω_R, such that

$$\nabla\mu = \mu_T \, \nabla T + \mu_p \, \nabla p + \mu_{\omega_R} \, \nabla\omega_R, \tag{6.6.3}$$

where a subscript indicates a partial derivative. Substituting this expression into (6.6.2) provides us with a corresponding expansion in terms of the mass fraction, temperature, and pressure,

$$j_R = -\rho \, \mathcal{D}_{RB} \left(\nabla\omega_R + k_T \, \nabla \ln T + k_p \nabla \ln p \right), \tag{6.6.4}$$

where k_T is a physical constant pertaining to the temperature and k_p is a physical constant pertaining to the pressure. The first term on the right-hand side expresses Fickian diffusion, the second term expresses a thermal mass flux according to the Soret effect, and the third term expresses baro-diffusion. The heat flux, $\mathbf{q}$, is given by a similar expression.

The heat flux is expressed by the corresponding form

$$\mathbf{q} = -\delta \, \nabla\mu - \gamma \, \nabla T + \mu \, j_R, \tag{6.6.5}$$

where γ and δ are pressure and temperature dependent coefficients. The last term on the right-hand side expresses diffusion-mediated heat transport. It can be shown that $\delta = \beta T$.

6.6.1 Comprehensive Laws

Comprehensive expressions for the diffusive mass flux in a binary fluid have been developed under the auspices of irreversible thermodynamics. For example,

$$\begin{bmatrix} \boldsymbol{j}_R \\ \boldsymbol{j}_B \end{bmatrix} = -\rho \begin{bmatrix} \mathcal{D}_{RR} & \mathcal{D}_{RB} \\ \mathcal{D}_{BR} & \mathcal{D}_{BB} \end{bmatrix} \cdot \begin{bmatrix} \mathbf{d}_R \\ \mathbf{d}_B \end{bmatrix} + \begin{bmatrix} \rho_R\theta_R \\ \rho_B\theta_B \end{bmatrix} \nabla \ln T, \tag{6.6.6}$$

where $\mathcal{D}_{XY}$ are diffusivities, $\mathbf{d}_X$ are diffusion driving forces, and θ_X is a species thermal diffusion coefficient (e.g., http://www.scholarpedia.org/article/Multicomponent_Flow). The diffusivity matrix on the right-hand side is symmetric and positive semi-definite.

To ensure that $\boldsymbol{j}_R + \boldsymbol{j}_B = \mathbf{0}$, we require the constraints

$$\mathcal{D}_{RR} + \mathcal{D}_{BR} = \mathcal{D}_{RB} + \mathcal{D}_{BB} = \rho_R\theta_R + \rho_B\theta_B = 0. \tag{6.6.7}$$

The accompanying expression for the thermal flux is

$$\mathbf{q} = -\mathbf{K} \cdot \nabla T - p\,(\theta_R \mathbf{d}_R + \theta_B \mathbf{d}_B) + h_R \boldsymbol{j}_R + h_B \boldsymbol{j}_B, \tag{6.6.8}$$

where $\mathbf{K}$ is the thermal conductivity matrix, and h_R and h_B are specific enthalpies.

6.6.2 Exercise

6.6.1 Explain the constraints shown in (6.6.7).

6.7 Fractional Diffusion

Fick's law discussed in Sect. 6.1 can be generalized into a fractional-diffusion law for the red mass flux in a binary mixture expressed by

$$\boldsymbol{j}_R = -\rho\, \mathcal{D}^{(\alpha)}_{RB}\, \nabla^{\alpha-1} \omega_R, \tag{6.7.1}$$

where $\mathcal{D}^{(\alpha)}_{RB}$ is a fractional diffusivity, $\nabla^{\alpha-1}\omega_R$ the fractional gradient of the mass fraction, ω_R, and α is the fractional order taking values in the interval (0, 2]. As the parameter α tends to 2, the fractional gradient reduces to the ordinary gradient, ∇, describing regular Fickian diffusion.

In Sect. 6.8, we will demonstrate that fractional diffusion is associated with the collective motion of particles that execute random flights to adjacent sites, as opposed to random walks to nearest neighbor sites.

6.7.1 Fractional Laplacian

The fractional Laplacian encountered in a transport equation involving a diffusive flux is the divergence of the fractional gradient, defined as

$$\nabla^{\alpha} \omega_R = \nabla \cdot \nabla^{\alpha-1} \omega_R\,, \tag{6.7.2}$$

where $\nabla\cdot$ is the ordinary divergence operator. As the fractional order, α, tends to 2, the fractional Laplacian reduces to the ordinary Laplacian, ∇^2, which is the sum of the second derivatives in Cartesian coordinates, x, y, and z or is given by other standard expressions in orthogonal or non-orthogonal curvilinear coordinates.

It is important to bear in mind that fractional diffusion and the associated Laplacian are defined in an infinite domain. Further assumptions must be made over the complement of a partially infinite or finite solution domain.

6.7.2 Fractional Diffusion in One Dimension

In one dimension along the x axis, the fractional gradient of a function, $f(x)$, is a fractional derivative given by an integral representation,

$$f^{(\alpha-1)}(x) \equiv \epsilon^{(1)}_{1,\alpha} \int_0^\infty \frac{f(x+v)-f(x-v)}{v^\alpha}\,\mathrm{d}v \tag{6.7.3}$$

for $0 \le \alpha < 2$, where x is an arbitrary evaluation point, v is an integration variable, and the coefficient

$$\epsilon^{(1)}_{1,\alpha} = \frac{1}{2\,\Gamma(1-\alpha)\,\cos(\dfrac{\alpha\pi}{2})} = \frac{2^{\alpha-1}}{\sqrt{\pi}}\,\frac{\Gamma(\frac{1+\alpha}{2})}{\Gamma(\frac{2-\alpha}{2})} \tag{6.7.4}$$

is defined in terms of the Gamma function, Γ. The integral representation is an unmistakable indicator of a nonlocal behavior.

6.7.3 Principal-Value Integral

An alternative representation of the fractional derivative in terms of a principal-value integral (pv) is

$$f^{(\alpha-1)}(x) = \epsilon^{(1)}_{1,\alpha} \int_{-\infty}^{\infty} \mathrm{pv}\ \frac{v}{|v|^{\alpha+1}}\, f(x+v)\,\mathrm{d}v. \tag{6.7.5}$$

The fraction inside the integral is positive when $v > 0$ and negative when $v < 0$. To illustrate the notion of the principal-value integral, we introduce a cut-off length, $\omega(x) > 0$, break up the infinite integration domain in (6.7.5) into three pieces, and rearrange as follows:

$$f^{(\alpha-1)}(x) = \epsilon^{(1)}_{1,\alpha}\big(f(x)\,\mathcal{K}_0 + \mathcal{K}_1(x) + \dot{f}(x)\,\mathcal{K}_2(x) + \mathcal{K}_3(x)\big), \tag{6.7.6}$$

where $\dot{f}(x) = \mathrm{d}f/\mathrm{d}x$ is the ordinary first derivative. We have introduced a principal-value integral,

$$\mathcal{K}_0 \equiv \int_{-\omega(x)}^{\omega(x)} \mathrm{pv}\ \frac{v}{|v|^{1+\alpha}}\,\mathrm{d}v, \tag{6.7.7}$$

a regularized integral,

$$\mathcal{K}_1(x) \equiv \int_{-\omega(x)}^{\omega(x)} \frac{v}{|v|^{1+\alpha}}\big(f(x+v) - f(x) - \dot{f}(x)\,v\big)\,\mathrm{d}v, \tag{6.7.8}$$

a proper integral,

$$\mathcal{K}_2(x) \equiv \int_{-\omega(x)}^{\omega(x)} \frac{1}{|v|^{-1+\alpha}} \, \mathrm{d}v, \tag{6.7.9}$$

and another proper integral,

$$\mathcal{K}_3(x) \equiv -\int_{-\infty}^{\omega(x)} \frac{1}{v^\alpha} f(x+v)\, \mathrm{d}v + \int_{\omega(x)}^{\infty} \frac{1}{v^\alpha} f(x+v)\, \mathrm{d}v. \tag{6.7.10}$$

The notion of a principal-value integral lies in the observation that, due to cancellation of positive and negative, finite or infinite contributions, the principal-value integral $\mathcal{K}_0$ is zero by convention. Rearranging the rest of the terms in (6.7.6), we obtain

$$f^{(\alpha-1)}(x) \simeq \epsilon_{1,\alpha}^{(1)} \big(\mathcal{J}_1(x) + 2\, \dot{f}(x)\, \mathcal{J}_2(x) + \mathcal{J}_3(x) \big), \tag{6.7.11}$$

where

$$\mathcal{J}_1(x) \equiv \int_0^{\omega(x)} \frac{f(x+v) - f(x-v) - 2\, \dot{f}(x)\, v}{v^\alpha} \, \mathrm{d}v \tag{6.7.12}$$

is a regularized integral,

$$\mathcal{J}_2(x) \equiv \int_0^{\omega(x)} v^{1-\alpha} \, \mathrm{d}v = \frac{1}{2-\alpha}\, \omega(x)^{2-\alpha} \tag{6.7.13}$$

is a standard integral, and

$$\mathcal{J}_3(x) \equiv \int_{\omega(x)}^{\infty} \frac{f(x+v) - f(x-v)}{v^\alpha} \, \mathrm{d}v \tag{6.7.14}$$

is a regular integral over a semi-infinite domain. Expression (6.7.11) also arises from the principal-value regularization of (6.7.3).

Since the integrand of the first integral, $\mathcal{J}_1(x)$, remains nonsingular as v tends to zero, this integral can be computed by standard numerical methods, including the trapezoidal rule. If we choose $\omega(x) \gg |x|$, the third integral, $\mathcal{J}_3(x)$, will be negligibly small, provided that the function $f(x)$ decays sufficiently fast far from the origin.

6.7.4 Second Moment of the Gaussian Distribution

As an example, we consider the second moment of the Gaussian distribution over the entire x axis,

$$f(x) = x^2\, \mathrm{e}^{-x^2}. \tag{6.7.15}$$

The fractional derivative computed using a numerical method based on the preceding regularization is shown in Fig. 6.4 for several fractional orders, α. As α tends to 2, the fractional derivative tends to the ordinary first derivative drawn with the solid bold line.

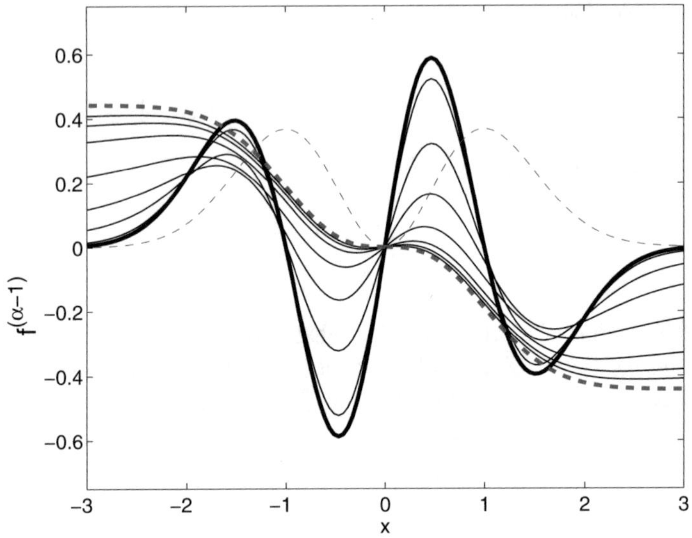

Fig. 6.4 Fractional derivative of the second moment of the Gaussian distribution, $f(x) = x^2\,\mathrm{e}^{-x^2}$ (broken blue line), for fractional order $\alpha = 2$ (solid bold line), 1.9, 1.5, 1.0, 0.5, 0.2, 0.1, and 0 (broken red bold line)

6.7.5 Fractional Laplacian in One Dimension

The fractional Laplacian associated with the fractional derivative in one dimension is given by

$$\nabla^\alpha f(x) \equiv \frac{\partial f^{(\alpha-1)}}{\partial x} = c_{1,\alpha} \int_{-\infty}^{\infty} \mathrm{pv} \; \frac{f(x+v) - f(x)}{|v|^{1+\alpha}} \, \mathrm{d}v, \tag{6.7.16}$$

where $c_{1,\alpha} \equiv \alpha\, \epsilon_{1,\alpha}^{(1)}$. The principal-value integral is interpreted in the sense described previously in this section for the fractional derivative. An association between this expression and random particle flights will be made in Sect. 6.8.

6.7.6 Population Evolution

In Sect. 4.12, we developed the evolution equation (4.12.3) for a particle density distribution, $f(x,t)$, repeated below for convenience,

$$\frac{f(x, t+\Delta t) - f(x,t)}{\Delta t} = \frac{1}{\Delta t} \int_{-\infty}^{\infty} \big(f(x-\xi, t) - f(x,t)\big)\, \phi(\xi)\, \mathrm{d}\xi, \tag{6.7.17}$$

where $\phi(\xi)$ is a probability density function for the particle jump lengths. Comparing the right-hand side with (6.7.16), we are tempted to set

$$\phi(\xi) \sim \frac{1}{|\xi|^{1+\alpha}} \tag{6.7.18}$$

for $0 \le \alpha < 2$. However, the mandatory normalization condition

$$\int_{-\infty}^{\infty} \phi(\xi)\, \mathrm{d}\xi = 1 \tag{6.7.19}$$

cannot be satisfied due to a strong non-integrable singularity at $\xi = 0$. Notwithstanding this essential difficulty, the probability density function shown in (6.7.18) can be accepted in the way that an improper prior probability density function is accepted in Bayesian analysis.

6.7.7 Green's Function

The Green's function of the unsteady fractional diffusion equation, denoted by $\mathcal{G}_\alpha$, satisfies the equation

$$\frac{\partial \mathcal{G}_\alpha}{\partial t} = \mathcal{D}_\alpha \nabla^{(\alpha)} \mathcal{G}_\alpha + \delta_1(x - x_0)\, \delta_1(t - t_0) \tag{6.7.20}$$

for $t > t_0$ and any x, where δ_1 is the Dirac delta function in one dimension, x_0 is a chosen singular source point, t_0 is a chosen application time, $\mathcal{D}_\alpha$ is the fractional diffusivity with units of ℓ^α/τ, ℓ is a characteristic length, and τ is a characteristic time. Using the method of Fourier transforms, we find that

$$\mathcal{G}_\alpha(\widehat{x}, \widehat{t}) = \frac{1}{(\mathcal{D}_\alpha \widehat{t}\,)^{1/\alpha}}\, \Phi_\alpha(\eta_\alpha), \tag{6.7.21}$$

where $\widehat{t} \equiv t - t_0$, $\widehat{x} \equiv x - x_0$,

$$\eta_\alpha \equiv \frac{\widehat{x}}{(\mathcal{D}_\alpha \widehat{t}\,)^{1/\alpha}} \tag{6.7.22}$$

is a dimensionless similarity variable,

$$\Phi_\alpha(\eta_\alpha) = \frac{1}{2\pi} \int_{-\infty}^{\infty} \mathrm{e}^{-w^\alpha} \exp(-\mathrm{i}\, \eta_\alpha w)\, \mathrm{d}w \tag{6.7.23}$$

is a dimensionless function, and i is the imaginary unit, $\mathrm{i}^2 = -1$. Resolving the complex exponential into its real and imaginary parts, we obtain

$$\Phi_\alpha(\eta_\alpha) = \frac{1}{\pi} \int_{0}^{\infty} \mathrm{e}^{-w^\alpha} \cos(\eta_\alpha w)\, \mathrm{d}w. \tag{6.7.24}$$

Analytical expressions for the definite integral on the right-hand side are available only for certain values of α.

For $\alpha = 2$, we obtain a Gaussian distribution associated with ordinary diffusion. For $\alpha = 1$, we find that

$$\mathcal{G}_1(\widehat{x}, \widehat{t}) = \frac{1}{\pi \mathcal{D}_1 \widehat{t}}\, \frac{1}{1 + \eta_1^2}, \tag{6.7.25}$$

where $\eta_1 \equiv \widehat{x}/(\mathcal{D}_1 \widehat{t})$ is a dimensionless similarity variable. We observe that the Green's function for $\alpha = 1$ describes an algebraic decay in $\widehat{x}$, which can be contrasted with the Gaussian decay arising when $\alpha = 2$.

6.7.8 Fractional Diffusion in Two Dimensions

In two dimensions, the fractional gradient is given by an integral representation over the entire plane,

$$\nabla^{\alpha-1} f(\mathbf{x}) = \epsilon_{2,\alpha}^{(1)} \iint \text{pv} \; \frac{1}{v^{1+\alpha}} \, f(\mathbf{x}+\mathbf{v}) \, \mathbf{e}_v \, \mathrm{d}A(\mathbf{v}) \tag{6.7.26}$$

for $0 \le \alpha < 2$, where $\mathbf{x}$ is an arbitrary evaluation point, $v = |\mathbf{v}|$, $\mathbf{e}_v = \frac{1}{v}\mathbf{v}$ is a unit vector pointing from the evaluation point, $\mathbf{x}$, to an arbitrary integration point, $\mathrm{d}A(\mathbf{v})$ is an areal element, the integration is carried over the entire $\mathbf{v}$ plane, and

$$\epsilon_{2,\alpha}^{(1)} \equiv \frac{2^{\alpha-1}}{\pi} \, \frac{\Gamma(\frac{2+\alpha}{2})}{\Gamma(\frac{2-\alpha}{2})} \tag{6.7.27}$$

is an appropriate coefficient defined in terms of the Gamma function, Γ. The fractional Laplacian is given by

$$\nabla^{\alpha} f(\mathbf{x}) \equiv c_{2,\alpha} \iint \text{pv} \; \frac{f(\mathbf{x}+\mathbf{v}) - f(\mathbf{x})}{v^{2+\alpha}} \, \mathrm{d}A(\mathbf{v}), \tag{6.7.28}$$

where $c_{2,\alpha} \equiv \alpha \, \epsilon_{2,\alpha}^{(1)}$ and the integration domain is the entire $\mathbf{v}$ plane.

To clarify the meaning of the principal-value integral in (6.7.28), we derive the following expansion of the integrand for small v,

$$\frac{f(\mathbf{x}+\mathbf{v}) - f(\mathbf{x})}{v^{2+\alpha}} = \frac{1}{v^{1+\alpha}} \, \mathbf{e}_v \cdot (\nabla f)_{\mathbf{x}} + \frac{1}{2} \, \frac{1}{v^{\alpha}} \, \mathbf{e}_v \otimes \mathbf{e}_v : (\nabla \otimes \nabla f)_{\mathbf{x}} + \cdots . \tag{6.7.29}$$

The concept of the principal-value integral hinges on the convention that the integral of the first term on the right-hand side of (6.7.29) over a circular disk centered at the origin of $\mathbf{v}$ is zero due to the cancellation of infinitely large positive and negative contributions.

6.7.9 Fractional Diffusion in Three Dimensions

In three dimensions, the fractional gradient is given by an integral representation over the entire three-dimensional space,

$$\nabla^{\alpha-1} f(\mathbf{x}) = \epsilon_{3,\alpha}^{(1)} \iiint \text{pv} \; \frac{1}{v^{2+\alpha}} \, f(\mathbf{x}+\mathbf{v}) \, \mathbf{e}_v \, \mathrm{d}V(\mathbf{v}) \tag{6.7.30}$$

for $0 \le \alpha < 2$, where the integration domain is the entire three-dimensional space,

$$\epsilon_{3,\alpha}^{(1)} \equiv -\frac{1}{4\pi} \, \frac{1+\alpha}{\Gamma(1-\alpha)\cos(\dfrac{\alpha\pi}{2})} = \frac{2^{\alpha-1}}{\pi^{3/2}} \, \frac{\Gamma(\frac{3+\alpha}{2})}{\Gamma(\frac{2-\alpha}{2})} \tag{6.7.31}$$

is a coefficient defined in terms of the Gamma function, Γ. The fractional Laplacian is given by

$$\nabla^{\alpha} f(\mathbf{x}) \equiv c_{3,\alpha} \iiint \text{pv} \; \frac{f(\mathbf{x}+\mathbf{v}) - f(\mathbf{x})}{v^{3+\alpha}} \, \mathrm{d}V(\mathbf{v}), \tag{6.7.32}$$

where $c_{3,\alpha} \equiv \alpha\, \epsilon_{3,\alpha}^{(1)}$ and the integration domain is the entire three-dimensional space. The notion of the principal-value integral is similar to that discussed earlier in this section for the fractional Laplacian in the plane.

6.7.10 Exercises

6.7.1 Verify that as α tends to 2, the fractional derivative defined in (6.7.5) reduces to the ordinary first derivative.

6.7.2 Discuss the interpretation of (6.7.26) in the context of a population evolution over a plane.

6.8 Random Jumpers

The physical interpretation of fractional diffusion can be illustrated by considering the collective motion of a large number of particles distributed over the x axis at positions $x_i = i\,\Delta x$ for $i = 0, \pm 1, \ldots$, where Δx is a specified interval. At a designated time, each particle starts making random jumps such that the probability of jumping by k intervals is denoted as π_k for $k = 0, \pm 1 \ldots$, subject to the constraint that the particles are not asleep,

$$\pi_0 = 0. \tag{6.8.1}$$

The mandatory probability normalization condition requires that

$$\sum_{k=-\infty}^{\infty} \pi_k = 1. \tag{6.8.2}$$

In the case of symmetric jumpers, $\pi_k = -\pi_{-k}$. If the particles are only able to step to their nearest neighboring sites, $\pi_k = 0$ for any k, except that $\pi_{-1} = \frac{1}{2}$ and $\pi_1 = \frac{1}{2}$.

6.8.1 Expected Displacement and Variance

The expected displacement of each random walker after each step is

$$\overline{d} = \Delta x \sum_{k=-\infty}^{\infty} k\, \pi_k, \tag{6.8.3}$$

and the associated variance is

$$\sigma_d^2 = \sum_{k=-\infty}^{\infty} \left(k\Delta x - \overline{d}\right)^2 \pi_k, \tag{6.8.4}$$

where σ_d is the standard deviation. The infinite sums on the right-hand sides of (6.8.3) and (6.8.4) converge only when π_k decays sufficiently fast with respect to the particle excursion index, k.

6.8.2 Population Balances

After a jump has been made, particles have arrived at the ith node from the $i-1$ node with probability π_1, from the $i-2$ node with probability π_2, and, more generally, from the $i-k$ node with probability π_k. A population balance at the ith node requires that

$$p_i^{(n+1)} = \sum_{k=-\infty}^{\infty} p_{i-k}^{(n)} \, \pi_k, \tag{6.8.5}$$

where p_i^n is the population at the ith node at the nth time level.

Using the mandatory normalization condition (6.8.2) on the discrete displacement probability distribution, we obtain

$$p_i^{(n+1)} - p_i^{(n)} = \sum_{k=-\infty}^{\infty} \left(p_{i-k}^{(n)} - p_i^{(n)} \right) \pi_k. \tag{6.8.6}$$

The left-hand side is the incremental population change at the ith node after one time step.

6.8.3 Power-Law Probability Distribution

Of particular interest are symmetric jumpers with a power-law discrete probability distribution involving a positive parameter α that takes values in the interval $(0, 2]$, where $\pi_0(\alpha) = 0$,

$$\pi_k(\alpha) = \frac{1}{2\,\zeta_{1+\alpha}} \frac{1}{|k|^{1+\alpha}} \tag{6.8.7}$$

for $k = \pm 1, \pm 2, \ldots$, and

$$\zeta_s \equiv \sum_{r=1}^{\infty} \frac{1}{r^s} \tag{6.8.8}$$

is the Riemann zeta function defined for $s > 1$ and undefined for $s \le 1$. The power-law distribution is a convenient prototype of a heavy-tailed discrete probability distribution.

To confirm the mandatory normalization condition, we compute the sum of all probabilities,

$$\sum_{k=-\infty}^{\infty} \pi_k = \frac{1}{2\,\zeta_{1+\alpha}} \sum_{k=-\infty}^{\infty}{}' \frac{1}{|k|^{1+\alpha}} = 1, \tag{6.8.9}$$

where the prime after the sum signifies the omission of the singular term corresponding to $k = 0$.

Special values of the zeta function are

$$\zeta_1 = \infty, \qquad \zeta_2 = \tfrac{1}{6}\,\pi^2, \qquad \zeta_3 = 1.2021\cdots, \qquad \zeta_4 = \tfrac{1}{90}\,\pi^4. \tag{6.8.10}$$

Using the second of these values, we find that when $\alpha = 1$, the discrete probability distribution is

$$\pi_0(\alpha = 1) = 0, \qquad \pi_k(\alpha = 1) = \frac{3}{\pi^2} \frac{1}{k^2} \tag{6.8.11}$$

for $k = \pm 1, \pm 2, \ldots$.

6.8.4 Population Balance for the Power-Law Distribution

We may introduce a time elapsed in each step, Δt, adopt the power-law distribution (6.8.7), and recast the balance equation (6.8.6) into the form

$$\frac{p_i^{(n+1)} - p_i^{(n)}}{\Delta t} = \kappa_\alpha \frac{1}{\Delta x^\alpha} \sum_{k=-\infty}^{\infty}{}' \frac{p_{i+k}^{(n)} - p_i^{(n)}}{|k|^{1+\alpha}}, \tag{6.8.12}$$

where the prime after the summation symbol signifies omission of the singular term $k = 0$,

$$\kappa_\alpha \equiv \frac{1}{2\,\zeta_{1+\alpha}} \frac{\Delta x^\alpha}{\Delta t} \tag{6.8.13}$$

is a diffusivity with units of ℓ^α/τ, ℓ is a characteristic length and τ is a characteristic time.

Note that

$$\kappa_2 \equiv \frac{1}{2\,\zeta_3} \frac{\Delta x^2}{\Delta t} \simeq \frac{1}{2.4042} \frac{\Delta x^2}{\Delta t}, \tag{6.8.14}$$

which is somewhat smaller than the classical Einstein diffusivity associated with nearest neighbor walkers where the denominator in the fraction is 2 instead of the approximate value of 2.4042 on the right-hand side, as shown in (4.10.5) for $q + s = 1$.

Equation (6.8.12) can be expressed in the deliberate form

$$\frac{p_i^{(n+1)} - p_i^{(n)}}{\Delta t} = \kappa_\alpha \Big(\sum_{k=-\infty}^{\infty}{}' \frac{p_{i+k}^{(n)} - p_i^{(n)}}{|v_k|^{1+\alpha}} \Delta x \Big), \tag{6.8.15}$$

where $v_k = k\Delta x$. The left-hand side is a discrete representation of the time derivative, $\partial p_i/\partial t$, at the ith node. The right-hand side is a discrete representation of an underlying linear differential operator whose precise form depends on α.

6.8.5 Fractional Diffusion Equation

To this end, we make an association between the difference equation (6.8.15) and the fractional diffusion equation

$$\frac{\partial p}{\partial t} = \mathcal{D}_\alpha \nabla^\alpha p, \tag{6.8.16}$$

where $\mathcal{D}_\alpha$ is a fractional diffusivity and $\nabla^\alpha p$ is the fractional Laplacian. Substituting expression (6.7.16) for the fractional Laplacian, we obtain

$$\frac{\partial p}{\partial t} = \mathcal{D}_\alpha c_{1,\alpha} \int_{-\infty}^{\infty} \mathrm{pv} \; \frac{p(x+v) - p(x)}{|v|^{1+\alpha}} \, dv. \tag{6.8.17}$$

Now comparing equation (6.8.17) with (6.8.15), we obtain

$$\kappa_\alpha \equiv \frac{1}{2\,\zeta_{1+\alpha}} \frac{\Delta x^\alpha}{\Delta t} \simeq \mathcal{D}_\alpha c_{1,\alpha}, \tag{6.8.18}$$

and therefore

$$\mathcal{D}_\alpha \simeq \frac{1}{2\,\zeta_{1+\alpha}}\,\frac{1}{c_{1,\alpha}}\,\frac{\Delta x^\alpha}{\Delta t} \tag{6.8.19}$$

within numerical discretization error, which can be substantial.

6.8.6 Numerical Simulation

The particle jumping activity can be implemented in a numerical simulation. As a preliminary, we consider a method of generating a sample of integers, i, consistent with a specified discrete probability distribution. In the context of fractional diffusion, we are particularly interested in the power-law distribution.

Assume that a random integer variable, i, takes values inside a specified interval $i_{\min} \le i \le i_{\max}$, with corresponding discrete probability distribution p_i (dpd), where $i_{\min}$ and $i_{\max}$ are specified lower and upper limits. Normalization requires that all probabilities add to unity,

$$\sum_{i=i_{\min}}^{i_{\max}} p_i = 1. \tag{6.8.20}$$

The associated cumulative discrete probability distribution (cdpd) is found by adding left-sided contributions,

$$\Pi_i \equiv \sum_{j=i_{\min}}^{i} p_j. \tag{6.8.21}$$

A graph of Π_i against i provides us with a generally uneven staircase function, as shown in Fig. 6.5.

To generate a sample of integers, i, consistent with a specified discrete probability distribution, we introduce a random variable, ϱ, with uniform probability distribution in the standard interval [0, 1], and assign the value of i as follows:

- If $\varrho \le \Pi_{i_{\min}}$, then $i = i_{\min}$.
- If $\Pi_{j-1} < \varrho \le \Pi_j$, then $i = j$ for $j = 2, \ldots, i_{\max}$, as illustrated with the horizontal arrow in Fig. 6.5.

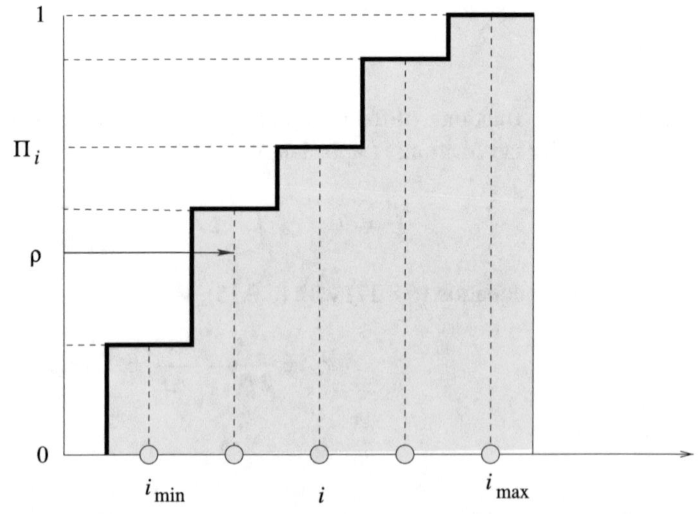

Fig. 6.5 Illustration of a cumulative discrete probability distribution (cdpd), Π_i

The method is implemented in the following Matlab function named *dpd* that receives a random deviate, r, representing ϱ, and returns the corresponding value of i consistent with the specified discrete probability distribution:

```
function i = dpd(imin,imax,p,r)

%=========
% Copyright by Constantine Pozrikidis,
                              2020
% All rights reserved
%
% return random integers consistent
                    with a dpd
%
% p:    probability distribution
% pcum: cumulative distribution
% r:    random deviate
%=========

pcum = p(imin);

%---
if(r <= pcum)
  i = imin;
  return
end
%---

for j=imin+1:imax
 pcum = pcum + p(j);
 if(pcum >= r)
   i=j;
   return;
 end
end

return
```

Heaps of random numbers in the range $i = 1, \ldots, 32$ described by the dpd

$$p_i = \frac{1}{c} \frac{1}{i^2} \tag{6.8.22}$$

are shown in Fig. 6.6a, where c is a normalization constant ensuring that the discrete probabilities in the specified range add to unity.

Heaps of random numbers defined in the range $i = 1, \ldots, 32$ corresponding to the Gaussian probability density function with unit standard deviation are shown in Fig. 6.6b.

6.8.7 Code

Now we consider n_c columns of particles sitting on the x axis, where n_c is a specified integer. When n_c is odd, the particles are situations at notched positions

$$x_i = i\,\Delta x \qquad \text{for} \qquad i = -\tfrac{1}{2}\,(n_c - 1), \ldots, \tfrac{1}{2}\,(n_c - 1). \tag{6.8.23}$$

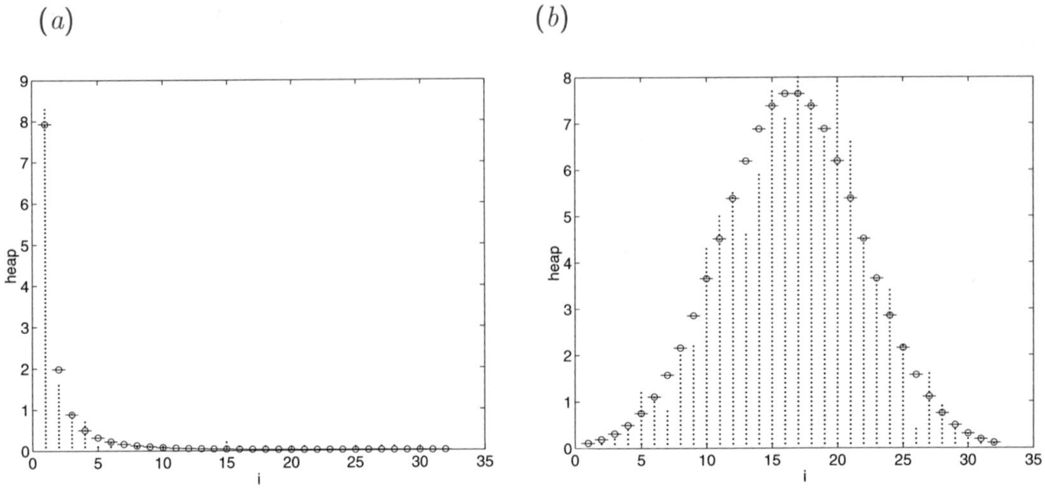

Fig. 6.6 Heaps of random numbers corresponding to (**a**) the discrete probability distribution described by (6.8.22) and (**b**) the Gaussian distribution with unit standard deviation

When n_c is even, the particles are situations at notched positions

$$x_i = (i - \tfrac{1}{2})\,\Delta x \qquad \text{for} \qquad i = -\tfrac{1}{2}\,(n_c - 2), \ldots, \tfrac{1}{2}\,n_c, \tag{6.8.24}$$

where Δx is a specified interval. When $n_c = 1$, we obtain one column corresponding to $i = 0$, located at $x_0 = 0$. When $n_c = 2$, we obtain two columns corresponding to $i = 0, 1$, located at $x_0 = -\frac{1}{2}\,\Delta x$ and $x_1 = \frac{1}{2}\,\Delta x$.

The following Matlab code named *jumpers* performs the simulation of the particle motion for a power-law distribution, p_i, associated with fractional diffusion:

```
%====================================
% Copyright by Constantine Pozrikidis, 2020
% All rights reserved
%
% random jumpers with a specified
% flight discrete probability distribution (dpd)
%====================================
%---
% Probability of jumping by i units (p_i)
% runs from i=imin to imax,

% and is zero otherwise
%---

Dx = 1.000;     % spatial step (arbitrary)
Dt = 1.000;     % time step step (arbitrary)

alpha = 1.9; % fractional order in (0, 2);

imin = 1;  % arbitrary
imax = 32; % arbitrary

nsteps = 2*32;

%---
% prepare
```

```
%---

ni = imax-imin+1;

%---
% number of walkers (arbitrary)
% walkers will be assigned
% in nc initial columns
%---

nc = 1;
nwc = 2*2^10;    % walkers per column (arbitrary)

nwalker = nc*nwc;

%---
% odd-even number of columns
%---

oddevenc = 1;
if(mod(nc,2)==0) oddevenc = 2; end



%---
% specify the dpd
%---

norm = 0.0;

for i=imin:imax
 p(i) = 1.0/i^(1.0+alpha);
 norm = norm + p(i);
end

%---
% normalize the dpd
%---

for i=imin:imax
 p(i) = p(i)/norm;
end

%---
% define notches over the x axis
% and initial population
%---

notch = 8*ni;  % arbitrary

ishift = notch+1;

if(oddevenc==1) xshift = 0.0; end
if(oddevenc==2) xshift = 0.5; end

for i=-notch:notch
 xnotch(i+ishift) = (i-xshift)*Dx;  % x position of notch
 pnotch(i+ishift) = 0;              % notch population
end
```

```
%---
% assign the initial x-position of the walkers
% in nc columns
% and compute the initial notch population
%---

if(oddevenc==1) klow =-(nc-1)/2; khigh = (nc-1)/2; end
if(oddevenc==2) klow =-nc/2+1;   khigh = nc/2;     end

Ic = 0; % walker counter


for i=1:nwc
  for k=klow:khigh
   Ic = Ic+1;
   xwalker(Ic) = k;
   pnotch(ishift+k) = pnotch(ishift+k)+1;
  end
end

%---
% plot the initial notch population
%---

figure(1)
hold on
xlabel('i','fontsize',15)
ylabel('population','fontsize',15)
set(gca,'fontsize',15)
box on
plot(xnotch,pnotch/nwalker,'ko')
axis([-20 20 0 1])

%---
% simulation
%---

time = 0.0;

%---
% run over walkers

% and update notches
%---

%---
for n=1:nsteps
%---

 for i=1:nwalker

  j = xwalker(i);          % current x position of walker

  flight = dpd(imin,imax,p,rand);
  if(rand>0.5)             % fly to the left or right
   flight = -flight;
  end

  % update the x position of walker:
```

```
    xwalker(i) = j + flight;

    % update the notch population:
    pnotch(j+ishift)         = pnotch(j+ishift)         - 1;
    pnotch(j+ishift+flight)  = pnotch(j+ishift+flight)  + 1;
   end

   time = time + Dt;

%---
% plot
%---

   clf
   hold on
   xlabel('i','fontsize',15)
   ylabel('population','fontsize',15)
   set(gca,'fontsize',15)
   box on
   plot(xnotch,pnotch/nwalker,'k.')
   pause(0.1)

end
```

The code makes a call to the function *dpd* discussed earlier in this section to compute random numbers consistent with the specified discrete probability distribution.

Population distributions after a certain evolution time for fractional diffusion orders $\alpha = 0.1$ and 1.9 are shown in Fig. 6.7. The results confirm that the lower the fractional order, α, the broader the particle distribution after a given evolution.

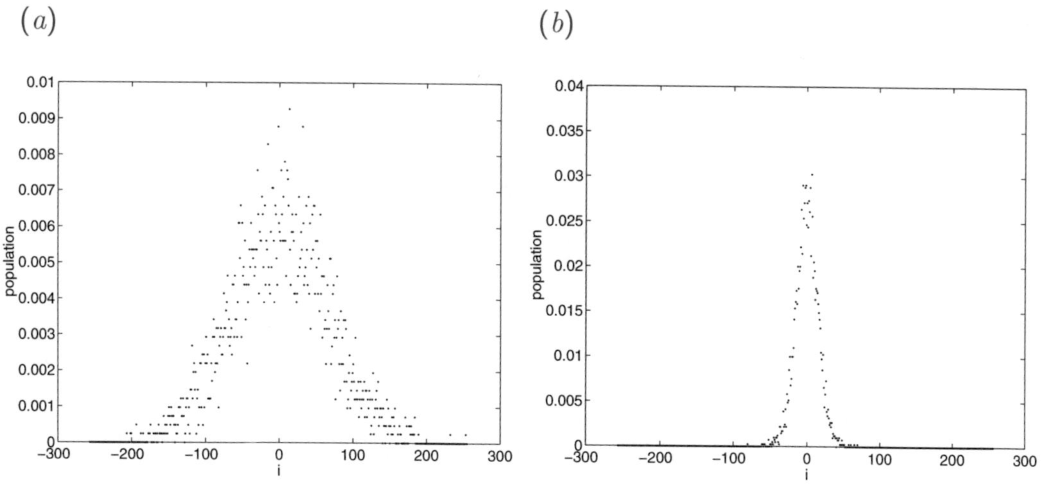

Fig. 6.7 Particle population distribution after a certain evolution time for energized jumpers with fractional order (**a**) $\alpha = 0.1$ and (**b**) 1.9

6.8.8 Exercises

6.8.1 Explain in physical terms why $\mathcal{D}_2$ shown in (6.8.14) is somewhat smaller than the classical Einstein diffusivity associated with nearest neighbor walkers.

6.8.2 Generate and plot heaps of random numbers described by the discrete probability distribution

$$p_i = \frac{1}{c} \frac{i^2}{1+i^4} \tag{6.8.25}$$

in the range $i = 1, \ldots, 32$, where c is a normalization constant.

6.8.3 Investigate the effect of initial jumper columns on the evolution of the particle distribution.

Species and Mixture Hydrodynamics

7

We have regarded a mixture as a superposition of coexisting, interacting, and possibly reacting species. We have discussed the motion of monochromatic species parcels and analyzed the mixture kinematics with reference to the species diffusion velocities defined as the differences between the individual species velocities and the mixture mass velocity. In this chapter, we turn our attention to the balance of forces and introduce species and mixture stresses that develop spontaneously in hydrostatics or as a result of the motion.

Species stresses serve as a vehicle for writing a complete set of governing equations, subject to assumptions and approximations. In this framework, the motion of the individual species and mixture is governed by corresponding equations of motion involving species and mixture stresses that are related by species diffusion fluxes. This viewpoint leads us to the notion of species and mixture hydrodynamics under the auspices of the continuum.

7.1 Species Stress Tensor

A fundamental concept underlying the hydrodynamics of mixtures is the notion of the Cauchy species stress tensor, also called the partial stress tensor corresponding to the partial vapor pressure employed in thermodynamics. The red stress tensor attributed to a red parcel, $\boldsymbol{\sigma}_{\mathrm{R}}$, is defined such that the traction exerted on a small surface of a red parcel is

$$\mathbf{f}_{\mathrm{R}} = \mathbf{n} \cdot \boldsymbol{\sigma}_{\mathrm{R}}, \tag{7.1.1}$$

where $\mathbf{n}$ is the unit vector normal to the red parcel surface. This definition is identical to that for a homogeneous fluid, as discussed in Sect. 2.1.

The traction can be resolved into a tangential component with an otherwise arbitrary direction representing a shear stress, and a normal component representing the normal stress.

Since the species stress tensor $\boldsymbol{\sigma}_{\mathrm{R}}$ is not necessarily symmetric, the order of multiplication in (7.1.1) may not generally be reversed, that is, $\mathbf{n} \cdot \boldsymbol{\sigma}_{\mathrm{R}} \neq \boldsymbol{\sigma}_{\mathrm{R}} \cdot \mathbf{n}$.

C. Pozrikidis, *Transport Processes Primer*,
https://doi.org/10.1007/978-1-4939-9909-5_7

7.1.1 Interpretation

The interpretation of the partial stress tensor requires careful consideration. One might expect that the traction exerted on an infinitesimal surface of a mixture, **f**, is the sum of the traction exerted on all species. However, this is not true in the presence of diffusion, that is,

$$\mathbf{f} \neq \sum_{X} \mathbf{f}_X, \tag{7.1.2}$$

and correspondingly

$$\boldsymbol{\sigma} \neq \sum_{X} \boldsymbol{\sigma}_X, \tag{7.1.3}$$

where $\boldsymbol{\sigma}$ is the mixture stress tensor. Instead, we will see that

$$\boldsymbol{\sigma} = \sum_{X} \Big(\boldsymbol{\sigma}_X - \frac{1}{\rho_X}\, \boldsymbol{j}_X \otimes \boldsymbol{j}_X\Big), \tag{7.1.4}$$

where $\boldsymbol{j}_X$ is the species mass flux. Moreover, we will see that

$$\boldsymbol{\sigma} - \rho\, \mathbf{u} \otimes \mathbf{u} = \sum_{X} \big(\boldsymbol{\sigma}_X - \rho_X\, \mathbf{u}_X \otimes \mathbf{u}_X\big), \tag{7.1.5}$$

where the left-hand side is the mixture stress–momentum tensor and the right-hand side is the sum of the species stress–momentum tensors. These relations suggest that, in transitioning from species to mixture, the partial stress tensors are not necessarily additive.

The reason for the inequality (7.1.2) is that the species tractions apply to different physical bodies, which are monochromatic parcels moving with different velocities. The superficial conceptual discrepancy is truly a matter of semantics.

7.1.2 Species Pressures

A red species pressure, also called the partial pressure, can be defined in terms of the trace of the red stress tensor as

$$p_R = -\tfrac{1}{3}\, \mathrm{trace}(\boldsymbol{\sigma}_R). \tag{7.1.6}$$

The species pressure consists of a hydrostatic or thermodynamic part, p_R^{stat}, and a hydrodynamic part, p_R^{dyn}, so that

$$p_R = p_R^{\mathrm{stat}} + p_R^{\mathrm{dyn}}. \tag{7.1.7}$$

Raoult's law states that the partial red species pressure above an ideal liquid solution is given by

$$p_R^{\mathrm{stat}} = p_R^{v}\, f_R^{\mathrm{liquid}}, \tag{7.1.8}$$

where p_R^{v} is the red species vapor pressure and f_R^{liquid} is the liquid molar fraction. The dynamic part vanishes in a stationary fluid.

The total static pressure in a multicomponent mixture is the sum of all species pressures,

$$p^{\text{stat}} = \sum_{\text{X}} p_{\text{X}}^{\text{stat}}. \tag{7.1.9}$$

Pressure additivity is a fundamental notion of thermodynamics.

7.1.3 Deviatoric Stress Tensor

The deviatoric part of the red stress tensor is defined in terms of the species pressure by the equation

$$\boldsymbol{\sigma}_{\text{R}} = -p_{\text{R}}\,\mathbf{I} + \boldsymbol{\sigma}_{\text{R}}^{\text{dev}}, \tag{7.1.10}$$

where $\mathbf{I}$ is the identity matrix. The mixture stress tensor is given

$$\boldsymbol{\sigma} = -p^{\text{stat}}\,\mathbf{I} + (-p^{\text{dyn}}\,\mathbf{I} + \boldsymbol{\sigma}^{\text{dev}}), \tag{7.1.11}$$

where $\boldsymbol{\sigma}^{\text{dev}}$ is the deviatoric part of the mixture stress tensor.

7.1.4 Exercise

7.1.1 Discuss the interpretation of Eq. (7.1.9) with reference to the additivity of partial vapor pressures.

7.2 Equation of Species Parcel Motion

Consider the motion of an arbitrary red or blue parcel in a binary fluid. We recall that the boundary of a monochromatic parcel remains sharp and well defined even in the presence of diffusion. Red point particles at the surface of a red parcel remain at the surface of the parcel at all times, and blue point particles at the surface of a blue parcel remain at the surface of the parcel at all times. Red and blue parcels generally expand or contract due to diffusion as they evolve under the action of an imposed flow.

7.2.1 Equation of Species Parcel Motion

The momentum of a red parcel was defined in Eq. (5.6.1) in terms of the red density, ρ_{R}, and velocity $\mathbf{u}_{\text{R}}$, repeated below for convenience,

$$\mathbf{M}_{\text{R}} = \iiint_{\text{red parcel}} \rho_{\text{R}}\mathbf{u}_{\text{R}}\,\mathrm{d}V. \tag{7.2.1}$$

Expressions for the rate of change of the parcel momentum in terms of the red velocity field and red point-particle acceleration were derived in Sect. 5.6.

Newton's second law of motion requires that the rate of change of momentum of a red parcel is balanced by the surface and volume forces acting on the parcel. Accounting for an interspecies interaction volume force and for momentum transfer due to species conversion, we obtain the following red parcel equation of motion:

$$\frac{\mathrm{d}}{\mathrm{d}t}\iiint_{\text{red parcel}} \rho_{\mathrm{R}}\mathbf{u}_{\mathrm{R}}\,\mathrm{d}V = \iint_{\text{red parcel}} \mathbf{n}\cdot\boldsymbol{\sigma}_{\mathrm{R}}\,\mathrm{d}S + \iiint_{\text{red parcel}} \rho_{\mathrm{R}}\,\mathbf{g}\,\mathrm{d}V$$
$$+ \iiint_{\text{red parcel}} \iota_{\mathrm{R}}\,\mathbf{b}_{\mathrm{R}}\,\mathrm{d}V + \iiint_{\text{red parcel}} \boldsymbol{\phi}_{\mathrm{R}}\,\mathrm{d}V + \iiint_{\text{red parcel}} \dot{\mathbf{m}}_{\mathrm{R}}\,\mathrm{d}V, \tag{7.2.2}$$

subject to the following definitions:

- $\boldsymbol{\sigma}_{\mathrm{R}}$ is the red stress tensor (partial stress tensor) defined in Sect. 7.1 in terms of the traction.
- $\mathbf{g}$ is the gravitational acceleration.
- $\mathbf{b}_{\mathrm{R}}$ is a red-specific force field with corresponding proportionality coefficient ι_{R}. This field would arise in the presence of an electrical field if the red species were ionic.
- $\boldsymbol{\phi}_{\mathrm{R}}$ is a volumetric interaction force acting on the red species due to the presence of the blue species, as discussed in Sect. 7.3. This interaction force is assumed to be independent of the rate of species conversion.
- $\dot{\mathbf{m}}_{\mathrm{R}}$ is the volumetric rate of generation of red momentum due to species conversion, as discussed in Sect. 7.4.

Equation (7.2.2) is a generalization of the familiar Cauchy equation of motion for a homogeneous fluid discussed in Sect. 2.3, where the last two terms on the right-hand side do not appear.

7.2.2 Constraints

A similar equation of motion can be written for a blue or any species parcel under the stipulation that

$$\sum_{\mathrm{X}} \boldsymbol{\phi}_{\mathrm{X}} = 0, \qquad \sum_{\mathrm{X}} \dot{\mathbf{m}}_{\mathrm{X}} = 0, \tag{7.2.3}$$

where the sum is over all species. The first constraint is a consequence of Newton's law of action and reaction, as discussed in Sect. 7.3. The second constraint originates from the principle of momentum conservation in the absence of external forces: a chemical reaction cannot generate momentum, as discussed in Sect. 7.4.

7.2.3 Species Momentum Transport

Applying the Reynolds transport equation (4.7.17) for the momentum of a red parcel by setting $\phi = \rho_{\mathrm{R}}\mathbf{u}_{\mathrm{R}}$, we obtain

$$\frac{\mathrm{d}}{\mathrm{d}t}\iiint_{\text{red parcel}} \rho_{\mathrm{R}}\mathbf{u}_{\mathrm{R}}\,\mathrm{d}V = \iiint_{\text{red parcel}} \frac{\partial(\rho_{\mathrm{R}}\mathbf{u}_{\mathrm{R}})}{\partial t}\,\mathrm{d}V + \iint_{\text{red parcel}} (\rho_{\mathrm{R}}\mathbf{u}_{\mathrm{R}})\,\mathbf{n}\cdot\mathbf{u}_{\mathrm{R}}\,\mathrm{d}S, \tag{7.2.4}$$

where $\mathbf{n}$ is the outward unit vector normal to instantaneous surface of the red parcel, as also shown in (5.6.12).

Now setting the right-hand side of (7.2.2) equal to the right-hand side of (7.2.4), we derive the Eulerian form of the equation of red-parcel motion,

$$\iiint_{\text{red parcel}} \frac{\partial(\rho_R \mathbf{u}_R)}{\partial t}\, dV + \iint_{\text{red parcel}} (\rho_R \mathbf{u}_R)\, \mathbf{n} \cdot \mathbf{u}_R\, dS = \iint_{\text{red parcel}} \mathbf{n} \cdot \boldsymbol{\sigma}_R\, dS$$
$$+ \iiint_{\text{red parcel}} \rho_R\, \mathbf{g}\, dV + \iiint_{\text{red parcel}} \iota_R\, \mathbf{b}_R\, dV + \iiint_{\text{red parcel}} \boldsymbol{\phi}_R\, dV + \iiint_{\text{red parcel}} \dot{\mathbf{m}}_R\, dV. \quad (7.2.5)$$

With the exception of the last two terms on the right-hand side, this equation is identical to that for a homogeneous fluid or mixture parcel.

Similar equations can be written for a blue parcel involving the blue stress tensor. The red and blue equations can be added only when a red and a blue parcel coincide at a particular instant in time. However, differential equations deriving from these balances can be added at any point.

7.2.4 Integral Red Momentum Balance

To derive an integral red momentum balance, we consider the red parcel occupying a stationary or moving control volume at a particular instant in time, $\mathcal{V}_c$, and restate equation (7.2.5) as

$$\iiint_{\mathcal{V}_c} \frac{\partial(\rho_R \mathbf{u}_R)}{\partial t}\, dV = \iint_{\mathcal{S}_c} (\rho_R\, \mathbf{u}_R)\, (\mathbf{n}^{\text{in}} \cdot \mathbf{u}_R)\, dS - \iint_{\mathcal{S}_c} \mathbf{n}^{\text{in}} \cdot \boldsymbol{\sigma}_R\, dS$$
$$+ \iiint_{\mathcal{V}_c} \rho_R\, \mathbf{g}\, dV + \iiint_{\mathcal{V}_c} \iota_R\, \mathbf{b}_R\, dV + \iiint_{\mathcal{V}_c} \boldsymbol{\phi}_R\, dV + \iiint_{\mathcal{V}_c} \dot{\mathbf{m}}_R\, dV, \quad (7.2.6)$$

where $\mathbf{n}^{\text{in}} = -\mathbf{n}$ is the unit normal vector pointing into the control volume and $\mathcal{S}_c$ is the boundary of the control volume.

The transport equation (7.2.6) states that the rate of accumulation of red momentum inside a control volume, expressed by the left-hand side, is determined by convection, boundary traction, and the body forces and momentum conversion represented by the last four terms on the right-hand side.

A similar equation can be written for the blue species on the same control volume, and the red and blue equations can be added to give an integral mixture momentum balance. Due to cancellation, the last two integrals on the right-hand side do not appear in the mixture balance.

7.2.5 Suspensions, Emulsions, and Granular Materials

The equations of red parcel motion for chemical solutions stated in this section also apply to the particulate flow of suspensions, emulsions, and granular materials. The only essential difference is that the mixture lag velocities arising in further analysis are not necessarily diffusion velocities.

7.2.6 Exercise

7.2.1 Discuss an example of a body force specific to the red or blue molecules.

7.3 Species Interaction Force

The volumetric interaction force acting on the red species due to the presence of the blue species in a binary solution, denoted by $\boldsymbol{\phi}_R$, may also be interpreted as a volumetric rate of exchange of momentum for reasons other than a chemical reaction.

7.3.1 Action and Reaction

To conform with Newton's law of action and reaction, we stipulate that the volumetric interaction force imparted by the blue species to the red species, $\boldsymbol{\phi}_R$, is equal in magnitude and opposite in direction to that imparted by the red species to the blue species, $\boldsymbol{\phi}_B$,

$$\boldsymbol{\phi}_R = -\boldsymbol{\phi}_B. \tag{7.3.1}$$

In the absence of diffusion, that is, when the red and blue velocities are the same, the red and blue molecules move in synch and an interaction force does not appear, $\boldsymbol{\phi}_R = \mathbf{0}$ and $\boldsymbol{\phi}_B = \mathbf{0}$.

7.3.2 Interaction Laws

It appears reasonable to stipulate that the interaction force is proportional to the difference in the local velocities between the two species, $\mathbf{u}_R - \mathbf{u}_B$. The reason is that if the velocities were the same, momentum would not be exchanged. Moreover, it appears reasonable to stipulate that the interaction force is proportional to the product of the corresponding molecular fractions, f_R and f_B, regarded as interception probabilities. We recall that

$$f_X = \omega_X \frac{M}{M_X}, \tag{7.3.2}$$

where M_X is the molar mass of species X and M is the molar mass of the mixture, and write

$$\boldsymbol{\phi}_R = -\kappa\, \rho\, \omega_R\, \omega_B\, (\mathbf{u}_R - \mathbf{u}_B), \tag{7.3.3}$$

where ρ is the mixture density and κ is an interaction coefficient with units of inverse time exhibiting an implicit dependence on f_R. An alternative representation of (7.3.3) in terms of the species fluxes is

$$\boldsymbol{\phi}_R = -\kappa\, (\omega_B\, \boldsymbol{j}_R - \omega_R\, \boldsymbol{j}_B). \tag{7.3.4}$$

Yet another expression in terms of the mixture molar concentration, c, and molar fractions is

$$\boldsymbol{\phi}_R = -\kappa\, c\, \frac{M_R M_B}{M}\, f_R\, f_B\, (\mathbf{u}_R - \mathbf{u}_B). \tag{7.3.5}$$

Similar expressions can be written for the blue species.

7.3.3 Scaling

Implicit in (7.3.3) is the assumption that the interaction force scales with the mixture density, and thus with the species momenta. For example, when the blue species is stationary, the red species is opposed by a resistive force due to red–blue collisions and other intermolecular interactions. An alternative scaling with respect to the mixture viscosity may also be employed. The two scalings are related by the mixture kinematic viscosity.

7.3.4 Darcy and Brinkmann Laws

The interaction law expressed by (7.3.3) is similar to Darcy's or Brinkmann's law for flow in porous media where the resistive force in flow through a host medium is proportional to the nominal fluid velocity relative to that of the host medium.

7.3.5 Binary Mixture

Equations (7.3.3) and (7.3.4) are also valid in a pairwise sense for a multicomponent mixture. In the case of a binary mixture, because $\omega_R + \omega_B = 1$ and $\boldsymbol{j}_R + \boldsymbol{j}_B = \mathbf{0}$, we obtain

$$\boldsymbol{\phi}_R = -\kappa\, \boldsymbol{j}_R, \qquad \boldsymbol{\phi}_B = -\kappa\, \boldsymbol{j}_B. \tag{7.3.6}$$

Thus, the interaction force in a binary mixture is proportional to the negative of the diffusive species flux.

7.3.6 Fick's Law

Substituting into (7.3.4) Fick's law expressed by (6.1.1), we obtain

$$\boldsymbol{\phi}_R = \kappa\, \rho\, \mathcal{D}_{RB}\, (\omega_B \nabla \omega_R - \omega_B \nabla \omega_R). \tag{7.3.7}$$

In the case of a binary mixture, we obtain

$$\boldsymbol{\phi}_R = \kappa\, \rho\, \mathcal{D}_{RB} \nabla \omega_R, \qquad \boldsymbol{\phi}_B = \kappa\, \rho\, \mathcal{D}_{BR} \nabla \omega_B. \tag{7.3.8}$$

The sum of these two interaction forces is confirmed to be zero, provided that $\mathcal{D}_{RB} = \mathcal{D}_{BR}$.

7.3.7 Generalization

Other empirical or theoretical expressions can be employed for the interaction force, as appropriate. A possible generalization of (7.3.3) is

$$\boldsymbol{\phi}_R = -\rho\, \omega_R \omega_B\, f(\omega_R)\, \kappa\, (\mathbf{u}_R - \mathbf{u}_B), \tag{7.3.9}$$

where $f(\omega_R)$ is an appropriate function.

Thermophoretic contributions to the interaction force associated with temperature variations are also possible. A general interaction law for a gas mixture involves three contributions,

$$\boldsymbol{\phi}_{\rm R} = -\alpha_{\rm RB}\,(\mathbf{u}_{\rm R} - \mathbf{u}_{\rm B}) + \beta_{\rm RB}\,\nabla T_{\rm B} - \beta_{\rm BR}\,\nabla T_{\rm R}, \tag{7.3.10}$$

where $T_{\rm R}$ and $T_{\rm B}$ are the temperatures of the red and blue species and $\alpha_{\rm RB}$, $\beta_{\rm RB}$, and $\beta_{\rm BR}$ are phenomenological coefficients (e.g., Ramshaw, J. D. (1993) Hydrodynamic theory of multicomponent diffusion and thermal diffusion in multitemperature gas mixtures. *J. Non-Equilib. Thermodyn.* **18**, 121–134.)

In the case of a red–blue mixture of ideal gases, the interaction coefficient is given by

$$\alpha_{\rm RB} = \frac{p}{\mathcal{D}_{\rm RB}}\,\frac{p_{\rm R}}{p}\,\frac{p_{\rm B}}{p}, \tag{7.3.11}$$

where $p_{\rm R}$ and $p_{\rm B}$ are partial pressures, p is the mixture pressure, and $\mathcal{D}_{\rm RB}$ is a binary diffusivity. We may use the ideal-gas law to set $p = cRT$ in the first numerator on the left-hand side. Comparing the resulting expression with (7.3.5) shows that

$$\kappa = \frac{M}{M_{\rm R} M_{\rm B}}\,\frac{RT}{\mathcal{D}_{\rm RB}}, \tag{7.3.12}$$

where the right-hand side has units of inverse time. Not surprisingly, the higher the diffusivity, the weaker the interaction coefficient.

7.3.8 Multicomponent Mixture

The interaction law (7.3.3), or any other interaction law, can be assumed to apply in an additive pairwise fashion in a multicomponent solution. The total interaction force exerted on the red species is given by

$$\boldsymbol{\phi}_{\rm R} = \sum_{\rm X} \boldsymbol{\phi}_{\rm RX}, \tag{7.3.13}$$

where the sum is over all other species. To conform with Newton's law of action and reaction, we require that $\boldsymbol{\phi}_{\rm RX} = -\boldsymbol{\phi}_{\rm XR}$, thereby ensuring that

$$\sum_{\rm X} \boldsymbol{\phi}_{\rm X} = \mathbf{0}. \tag{7.3.14}$$

We may assume that the pairwise interaction force is given by

$$\boldsymbol{\phi}_{\rm XY} = -\kappa_{\rm XY}\,\rho\,\omega_{\rm X}\,\omega_{\rm Y}\,(\mathbf{u}_{\rm X} - \mathbf{u}_{\rm Y}), \tag{7.3.15}$$

where $\kappa_{\rm XY}$ is an interspecies interaction coefficient satisfying the symmetry condition $\kappa_{\rm XY} = \kappa_{\rm YX}$. In terms of the diffusive fluxes,

$$\boldsymbol{\phi}_{\rm XY} = -\kappa_{\rm XY}\,(\omega_{\rm Y}\,\boldsymbol{j}_{\rm X} - \omega_{\rm X}\,\boldsymbol{j}_{\rm Y}). \tag{7.3.16}$$

Other pairwise interaction laws can be developed.

7.3.9 Exercise

7.3.1 Derive (7.3.3) from (7.3.4).

7.4 Exchange of Momentum Due to Species Conversion

Consider a reversible chemical reaction between blue and red molecules where two blue molecules combine into one red molecule and one red molecule disintegrates into two molecules,

$$2\,\mathrm{B} \rightleftharpoons \mathrm{R}, \tag{7.4.1}$$

where the molar masses are related by $M_\mathrm{R} = 2M_\mathrm{B}$. The reaction occurs forward and backward at different rates. At dynamic equilibrium, the forward and backward rates are the same.

Let $\dot{r}_\mathrm{BR}$ be the rate of red mass generation according to the forward reaction $2\mathrm{B} \rightarrow \mathrm{R}$, and $\dot{r}_\mathrm{RB}$ be the rate of blue mass generation according to the reverse reaction $\mathrm{R} \rightarrow 2\mathrm{B}$. The net rate of red mass generation is

$$\dot{r}_\mathrm{R} = \dot{r}_\mathrm{BR} - \dot{r}_\mathrm{RB} \tag{7.4.2}$$

and the net rate of blue mass generation is

$$\dot{r}_\mathrm{B} = -\dot{r}_\mathrm{R} = \dot{r}_\mathrm{RB} - \dot{r}_\mathrm{BR}. \tag{7.4.3}$$

We observe that

$$\dot{r}_\mathrm{B} = -\dot{r}_\mathrm{R}, \tag{7.4.4}$$

as required for mass conservation. At equilibrium, $\dot{r}_\mathrm{BR} = \dot{r}_\mathrm{RB}$, that is, the forward and reverse mass generation rates are the same.

7.4.1 Kinetics

We may assume that the rates of forward and backward reaction follow the stoichiometry, that is,

$$\dot{r}_\mathrm{BR} = \lambda_\mathrm{BR}\, f_\mathrm{B}^2, \qquad \dot{r}_\mathrm{RB} = \lambda_\mathrm{RB}\, f_\mathrm{R}, \tag{7.4.5}$$

where λ_BR and λ_RB are reaction-rate constants. In this framework, the molecular fractions are regarded as scaled probabilities of interception or presence in a small volume.

7.4.2 Molecules Carrying Momentum

When two blue molecules combine to form one red molecule, they carry with them blue momentum, which then becomes red momentum. Conversely, when one red molecule breaks up into two blue molecules, it disseminates red momentum, which then becomes blue momentum.

The net rate of red momentum conversion due to a reversible reaction is then

$$\dot{\mathbf{m}}_\mathrm{R} = \dot{r}_\mathrm{BR}\, \mathbf{u}_\mathrm{B} - \dot{r}_\mathrm{RB}\, \mathbf{u}_\mathrm{R}, \tag{7.4.6}$$

and the corresponding rate of net blue momentum generation is

$$\dot{\mathbf{m}}_\mathrm{B} = -\dot{\mathbf{m}}_\mathrm{R} = \dot{r}_\mathrm{RB}\, \mathbf{u}_\mathrm{R} - \dot{r}_\mathrm{BR}\, \mathbf{u}_\mathrm{B}. \tag{7.4.7}$$

Note that species momentum generation takes place even when the net species conversion rate is zero, that is, at chemical equilibrium, provided that the species velocities are not the same.

Expressions (7.4.6) and (7.4.7) ensure that

$$\dot{\mathbf{m}}_\mathrm{R} + \dot{\mathbf{m}}_\mathrm{B} = \mathbf{0}, \tag{7.4.8}$$

which guarantees that conversion neither generates nor absorbs momentum at any instant.

To confirm the last assertion, we consider the conversion of two blue molecules with velocities v_B1 and v_B2 into one red molecule with resulting velocity v_R, or *vice versa*, and introduce a kinetic energy balance in one dimension,

$$\tfrac{1}{2}\, M_\mathrm{B}\, (v_\mathrm{B1} - V)^2 + \tfrac{1}{2}\, M_\mathrm{B}\, (v_\mathrm{B2} - V)^2 = \tfrac{1}{2}\, M_\mathrm{R}\, (v_\mathrm{R} - V)^2 + \Delta E, \tag{7.4.9}$$

where ΔE is the molecular energy absorption or release and V is an arbitrary frame-of-reference velocity. Setting $M_\mathrm{R} = 2M_\mathrm{B}$, and requiring that the energy balance is independent of V, we find that

$$M_\mathrm{B}\, v_\mathrm{B1} + M_\mathrm{B}\, v_\mathrm{B2} = M_\mathrm{R}\, v_\mathrm{R}, \tag{7.4.10}$$

which is a statement of momentum conservation.

Expressions (7.4.6) and (7.4.7) can be restated as

$$\dot{\mathbf{m}}_\mathrm{R} = \dot{r}_\mathrm{R}\, \frac{\dot{r}_\mathrm{BR}\, \mathbf{u}_\mathrm{B} - \dot{r}_\mathrm{RB}\, \mathbf{u}_\mathrm{R}}{\dot{r}_\mathrm{BR} - \dot{r}_\mathrm{RB}}, \qquad \dot{\mathbf{m}}_\mathrm{B} = \dot{r}_\mathrm{B}\, \frac{\dot{r}_\mathrm{RB}\, \mathbf{u}_\mathrm{R} - \dot{r}_\mathrm{BR}\, \mathbf{u}_\mathrm{B}}{\dot{r}_\mathrm{RB} - \dot{r}_\mathrm{BR}}. \tag{7.4.11}$$

When $\mathbf{u}_\mathrm{R} = \mathbf{u}_\mathrm{B} = \mathbf{u}$, we obtain $\dot{\mathbf{m}}_\mathrm{R} = \dot{r}_\mathrm{R}\, \mathbf{u}$ and $\dot{\mathbf{m}}_\mathrm{B} = \dot{r}_\mathrm{B}\, \mathbf{u}$. A general expression for the species X interaction force in a multicomponent solution is

$$\dot{\mathbf{m}}_\mathrm{X} = \dot{r}_\mathrm{X}\, \mathbf{J}_\mathrm{X}, \tag{7.4.12}$$

where $\mathbf{J}_\mathrm{X}$ has a suitable form, such as that displayed in (7.4.11) for a binary mixture.

7.4.3 Multicomponent Mixtures

Similar expressions can be written for any species in a multicomponent solution. To ensure that chemical conversion does not generate momentum, we require that

$$\sum_\mathrm{X} \dot{\mathbf{m}}_\mathrm{X} = \mathbf{0}, \tag{7.4.13}$$

where the sum is over all species.

7.4.4 Other Interspecies Momentum Transfer Laws

The following alternative expressions are sometimes used for the rate of momentum production due to a chemical reaction,

$$\dot{\mathbf{m}}_{\rm R} = \dot{r}_{\rm R}\mathbf{u}_{\rm R}, \qquad \dot{\mathbf{m}}_{\rm B} = \dot{r}_{\rm B}\mathbf{u}_{\rm B}. \tag{7.4.14}$$

These expressions arise by replacing $\mathbf{u}_{\rm B}$ by $\mathbf{u}_{\rm R}$ on the right-hand side of (7.4.6), and $\mathbf{u}_{\rm R}$ by $\mathbf{u}_{\rm B}$ on the right-hand side of (7.4.7).

These alternative expressions do not satisfy (7.4.8). Instead, the following condition involving the species interaction forces is required in the mathematical formulation,

$$\boldsymbol{\phi}_{\rm R} + \dot{\mathbf{m}}_{\rm R} + \boldsymbol{\phi}_{\rm B} + \dot{\mathbf{m}}_{\rm B} = \mathbf{0}, \tag{7.4.15}$$

where the sum $\boldsymbol{\phi}_{\rm R} + \boldsymbol{\phi}_{\rm B}$ is not necessarily zero. It appears then that the species interaction forces account in part for the effect of a chemical reaction.

7.4.5 Exercise

7.4.1 Write the species and mixture equation of motion in terms of the corresponding stress–momentum tensors.

7.5 Cauchy Species Equation of Motion

To derive Cauchy's equation of motion for the red species, we use the divergence theorem to convert surface integrals into volume integrals in (7.2.5), and discard the integral signs in all terms to obtain

$$\frac{\partial(\rho_{\rm R}\mathbf{u}_{\rm R})}{\partial t} + \nabla\cdot(\rho_{\rm R}\,\mathbf{u}_{\rm R}\otimes\mathbf{u}_{\rm R}) = \nabla\cdot\boldsymbol{\sigma}_{\rm R} + \rho_{\rm R}\,\mathbf{g} + \iota_{\rm R}\,\mathbf{b}_{\rm R} + \boldsymbol{\phi}_{\rm R} + \dot{\mathbf{m}}_{\rm R}. \tag{7.5.1}$$

Equation (7.5.1) has the typical conservative form of an evolution equation for a vector field.

7.5.1 Lagrangian Form

Expanding the derivatives on the left-hand side of (7.5.1), and combining the result with the red mass density evolution equation

$$\frac{\partial\rho_{\rm R}}{\partial t} + \nabla\cdot(\rho_{\rm R}\mathbf{u}_{\rm R}) = \dot{r}_{\rm R}, \tag{7.5.2}$$

we obtain

$$\rho_{\rm R}\Big(\frac{\partial\mathbf{u}_{\rm R}}{\partial t} + \mathbf{u}_{\rm R}\cdot\nabla\mathbf{u}_{\rm R}\Big) = \nabla\cdot\boldsymbol{\sigma}_{\rm R} + \rho_{\rm R}\,\mathbf{g} + \iota_{\rm R}\,\mathbf{b}_{\rm R} + \boldsymbol{\phi}_{\rm R} + \boldsymbol{\eta}_{\rm R}, \tag{7.5.3}$$

where

$$\eta_R \equiv \dot{\mathbf{m}}_R - \dot{r}_R\,\mathbf{u}_R = \dot{r}_R\,(\mathbf{J}_R - \mathbf{u}_R) \tag{7.5.4}$$

and $\mathbf{J}_R$ is defined in (7.4.12).

In terms of the red material derivative of the red velocity, which is the red point-particle acceleration, the red species equation of motion reads

$$\rho_R \left(\frac{D\mathbf{u}_R}{Dt}\right)_R = \nabla \cdot \boldsymbol{\sigma}_R + \rho_R\,\mathbf{g} + \iota_R\,\mathbf{b}_R + \boldsymbol{\phi}_R + \eta_R. \tag{7.5.5}$$

The last two terms on the right-hand side do not appear in the case of equal species velocities, that is, in the absence of diffusion and chemical reaction.

7.5.2 Blue Species

Similar equations can be written for blue point particles. The counterpart of Eq. (7.5.1) is

$$\frac{\partial(\rho_B \mathbf{u}_B)}{\partial t} + \nabla \cdot (\rho_B\,\mathbf{u}_B \otimes \mathbf{u}_B) = \nabla \cdot \boldsymbol{\sigma}_B + \rho_B\,\mathbf{g} + \iota_B\,\mathbf{b}_B + \boldsymbol{\phi}_B + \dot{\mathbf{m}}_B \tag{7.5.6}$$

and the counterpart of Eq. (7.5.5) is

$$\rho_B \left(\frac{D\mathbf{u}_B}{Dt}\right)_B = \nabla \cdot \boldsymbol{\sigma}_B + \rho_B\,\mathbf{g} + \iota_B\,\mathbf{b}_B + \boldsymbol{\phi}_B + \eta_B, \tag{7.5.7}$$

where

$$\eta_B \equiv \dot{\mathbf{m}}_B - \dot{r}_B\,\mathbf{u}_B = \dot{r}_B\,(\mathbf{J}_B - \mathbf{u}_B). \tag{7.5.8}$$

We recall that $\dot{\mathbf{m}}_R$ and $\dot{\mathbf{m}}_B$ express the rate of momentum production due to a chemical reaction.

7.5.3 Momentum Exchange

Using expression (7.4.6) for $\dot{\mathbf{m}}_R$ and setting $\dot{r}_R = \dot{r}_{BR} - \dot{r}_{RB}$, we find that

$$\eta_R = \dot{r}_{BR}\,\mathbf{u}_B - \dot{r}_{RB}\,\mathbf{u}_R - (\dot{r}_{BR} - \dot{r}_{RB})\,\mathbf{u}_R. \tag{7.5.9}$$

Simplifying, we find

$$\eta_R = \dot{r}_{BR}\,(\mathbf{u}_B - \mathbf{u}_R). \tag{7.5.10}$$

For the blue species, we derive the analogous expression

$$\eta_B = \dot{r}_{RB}\,(\mathbf{u}_R - \mathbf{u}_B). \tag{7.5.11}$$

Adding these expressions, we find that

$$\eta_R + \eta_B = (\dot{r}_{BR} - \dot{r}_{RB})\,(\mathbf{u}_B - \mathbf{u}_R), \tag{7.5.12}$$

which is not necessarily zero. We recall that $\boldsymbol{\phi}_R + \boldsymbol{\phi}_B = \mathbf{0}$ and $\dot{\mathbf{m}}_R + \dot{\mathbf{m}}_B = \mathbf{0}$.

If we had used instead the expression for $\dot{\mathbf{m}}_R$ given by (7.4.14), we would have found that $\eta_R = \mathbf{0}$ and $\eta_B = \mathbf{0}$.

7.5.4 Stationary Species

In the case of a stationary red species, $\mathbf{u}_R = \mathbf{0}$, we obtain the simplified equation of motion

$$\nabla \cdot \boldsymbol{\sigma}_R + \rho_R\,\mathbf{g} + \iota_R\,\mathbf{b}_R + \boldsymbol{\phi}_R + \dot{\mathbf{m}}_R = \mathbf{0}, \tag{7.5.13}$$

which is analogous to the equation of hydrostatics. However, the term $\boldsymbol{\phi}_R$ is not necessarily zero due to the motion of the blue species through the stationary red species. A similar equation can be written for the blue species.

7.5.5 From Species to Mixture

Adding the red and blue species equations of motion (7.5.1) and (7.5.6), and recalling that the mixture density is $\rho = \rho_R + \rho_B$, and the mixture velocity is defined by $\rho\,\mathbf{u} = \rho_R\mathbf{u}_R + \rho_B\mathbf{u}_B$, we obtain

$$\frac{\partial(\rho\,\mathbf{u})}{\partial t} + \nabla\cdot(\rho_R\,\mathbf{u}_R \otimes \mathbf{u}_R + \rho_B\,\mathbf{u}_B \otimes \mathbf{u}_B) = \nabla \cdot (\boldsymbol{\sigma}_R + \boldsymbol{\sigma}_B) + \rho\,\mathbf{g} + \iota_R\,\mathbf{b}_R + \iota_B\,\mathbf{b}_B. \tag{7.5.14}$$

This equation can be compared with Cauchy's equation of motion for the mixture as a whole,

$$\frac{\partial(\rho\,\mathbf{u})}{\partial t} + \nabla(\rho\,\mathbf{u} \otimes \mathbf{u}) = \nabla \cdot \boldsymbol{\sigma} + \rho\,\mathbf{g} + \iota\,\mathbf{b}, \tag{7.5.15}$$

where $\boldsymbol{\sigma}$ is the mixture stress tensor and $\iota\mathbf{b} \equiv \iota_R\,\mathbf{b}_R + \iota_B\,\mathbf{b}_B$. The comparison shows that

$$\boldsymbol{\tau} = \boldsymbol{\tau}_R + \boldsymbol{\tau}_B, \tag{7.5.16}$$

where

$$\boldsymbol{\tau}_R \equiv \boldsymbol{\sigma}_R - \rho_R\,\mathbf{u}_R \otimes \mathbf{u}_R, \qquad \boldsymbol{\tau}_B \equiv \boldsymbol{\sigma}_B - \rho_B\,\mathbf{u}_B \otimes \mathbf{u}_B, \qquad \boldsymbol{\tau} \equiv \boldsymbol{\sigma} - \rho\,\mathbf{u} \otimes \mathbf{u} \tag{7.5.17}$$

are the species and mixture stress–momentum tensors. The relation between species and mixture stress tensors will be discussed further in Sects. 7.8 and 7.9.

7.5.6 Exercise

7.5.1 Discuss whether the species deviatoric stresses are zero in a stationary species.

7.6 Rankin–Hugoniot Momentum Condition

The rate of a change of the red momentum recorded by an observer moving with velocity $\mathbf{v}_R$ is given by

$$\frac{d(\rho_R \mathbf{u}_R)}{dt} = \frac{\partial(\rho_R \mathbf{u}_R)}{\partial t} + \mathbf{v}_R \cdot \nabla(\rho_R \mathbf{u}_R). \tag{7.6.1}$$

Combining this equation with the equation of motion (7.5.1), we obtain

$$\begin{aligned}\frac{d(\rho_R \mathbf{u}_R)}{dt} &= \mathbf{v}_R \cdot \nabla(\rho_R \mathbf{u}_R) - \nabla \cdot (\rho_R\, \mathbf{u}_R \otimes \mathbf{u}_R) \\ &\quad + \nabla \cdot \boldsymbol{\sigma}_R + \rho_R\, \mathbf{g} + \iota_R\, \mathbf{b}_R + \boldsymbol{\phi}_R + \dot{\mathbf{m}}_R. \end{aligned} \tag{7.6.2}$$

Consequently, the rate of change of the red momentum jump across a moving front of red momentum discontinuity moving with velocity $\mathbf{v}_R$ is

$$\begin{aligned}\frac{d(\Delta(\rho_R \mathbf{u}_R))}{dt} &= \mathbf{v}_R \cdot \nabla(\Delta(\rho_R \mathbf{u}_R)) - \nabla \cdot (\Delta(\rho_R\, \mathbf{u}_R \otimes \mathbf{u}_R)) \\ &\quad + \nabla \cdot \Delta\boldsymbol{\sigma}_R + \Delta\rho_R\, \mathbf{g} + \Delta\iota_R\, \mathbf{b}_R + \Delta\boldsymbol{\phi}_R + \Delta\dot{\mathbf{m}}_R, \end{aligned} \tag{7.6.3}$$

where Δ denotes the discontinuity or jump. All terms in this equation are finite, except for the two terms involving spatial derivatives normal to the evolving front.

To ensure regularity, we concentrate on these annoying terms and require that, to leading order,

$$\begin{aligned}\mathbf{0} &= (\mathbf{v}_R \cdot \mathbf{n})\big(\mathbf{n} \cdot \nabla(\Delta(\rho_R \mathbf{u}_R))\big) \\ &\quad -\mathbf{n} \cdot \nabla\big(\Delta(\rho_R(\mathbf{n} \cdot \mathbf{u}_R)\, \mathbf{u}_R)\big) + \mathbf{n} \cdot \nabla(\mathbf{n} \cdot \boldsymbol{\sigma}_R), \end{aligned} \tag{7.6.4}$$

where $\mathbf{n}_R$ is the unit vector normal to the red front, and obtain the Rankin–Hugoniot red momentum condition

$$(\mathbf{v}_R \cdot \mathbf{n})\, \Delta(\rho_R \mathbf{u}_R) = \Delta(\rho_R(\mathbf{n} \cdot \mathbf{u}_R)\, \mathbf{u}_R) - \Delta(\mathbf{n} \cdot \boldsymbol{\sigma}_R). \tag{7.6.5}$$

This expression relates the jump in red momentum to the jump in rate of transport of the red momentum to the jump in red fraction across a discontinuous front. Similar equations can be written for the blue species.

7.6.1 Exercise

7.6.1 Formulate and discuss the Rankin–Hugoniot momentum condition (7.6.5) for one-dimensional flow.

7.7 Species Angular Momentum Balance

The angular momentum of a red parcel with respect to a chosen reference point, $\mathbf{x}_\mathrm{R}$, is given by

$$\mathbf{A}_\mathrm{R} \equiv \iiint_{\text{red parcel}} \rho_\mathrm{R}\,(\mathbf{x}-\mathbf{x}_\mathrm{R}) \times \mathbf{u}_\mathrm{R}\,\mathrm{d}V, \tag{7.7.1}$$

as discussed in Sect. 5.7. Invoking Newton's second law of motion, we balance the rate of change of angular momentum with the torque due to surface and volume forces, and write

$$\begin{aligned}\frac{\mathrm{d}\mathbf{A}_\mathrm{R}}{\mathrm{d}t} &= \iint_{\text{red parcel}} (\mathbf{x}-\mathbf{x}_\mathrm{R}) \times (\mathbf{n}\cdot\boldsymbol{\sigma}_\mathrm{R})\,\mathrm{d}S + \iiint_{\text{red parcel}} \rho_\mathrm{R}\,(\mathbf{x}-\mathbf{x}_\mathrm{R}) \times \mathbf{g}\,\mathrm{d}V \\ &+ \iiint_{\text{red parcel}} \iota_\mathrm{R}\,(\mathbf{x}-\mathbf{x}_\mathrm{R}) \times \mathbf{b}_\mathrm{R}\,\mathrm{d}V + \iiint_{\text{red parcel}} \lambda_\mathrm{R}\,\mathbf{c}_\mathrm{R}\,\mathrm{d}V \\ &+ \iiint_{\text{red parcel}} (\mathbf{x}-\mathbf{x}_\mathrm{R}) \times \boldsymbol{\phi}_\mathrm{R}\,\mathrm{d}V + \iiint_{\text{red parcel}} (\mathbf{x}-\mathbf{x}_\mathrm{R}) \times \dot{\mathbf{m}}_\mathrm{R}\,\mathrm{d}V + \iiint_{\text{red parcel}} \boldsymbol{\psi}_\mathrm{R}\,\mathrm{d}V.\end{aligned} \tag{7.7.2}$$

The term $\mathbf{c}_\mathrm{R}$ inside the fourth integral on the right-hand side is a torque-inducing field with associated physical constant, λ_R. A torque-inducing field can be an electrical field applied to a ferrofluid.

The last term on the right-hand side of (7.7.2) expresses the volumetric rate of torque imparted on red molecules by blue molecules, under the restriction that

$$\boldsymbol{\psi}_\mathrm{R} + \boldsymbol{\psi}_\mathrm{B} = \mathbf{0}. \tag{7.7.3}$$

Physically, the rate of torque exerted on the red species by the blue species must be equal in magnitude and opposite in direction to that imparted to the blue species by the red species.

A similar equation can be written for a blue parcel. Evolution equations and integral balances can be written in the usual way.

7.7.1 Stress Tensor Spin

Considering the first term on the right-hand side of (7.7.2), we apply the divergence theorem to write

$$\iint_{\text{red parcel}} (\mathbf{x}-\mathbf{x}_\mathrm{R}) \times (\mathbf{n}\cdot\boldsymbol{\sigma}_\mathrm{R})\,\mathrm{d}S = \iiint_{\text{red parcel}} \big(\mathbf{s}_\mathrm{R} + (\mathbf{x}-\mathbf{x}_\mathrm{R}) \times \nabla\cdot\boldsymbol{\sigma}_\mathrm{R}\big)\,\mathrm{d}V, \tag{7.7.4}$$

where $\mathbf{s}_\mathrm{R}$ is a torque-like vector called the stress tensor spin, defined as

$$s_{\mathrm{R}_i} \equiv \epsilon_{ijk}\,\sigma_{\mathrm{R}\,jk} = \epsilon_{ijk}\,\widetilde{\sigma}_{\mathrm{R}_{jk}}, \tag{7.7.5}$$

and

$$\widetilde{\boldsymbol{\sigma}}_\mathrm{R} \equiv \tfrac{1}{2}\,(\boldsymbol{\sigma}_\mathrm{R} - \boldsymbol{\sigma}_\mathrm{R}^\mathrm{T}) \tag{7.7.6}$$

is the skew-symmetric part of $\boldsymbol{\sigma}_\mathrm{R}$. We note that only the skew-symmetric part of $\boldsymbol{\sigma}_\mathrm{R}$ contributes to the red spin, $\mathbf{s}_\mathrm{R}$. Conversely,

$$\widetilde{\sigma}_{\mathrm{R}_{ij}} = \frac{1}{2}\,\epsilon_{ijk}\, s_{\mathrm{R}_k}, \tag{7.7.7}$$

where summation is implied over the repeated index, k.

The right-hand side of (7.7.4) can be substituted for the first term on the right-hand side of (7.7.2) to yield a balance equation in terms of volume integrals alone.

7.7.2 Cauchy Angular Momentum Equation

Using expression (5.7.5) for the parcel rate of change of angular momentum, repeated below for convenience,

$$\frac{\mathrm{d}\mathbf{A}_{\mathrm{R}}}{\mathrm{d}t} = \iiint_{\text{red parcel}} (\mathbf{x}-\mathbf{x}_{\mathrm{R}}) \times \Big(\rho_{\mathrm{R}} \Big(\frac{\mathrm{D}\mathbf{u}_{\mathrm{R}}}{\mathrm{D}t}\Big)_{\mathrm{R}} + \dot{r}_{\mathrm{R}}\mathbf{u}_{\mathrm{R}}\Big)\, \mathrm{d}V, \tag{7.7.8}$$

and discarding the integral signs in (7.7.2), we derive the Cauchy angular-momentum equation for the red species,

$$\rho_{\mathrm{R}}\,(\mathbf{x}-\mathbf{x}_{\mathrm{R}}) \times \Big(\frac{\mathrm{D}\mathbf{u}_{\mathrm{R}}}{\mathrm{D}t}\Big)_{\mathrm{R}} + \dot{r}_{\mathrm{R}}\,(\mathbf{x}-\mathbf{x}_{\mathrm{R}}) \times \mathbf{u}_{\mathrm{R}} = \mathbf{s}_{\mathrm{R}} + (\mathbf{x}-\mathbf{x}_{\mathrm{R}}) \times \nabla\cdot\boldsymbol{\sigma}_{\mathrm{R}} + (\mathbf{x}-\mathbf{x}_{\mathrm{R}}) \times (\rho_{\mathrm{R}}\mathbf{g} + \iota_{\mathrm{R}}\mathbf{b}_{\mathrm{R}})$$
$$+\,\lambda_{\mathrm{R}}\,\mathbf{c}_{\mathrm{R}} + (\mathbf{x}-\mathbf{x}_{\mathrm{R}}) \times (\boldsymbol{\phi}_{\mathrm{R}} + \dot{\mathbf{m}}_{\mathrm{R}}) + \boldsymbol{\psi}_{\mathrm{R}}, \tag{7.7.9}$$

where $(\mathrm{D}\mathbf{u}_{\mathrm{R}}/\mathrm{D}t)_{\mathrm{R}}$ is the red point particle acceleration. Combining the second term of the left-hand side with the penultimate term on the right-hand side, we obtain

$$\rho_{\mathrm{R}}\,(\mathbf{x}-\mathbf{x}_{\mathrm{R}}) \times \Big(\frac{\mathrm{D}\mathbf{u}_{\mathrm{R}}}{\mathrm{D}t}\Big)_{\mathrm{R}} = \mathbf{s}_{\mathrm{R}} + (\mathbf{x}-\mathbf{x}_{\mathrm{R}}) \times \nabla\cdot\boldsymbol{\sigma}_{\mathrm{R}}$$
$$+\,(\mathbf{x}-\mathbf{x}_{\mathrm{R}}) \times (\rho_{\mathrm{R}}\mathbf{g} + \iota_{\mathrm{R}}\mathbf{b}_{\mathrm{R}}) + \lambda_{\mathrm{R}}\,\mathbf{c}_{\mathrm{R}} + (\mathbf{x}-\mathbf{x}_{\mathrm{R}}) \times (\boldsymbol{\phi}_{\mathrm{R}} + \boldsymbol{\eta}_{\mathrm{R}}) + \boldsymbol{\psi}_{\mathrm{R}}, \tag{7.7.10}$$

where $\boldsymbol{\eta}_{\mathrm{R}} \equiv \dot{\mathbf{m}}_{\mathrm{R}} - \dot{r}_{\mathrm{B}}\,\mathbf{u}_{\mathrm{R}}$.

Now formulating the cross product of each term in the red Cauchy's equation of motion (7.5.5) with the distance $\mathbf{x}-\mathbf{x}_{\mathrm{R}}$, and comparing the result with (7.7.10), we conclude that

$$\mathbf{s}_{\mathrm{R}} = -\lambda_{\mathrm{R}}\,\mathbf{c}_{\mathrm{R}} - \boldsymbol{\psi}_{\mathrm{R}} \tag{7.7.11}$$

The angular momentum equation (7.7.10) then simplifies to

$$\rho_{\mathrm{R}}\,(\mathbf{x}-\mathbf{x}_{\mathrm{R}}) \times \Big(\frac{\mathrm{D}\mathbf{u}_{\mathrm{R}}}{\mathrm{D}t}\Big)_{\mathrm{R}} = (\mathbf{x}-\mathbf{x}_{\mathrm{R}}) \times \nabla\cdot\boldsymbol{\sigma}_{\mathrm{R}}$$
$$+\,(\mathbf{x}-\mathbf{x}_{\mathrm{R}}) \times (\rho_{\mathrm{R}}\mathbf{g} + \iota_{\mathrm{R}}\mathbf{b}_{\mathrm{R}}) + (\mathbf{x}-\mathbf{x}_{\mathrm{R}}) \times (\boldsymbol{\phi}_{\mathrm{R}} + \boldsymbol{\eta}_{\mathrm{R}}). \tag{7.7.12}$$

The effect of species interaction is represented by the last term on the right-hand side.

7.7.3 Implicit Effect of the Torque Field

The torque fields associated with $\mathbf{c}_R$ and $\boldsymbol{\psi}_R$ appear in Eq. (7.7.12) implicitly by determining the antisymmetric part of the red stress tensor. Substituting (7.7.11) into (7.7.7), we find that

$$(\sigma_{ij} - \sigma_{ji})_R = -\epsilon_{ijk}\,(\lambda_R\,\mathbf{c}_R + \boldsymbol{\psi}_R)_k, \tag{7.7.13}$$

where summation is implied over the repeated index, k.

7.7.4 Symmetry of the Species Stress Tensor

We have found that in the absence of a torque-inducing field and torque-inducing molecular interactions, the species stress tensor is symmetric.

7.7.5 Exercise

7.7.1 Describe a mechanism by which a body can impart a torque on another body by means of an interaction force potential.

7.8 Parcel Superposition

Assume that a red and a blue parcel are superposed at some time instant, that is, they are regarded as constituents of the same mixture parcel. As time progresses, the monochromatic parcels and the mixture parcel will separate due to diffusion.

7.8.1 Species and Mixture Parcels

Applying Eq. (4.7.20) for the species X momentum by setting $\phi = \rho_X \mathbf{u}_X$, we obtain

$$\frac{\mathrm{d}}{\mathrm{d}t}\iiint_{\mathrm{X\ parcel}} \rho_X \mathbf{u}_X\,\mathrm{d}V = \frac{\mathrm{d}}{\mathrm{d}t}\iiint_{\mathrm{mixture\ parcel}} \rho_X \mathbf{u}_X\,\mathrm{d}V + \iint_{\mathrm{parcel}} (\rho_X \mathbf{u}_X)\,\mathbf{n}\cdot(\mathbf{u}_X - \mathbf{u})\,\mathrm{d}S, \tag{7.8.1}$$

where X can be R, B, or any other species in a multicomponent mixture. Summing over all species and invoking the definition of the diffusive mass flux, we obtain

$$\sum_X \frac{\mathrm{d}}{\mathrm{d}t}\iiint_{\mathrm{X\ parcel}} \rho_X \mathbf{u}_X\,\mathrm{d}V = \frac{\mathrm{d}}{\mathrm{d}t}\iiint_{\mathrm{mixture\ parcel}} \rho\,\mathbf{u}\,\mathrm{d}V + \sum_X \iint_{\mathrm{parcel}} \mathbf{u}_X\,\mathbf{n}\cdot \boldsymbol{j}_X\,\mathrm{d}S. \tag{7.8.2}$$

Rearranging, we obtain

$$\frac{\mathrm{d}}{\mathrm{d}t}\iiint_{\mathrm{mixture\ parcel}} \rho\,\mathbf{u}\,\mathrm{d}V = \sum_X \frac{\mathrm{d}}{\mathrm{d}t}\iiint_{\mathrm{X\ parcel}} \rho_X \mathbf{u}_X\,\mathrm{d}V - \sum_X \iint_{\mathrm{parcel}} \mathbf{u}_X\,\mathbf{n}\cdot \boldsymbol{j}_X\,\mathrm{d}S, \tag{7.8.3}$$

which provides us with the rate of change of momentum of a mixture parcel in terms of the rates of change of momentum of the constituent species parcels and the diffusive fluxes.

Because of parcel separation, the rate of change of momentum of a mixture parcel is not equal to the sum of the rates of change of momentum of the constituent species parcels.

7.8.2 Mixture Momentum of Parcels

Also applying Eq. (4.7.20) for the mixture momentum by setting $\phi = \rho\,\mathbf{u}$, we obtain

$$\frac{\mathrm{d}}{\mathrm{d}t}\iiint_{\text{X parcel}} \rho\,\mathbf{u}\,\mathrm{d}V = \frac{\mathrm{d}}{\mathrm{d}t}\iiint_{\text{mixture parcel}} \rho\,\mathbf{u}\,\mathrm{d}V + \iint_{\text{parcel}} (\rho\,\mathbf{u})\,\mathbf{n}\cdot(\mathbf{u}_{\mathrm{X}} - \mathbf{u})\,\mathrm{d}S. \tag{7.8.4}$$

Rearranging, we obtain

$$\frac{\mathrm{d}}{\mathrm{d}t}\iiint_{\text{mixture parcel}} \rho\,\mathbf{u}\,\mathrm{d}V = \frac{\mathrm{d}}{\mathrm{d}t}\iiint_{\text{X parcel}} \rho\,\mathbf{u}\,\mathrm{d}V - \iint_{\text{parcel}} \frac{1}{\omega_{\mathrm{X}}}\,\mathbf{n}\cdot\boldsymbol{j}_{\mathrm{X}}\,\mathrm{d}S. \tag{7.8.5}$$

From Eq. (7.8.4), we find that

$$\frac{\mathrm{d}}{\mathrm{d}t}\iiint_{\text{X parcel}} \rho\,\mathbf{u}\,\mathrm{d}V = \frac{\mathrm{d}}{\mathrm{d}t}\iiint_{\text{Y parcel}} \rho\,\mathbf{u}\,\mathrm{d}V + \iint_{\text{parcel}} (\rho\,\mathbf{u})\,\mathbf{n}\cdot(\mathbf{u}_{\mathrm{X}} - \mathbf{u}_{\mathrm{Y}})\,\mathrm{d}S \tag{7.8.6}$$

for any pair of species, X and Y.

7.8.3 From Species to Mixture

The motion of a mixture parcel is governed by Newton's second law of motion expressed by the familiar form

$$\frac{\mathrm{d}}{\mathrm{d}t}\iiint_{\text{mixture parcel}} \rho\,\mathbf{u}\,\mathrm{d}V = \iint_{\text{parcel}} \mathbf{n}\cdot\boldsymbol{\sigma}\,\mathrm{d}S + \iiint_{\text{parcel}} (\rho\,\mathbf{g} + \iota\mathbf{b})\,\mathrm{d}V, \tag{7.8.7}$$

where $\boldsymbol{\sigma}$ is the mixture Cauchy stress tensor and $\iota\mathbf{b} \equiv \sum \iota_{\mathrm{X}}\,\mathbf{b}_{\mathrm{X}}$. Combining this equation with (7.8.3), we obtain

$$\sum_{\mathrm{X}} \frac{\mathrm{d}}{\mathrm{d}t}\iiint_{\text{X parcel}} \rho_{\mathrm{X}}\mathbf{u}_{\mathrm{X}}\,\mathrm{d}V = \sum_{\mathrm{X}} \iint_{\text{parcel}} \mathbf{u}_{\mathrm{X}}\,\mathbf{n}\cdot\boldsymbol{j}_{\mathrm{X}}\,\mathrm{d}S + \iint_{\text{parcel}} \mathbf{n}\cdot\boldsymbol{\sigma}\,\mathrm{d}S + \iiint_{\text{parcel}} (\rho\,\mathbf{g} + \iota\mathbf{b})\,\mathrm{d}V. \tag{7.8.8}$$

Next, we add the species parcel equation of motion (7.2.2) over all species for superposed parcels. Taking into consideration the constraints imposed by (7.2.3), we obtain

$$\sum_{\mathrm{X}} \frac{\mathrm{d}}{\mathrm{d}t}\iiint_{\text{X parcel}} \rho_{\mathrm{X}}\mathbf{u}_{\mathrm{X}}\,\mathrm{d}V = \sum_{\mathrm{X}} \iint_{\text{X parcel}} \mathbf{n}\cdot\boldsymbol{\sigma}_{\mathrm{X}}\,\mathrm{d}S + \iiint_{\text{X parcel}} (\rho\,\mathbf{g} + \iota\mathbf{b})\,\mathrm{d}V. \tag{7.8.9}$$

Comparing the last two equations, we find that

$$\boldsymbol{\sigma} = \sum_{\mathrm{X}} \left(\boldsymbol{\sigma}_{\mathrm{X}} - \mathbf{u}_{\mathrm{X}} \otimes \boldsymbol{j}_{\mathrm{X}} \right), \tag{7.8.10}$$

which is equivalent to (7.1.4), as discussed in Sect. 7.9.

7.8.4 Y Momentum of X Parcel

We may apply Eq. (4.7.19) for the Y momentum by setting $\phi = \rho_{\mathrm{Y}} \mathbf{u}_{\mathrm{Y}}$, where Y is any species, and obtain

$$\frac{\mathrm{d}}{\mathrm{d}t} \iiint_{\mathrm{X\ parcel}} \rho_{\mathrm{Y}} \mathbf{u}_{\mathrm{Y}}\, \mathrm{d}V = \frac{\mathrm{d}}{\mathrm{d}t} \iiint_{\mathrm{Y\ parcel}} \rho_{\mathrm{Y}} \mathbf{u}_{\mathrm{Y}}\, \mathrm{d}V + \iint_{\mathrm{parcel}} (\rho_{\mathrm{Y}} \mathbf{u}_{\mathrm{Y}})\, \mathbf{n} \cdot (\mathbf{u}_{\mathrm{X}} - \mathbf{u}_{\mathrm{Y}})\, \mathrm{d}S. \tag{7.8.11}$$

The first term on the right-hand side can be evaluated using the Y parcel equation of motion, providing us with an expression for the rate of a change of the Y momentum encapsulated in an X parcel,

$$\frac{\mathrm{d}}{\mathrm{d}t} \iiint_{\mathrm{X\ parcel}} \rho_{\mathrm{Y}} \mathbf{u}_{\mathrm{Y}}\, \mathrm{d}V = \iiint_{\mathrm{X\ parcel}} \dot{r}_{\mathrm{Y}}\, \mathrm{d}V + \iint_{\mathrm{parcel}} (\rho_{\mathrm{Y}} \mathbf{u}_{\mathrm{Y}})\, \mathbf{n} \cdot (\mathbf{u}_{\mathrm{X}} - \mathbf{u}_{\mathrm{Y}})\, \mathrm{d}S. \tag{7.8.12}$$

Summing over Y, and recalling that $\sum \dot{r}_{\mathrm{Y}} = 0$, we find that

$$\sum_{\mathrm{Y}} \frac{\mathrm{d}}{\mathrm{d}t} \iiint_{\mathrm{X\ parcel}} \rho_{\mathrm{Y}} \mathbf{u}_{\mathrm{Y}}\, \mathrm{d}V = \iint_{\mathrm{parcel}} (\rho \mathbf{u})\, \mathbf{n} \cdot \mathbf{u}_{\mathrm{X}}\, \mathrm{d}S - \sum_{\mathrm{Y}} \iint_{\mathrm{parcel}} (\rho_{\mathrm{Y}} \mathbf{u}_{\mathrm{Y}})\, \mathbf{n} \cdot \mathbf{u}_{\mathrm{Y}}\, \mathrm{d}S. \tag{7.8.13}$$

The last integrand can be expressed in terms of the species momentum tensor, $\rho_{\mathrm{Y}} \mathbf{u}_{\mathrm{Y}} \otimes \mathbf{u}_{\mathrm{Y}}$.

7.8.5 Exercise

7.8.1 Express the last integral in (7.8.13) in terms of the species momentum tensor, $\rho_{\mathrm{Y}} \mathbf{u}_{\mathrm{Y}} \otimes \mathbf{u}_{\mathrm{Y}}$.

7.9 From Species to Mixture Stresses

The red-species equation of motion was derived in (7.5.1) in terms of the red-species density and velocity as

$$\frac{\partial (\rho_{\mathrm{R}} \mathbf{u}_{\mathrm{R}})}{\partial t} + \boldsymbol{\nabla} \cdot (\rho_{\mathrm{R}}\, \mathbf{u}_{\mathrm{R}} \otimes \mathbf{u}_{\mathrm{R}}) = \boldsymbol{\nabla} \cdot \boldsymbol{\sigma}_{\mathrm{R}} + \rho_{\mathrm{R}}\, \mathbf{g} + \iota_{\mathrm{R}}\, \mathbf{b}_{\mathrm{R}} + \boldsymbol{\phi}_{\mathrm{R}} + \dot{\mathbf{m}}_{\mathrm{R}}, \tag{7.9.1}$$

where $\mathbf{b}_{\mathrm{R}}$ is a red body force with associated physical constant ι_{R}, $\boldsymbol{\phi}_{\mathrm{R}}$ is a red species interaction force, and $\dot{\mathbf{m}}_{\mathrm{R}}$ is the rate of red momentum generation due to a chemical reaction.

The blue species equation of motion is given by a corresponding equation, subject to the restrictions that

$$\boldsymbol{\phi}_{\mathrm{R}} + \boldsymbol{\phi}_{\mathrm{B}} = \mathbf{0}, \qquad \dot{\mathbf{m}}_{\mathrm{R}} + \dot{\mathbf{m}}_{\mathrm{B}} = \mathbf{0} \tag{7.9.2}$$

in a binary mixture. Similar equations can be written for a multicomponent mixture.

7.9.1 Species Modified Stress Tensors

In preparation for transitioning from species to mixture, we recall the definition of the mixture mass velocity, $\mathbf{u} = \omega_R \mathbf{u}_R + \omega_B \mathbf{u}_B$, and rearrange the second term on the left-hand side of (7.9.1) to obtain

$$\frac{\partial(\rho_R \mathbf{u}_R)}{\partial t} + \nabla \cdot (\rho_R\, \mathbf{u} \otimes \mathbf{u}_R) = \nabla \cdot \widehat{\boldsymbol{\sigma}}_R + \rho_R\, \mathbf{g} + \iota_R\, \mathbf{b}_R + \boldsymbol{\phi}_R + \dot{\mathbf{m}}_R, \tag{7.9.3}$$

where

$$\widehat{\boldsymbol{\sigma}}_R \equiv \boldsymbol{\sigma}_R - \boldsymbol{j}_R \otimes \mathbf{u}_R \tag{7.9.4}$$

is a modified stress tensor and $\boldsymbol{j}_R \equiv \rho_R\,(\mathbf{u}_R - \mathbf{u})$ is the diffusive red species mass flux. In the absence of diffusion, $\widehat{\boldsymbol{\sigma}}_B = \boldsymbol{\sigma}_B$.

Alternative expressions for the modified red stress tensor are

$$\widehat{\boldsymbol{\sigma}}_R \equiv \boldsymbol{\sigma}_R - \frac{1}{\rho_R}\, \boldsymbol{j}_R \otimes \boldsymbol{j}_R - \boldsymbol{j}_R \otimes \mathbf{u} \tag{7.9.5}$$

and

$$\widehat{\boldsymbol{\sigma}}_R \equiv \boldsymbol{\sigma}_R - \rho_R\, \mathbf{u}_R \otimes \mathbf{u}_R + \rho_R\, \mathbf{u} \otimes \mathbf{u}_R, \tag{7.9.6}$$

and also

$$\widehat{\boldsymbol{\sigma}}_R \equiv \boldsymbol{\sigma}_R - \rho_R\, \mathbf{u}_R^{\text{diff}} \otimes \mathbf{u}_R^{\text{diff}} - \boldsymbol{j}_R \otimes \mathbf{u}. \tag{7.9.7}$$

These expressions admit a direct generalization to an arbitrary number of species in a multicomponent solution.

7.9.2 Mixture Stress and Equation of Motion

Adding the red equation of motion (7.9.3) to its counterpart for the red species, and recalling the restrictions (7.9.2), we obtain the familiar Cauchy equation of motion for the mixture,

$$\frac{\partial(\rho\, \mathbf{u})}{\partial t} + \nabla \cdot (\rho\, \mathbf{u} \otimes \mathbf{u}) = \nabla \cdot \boldsymbol{\sigma} + \rho\, \mathbf{g} + \iota\, \mathbf{b}, \tag{7.9.8}$$

where $\iota\, \mathbf{b} \equiv \iota_R\, \mathbf{b}_R + \iota_B\, \mathbf{b}_B$ is the combined species-specific body force, and

$$\boldsymbol{\sigma} = \widehat{\boldsymbol{\sigma}}_R + \widehat{\boldsymbol{\sigma}}_B \tag{7.9.9}$$

is the mixture Cauchy stress tensor.

Substituting the definitions of $\widehat{\boldsymbol{\sigma}}_R$ and $\widehat{\boldsymbol{\sigma}}_B$ from (7.9.4) and its counterpart for the blue species, we obtain the explicit expression

$$\boldsymbol{\sigma} = (\boldsymbol{\sigma}_R - \boldsymbol{j}_R \otimes \mathbf{u}_R) + (\boldsymbol{\sigma}_B - \boldsymbol{j}_B \otimes \mathbf{u}_B), \tag{7.9.10}$$

which is equivalent to

$$\boldsymbol{\sigma} = (\boldsymbol{\sigma}_R - \frac{1}{\rho_R}\, \boldsymbol{j}_R \otimes \boldsymbol{j}_R) + (\boldsymbol{\sigma}_B - \frac{1}{\rho_B}\, \boldsymbol{j}_B \otimes \boldsymbol{j}_B). \tag{7.9.11}$$

An alternative expression in terms of the diffusion velocities is

$$\boldsymbol{\sigma} = (\boldsymbol{\sigma}_{\rm R} - \rho_{\rm R}\, \mathbf{u}_{\rm R}^{\rm diff} \otimes \mathbf{u}_{\rm R}^{\rm diff}) + (\boldsymbol{\sigma}_{\rm B} - \rho_{\rm B}\, \mathbf{u}_{\rm B}^{\rm diff} \otimes \mathbf{u}_{\rm B}^{\rm diff}). \tag{7.9.12}$$

The diffusion velocities appear as an inertial correction in this expression.

7.9.3 Multicomponent Mixture

The sum on the right-hand sides of (7.9.11) can be extended to an arbitrary number of species in a multicomponent mixture,

$$\sigma = \sum_{\rm X} \big(\sigma_{\rm X} - \frac{1}{\rho_{\rm X}}\, \boldsymbol{j}_{\rm X} \otimes \boldsymbol{j}_{\rm X}\big) = \sum_{\rm X} \big(\sigma_{\rm X} - \rho_{\rm X}\, \mathbf{u}_{\rm X}^{\rm diff} \otimes \mathbf{u}_{\rm X}^{\rm diff}\big). \tag{7.9.13}$$

Note that the mixture stress tensor, $\boldsymbol{\sigma}$, is equal to the sum of the species stress tensors only when the diffusive species mass fluxes are zero.

7.9.4 Binary Mixture

Since $\boldsymbol{j}_{\rm B} = -\boldsymbol{j}_{\rm R}$ in a binary mixture, the mixture stress tensor is given by

$$\boldsymbol{\sigma} = \boldsymbol{\sigma}_{\rm R} + \boldsymbol{\sigma}_{\rm B} - (\frac{1}{\rho_{\rm R}} + \frac{1}{\rho_{\rm B}})\, \boldsymbol{j}_{\rm R} \otimes \boldsymbol{j}_{\rm R}. \tag{7.9.14}$$

The red flux can be replaced by the negative of the blue flux on the right-hand side.

7.9.5 Stress–Momentum Tensor

Substituting into (7.9.9) the alternative definition of $\widehat{\boldsymbol{\sigma}}_{\rm R}$ given in (7.9.5), and its counterpart for the blue species, we obtain

$$\boldsymbol{\sigma} - \rho\, \mathbf{u} \otimes \mathbf{u} = (\boldsymbol{\sigma}_{\rm R} - \rho_{\rm R} \mathbf{u}_{\rm R} \otimes \mathbf{u}_{\rm R}) + (\boldsymbol{\sigma}_{\rm B} - \rho_{\rm B} \mathbf{u}_{\rm B} \otimes \mathbf{u}_{\rm B}). \tag{7.9.15}$$

In terms of the stress–momentum tensors of the species and mixture, defined as

$$\boldsymbol{\tau}_{\rm R} \equiv \boldsymbol{\sigma}_{\rm R} - \rho_{\rm R}\, \mathbf{u}_{\rm R} \otimes \mathbf{u}_{\rm R}, \qquad \boldsymbol{\tau}_{\rm B} \equiv \boldsymbol{\sigma}_{\rm B} - \rho_{\rm B}\, \mathbf{u}_{\rm B} \otimes \mathbf{u}_{\rm B}, \qquad \boldsymbol{\tau} \equiv \boldsymbol{\sigma} - \rho\, \mathbf{u} \otimes \mathbf{u}. \tag{7.9.16}$$

Equation (7.9.15) states that

$$\boldsymbol{\tau} = \boldsymbol{\tau}_{\rm R} + \boldsymbol{\tau}_{\rm B}, \tag{7.9.17}$$

as shown in (7.5.16).

The generalization of this equation to multicomponent mixtures is straightforward. The sum of the stress–momentum tensor of the species is the stress–momentum tensor of the mixture.

$$\boldsymbol{\tau} = \sum_{\mathrm{X}} \boldsymbol{\tau}_{\mathrm{X}}. \tag{7.9.18}$$

Though defining the mixture stress in terms of species stresses is straightforward, a procedure for apportioning the mixture stresses into species stresses is unclear.

7.9.6 Symmetry of the Stress Tensor

Based on the preceding expressions, we confirm that if $\boldsymbol{\sigma}_{\mathrm{R}}$ and $\boldsymbol{\sigma}_{\mathrm{B}}$ are symmetric, $\boldsymbol{\sigma}$ will also be symmetric. Conversely, if $\boldsymbol{\sigma}$ is symmetric, then the sum $\boldsymbol{\sigma}_{\mathrm{R}} + \boldsymbol{\sigma}_{\mathrm{B}}$ is also symmetric, although the individual red and blue stress tensors are not necessarily symmetric.

7.9.7 Pressure

The pressure is defined as the negative of one-third of the trace of the stress tensor. From (7.9.11), we find that the mixture pressure is given by

$$p = p_{\mathrm{R}} + p_{\mathrm{B}} + \tfrac{1}{3}\,\frac{1}{\rho_{\mathrm{R}}}\,|\boldsymbol{j}_{\mathrm{R}}|^2 + \tfrac{1}{3}\,\frac{1}{\rho_{\mathrm{B}}}\,|\boldsymbol{j}_{\mathrm{B}}|^2, \tag{7.9.19}$$

which shows that the mixture pressure is no less than the sum of the species pressures. An alternative expression originating from Eq. (7.9.15) is

$$p = p_{\mathrm{R}} + p_{\mathrm{B}} + \tfrac{1}{3}\,\rho_{\mathrm{R}}\,|\mathbf{u}_{\mathrm{R}}|^2 + \tfrac{1}{3}\,\rho_{\mathrm{B}}\,|\mathbf{u}_{\mathrm{B}}|^2 - \tfrac{1}{3}\,\rho\,|\mathbf{u}|^2. \tag{7.9.20}$$

When the species velocities are the same, the mixture pressure is the sum of the species pressures irrespective of the species densities.

Using expression (4.5.4) for a binary mixture, repeated below for convenience,

$$\tfrac{1}{2}\,\rho_{\mathrm{R}}\,|\mathbf{u}_{\mathrm{R}}|^2 + \tfrac{1}{2}\,\rho_{\mathrm{B}}\,|\mathbf{u}_{\mathrm{B}}|^2 - \tfrac{1}{2}\,\rho\,|\mathbf{u}|^2 = \tfrac{1}{2}\,\rho\,\omega_{\mathrm{R}}\,\omega_{\mathrm{B}}\,|\mathbf{u}_{\mathrm{R}} - \mathbf{u}_{\mathrm{B}}|^2, \tag{7.9.21}$$

we obtain

$$p = p_{\mathrm{R}} + p_{\mathrm{B}} + \tfrac{1}{3}\,\rho\,\omega_{\mathrm{R}}\,\omega_{\mathrm{B}}\,|\mathbf{u}_{\mathrm{R}} - \mathbf{u}_{\mathrm{B}}|^2, \tag{7.9.22}$$

which confirms that the mixture pressure is no less than the sum of the species pressures.

7.9.8 Mechanical Energy Loss or Gain

Recalling expression (7.9.13) for the mixture stress tensor, we find that the rate of mechanical energy loss or gain due to the fluid motion is given by

$$\boldsymbol{\sigma} : \nabla\mathbf{u} = \sum_{\mathrm{X}} \boldsymbol{\sigma}_{\mathrm{X}} : \nabla\mathbf{u} - \sum_{\mathrm{X}} \frac{1}{\rho_{\mathrm{X}}}\,\boldsymbol{j}_{\mathrm{X}} \otimes \boldsymbol{j}_{\mathrm{X}} : \nabla\mathbf{u}. \tag{7.9.23}$$

Resolving the stress into an isotropic and a deviatoric component, and using expression (7.9.19) for the pressure, we find that

$$\boldsymbol{\sigma}^{\text{dev}} : \nabla \mathbf{u} = \sum_{\text{X}} \boldsymbol{\sigma}_{\text{X}}^{\text{dev}} : \nabla \mathbf{u} - \Psi, \tag{7.9.24}$$

where the superscript *dev* denotes the deviatoric part of the stress tensor excluding the pressure,

$$\Psi \equiv \sum_{\text{X}} \frac{1}{\rho_{\text{X}}} (\boldsymbol{j}_{\text{X}} \otimes \boldsymbol{j}_{\text{X}} - \tfrac{1}{3} |\boldsymbol{j}_{\text{X}}|^2 \mathbf{I}) : \mathbf{E}, \tag{7.9.25}$$

and

$$\mathbf{E} \equiv \tfrac{1}{2} (\nabla \mathbf{u} + (\nabla \mathbf{u})^{\text{T}}) - \tfrac{1}{3} (\nabla \cdot \mathbf{u}) \mathbf{I} \tag{7.9.26}$$

is the mixture rate-of-deformation tensor. Requiring that

$$\boldsymbol{\sigma}^{\text{dev}} : \nabla \mathbf{u} \geq 0, \tag{7.9.27}$$

we derive the constraint

$$\sum_{\text{X}} \boldsymbol{\sigma}_{\text{X}}^{\text{dev}} : \nabla \mathbf{u} \geq \Psi. \tag{7.9.28}$$

In developing constitutive equations for species stresses, this inequality must be satisfied. In the absence of diffusion, we recover the second Clausius–Duhem inequality stated in (3.9.8).

7.9.9 Constitutive Equations

When dotted with the unit normal vector, $\mathbf{n}$, the red stress tensor $\boldsymbol{\sigma}_{\text{R}}$ provides us with the traction exerted on the surface of a red parcel, defined as the force divided by an infinitesimal surface area, as discussed in Sect. 7.1. Since this force involves interspecies interactions, a constitutive equation $\boldsymbol{\sigma}_{\text{R}}$ involving the red velocity field and red properties alone can be developed only under idealized conditions (e.g., Atkin, R. J. & Crain, R. E. (1976) Continuum theories of mixtures: Basic theory and historical development. *Q. J. Mech. Appl. Math.* **29**, 209–244.)

7.9.10 Exercises

7.9.1 Simplify the red equation of motion in the absence of diffusion.

7.9.2 Derive expressions (7.9.5), (7.9.6), and (7.9.7).

7.9.3 Express (7.9.11) in terms of the diffusion velocities.

7.10 Stationary Mixtures

In the absence of global motion, $\mathbf{u} = \mathbf{0}$, we invoke the definition of the red and blue diffusive mass fluxes and obtain

$$\boldsymbol{j}_{\mathrm{R}} = \rho_{\mathrm{R}} \mathbf{u}_{\mathrm{R}}, \qquad \boldsymbol{j}_{\mathrm{B}} = \rho_{\mathrm{B}} \mathbf{u}_{\mathrm{B}}. \tag{7.10.1}$$

These expressions can be inverted to give the species velocities in terms of the corresponding fluxes, and then substituted into transport equations.

7.10.1 Species Equations of Motion

In a stationary mixture, the species velocities are the diffusion velocities. The general equation of motion for the red species, Eq. (7.9.3), is repeated below for convenience,

$$\frac{\partial(\rho_{\mathrm{R}} \mathbf{u}_{\mathrm{R}})}{\partial t} + \nabla \cdot (\rho_{\mathrm{R}} \, \mathbf{u} \otimes \mathbf{u}_{\mathrm{R}}) = \nabla \cdot \widehat{\boldsymbol{\sigma}}_{\mathrm{R}} + \rho_{\mathrm{R}} \, \mathbf{g} + \iota_{\mathrm{R}} \, \mathbf{b}_{\mathrm{R}} + \boldsymbol{\phi}_{\mathrm{R}} + \dot{\mathbf{m}}_{\mathrm{R}}, \tag{7.10.2}$$

where

$$\widehat{\boldsymbol{\sigma}}_{\mathrm{R}} \equiv \boldsymbol{\sigma}_{\mathrm{R}} - \boldsymbol{j}_{\mathrm{R}} \otimes \mathbf{u}_{\mathrm{R}} \tag{7.10.3}$$

is modified red stress tensor. In the absence of global motion, $\mathbf{u} = \mathbf{0}$, we obtain the simplified equation

$$\frac{\partial \, \boldsymbol{j}_{\mathrm{R}}}{\partial t} = \nabla \cdot \boldsymbol{\sigma}_{\mathrm{R}} - \nabla \cdot (\boldsymbol{j}_{\mathrm{R}} \otimes \mathbf{u}_{\mathrm{R}}) + \rho_{\mathrm{R}} \, \mathbf{g} + \iota_{\mathrm{R}} \, \mathbf{b}_{\mathrm{R}} + \boldsymbol{\phi}_{\mathrm{R}} + \dot{\mathbf{m}}_{\mathrm{R}}. \tag{7.10.4}$$

We will adopt Eq. (7.4.6) for the species rate of momentum exchange,

$$\dot{\mathbf{m}}_{\mathrm{R}} = \dot{r}_{\mathrm{BR}} \, \mathbf{u}_{\mathrm{B}} - \dot{r}_{\mathrm{RB}} \, \mathbf{u}_{\mathrm{R}}. \tag{7.10.5}$$

Similar equations can be written for the blue species.

We will assume that the density of the mixture, ρ, is uniform, as in the case of self-diffusion of a mixture of tagged species or isotopes.

7.10.2 Fick's Law

Substituting into (7.10.4) Fick's law,

$$\boldsymbol{j}_{\mathrm{R}} = -\rho \, \mathcal{D} \, \nabla \omega_{\mathrm{R}}, \tag{7.10.6}$$

and setting

$$\mathbf{u}_{\mathrm{R}} = -\mathcal{D} \, \nabla \ln \omega_{\mathrm{R}}, \qquad \mathbf{u}_{\mathrm{B}} = -\mathcal{D} \, \nabla \ln \omega_{\mathrm{B}}, \tag{7.10.7}$$

we obtain

$$\begin{aligned} -\rho \mathcal{D} \frac{\partial \, \nabla \omega_{\mathrm{R}}}{\partial t} &= \nabla \cdot \boldsymbol{\sigma}_{\mathrm{R}} - \rho \mathcal{D}^2 \nabla \cdot \big(\nabla \omega_{\mathrm{R}} \otimes \nabla \ln \omega_{\mathrm{R}} \big) \\ &\quad + \rho_{\mathrm{R}} \, \mathbf{g} + \iota_{\mathrm{R}} \, \mathbf{b}_{\mathrm{R}} + \boldsymbol{\phi}_{\mathrm{R}} - \mathcal{D} \big(\dot{r}_{\mathrm{BR}} \, \nabla \ln \omega_{\mathrm{B}} - \dot{r}_{\mathrm{RB}} \, \nabla \ln \omega_{\mathrm{R}} \big), \end{aligned} \tag{7.10.8}$$

where $\mathcal{D}$ is the binary diffusivity. The gradient and time differentiation operators can be transposed on the left-hand side. Solving for the first term on the right-hand side, we obtain

$$\begin{aligned}\nabla\cdot\boldsymbol{\sigma}_R &= -\rho\mathcal{D}\,\nabla\frac{\partial\omega_R}{\partial t} + \rho\mathcal{D}^2\nabla\cdot\big(\nabla\omega_R\otimes\nabla\ln\omega_R\big)\\ &\quad -\rho_R\,\mathbf{g} - \iota_R\,\mathbf{b}_R - \boldsymbol{\phi}_R + \mathcal{D}\big(\dot{r}_{RB}\,\nabla\ln\omega_B - \dot{r}_{BR}\,\nabla\ln\omega_R\big).\end{aligned} \tag{7.10.9}$$

The time derivative of the red mass fraction on the right-hand side can be computed from the red mass fraction evolution equation (6.3.5), repeated below for convenience,

$$\rho\,\frac{\partial\,\omega_R}{\partial t} = \rho\,\mathcal{D}\,\nabla^2\omega_R + \dot{r}_R. \tag{7.10.10}$$

Rearranging, we obtain

$$\begin{aligned}\nabla\cdot\boldsymbol{\sigma}_R &= -\mathcal{D}\big(\rho\,\mathcal{D}\,\nabla^2\nabla\omega_R + \nabla\dot{r}_R\big) + \rho\,\mathcal{D}^2\nabla\cdot\big(\nabla\omega_R\otimes\nabla\ln\omega_R\big)\\ &\quad -\rho_R\,\mathbf{g} - \iota_R\,\mathbf{b}_R - \boldsymbol{\phi}_R + \mathcal{D}\big(\dot{r}_{RB}\,\nabla\ln\omega_B - \dot{r}_{BR}\,\nabla\ln\omega_R\big).\end{aligned} \tag{7.10.11}$$

Consolidating two terms and recalling that the Laplacian is the divergence of the gradient, $\nabla^2 = \nabla\cdot\nabla$, and therefore

$$\nabla^2\nabla = \nabla\cdot(\nabla\otimes\nabla), \tag{7.10.12}$$

we obtain

$$\nabla\cdot\boldsymbol{\sigma}_R = -\rho\,\mathcal{D}^2\,\nabla\cdot\boldsymbol{\Lambda}_R - \rho_R\,\mathbf{g} - \iota_R\,\mathbf{b}_R - \boldsymbol{\phi}_R - \mathcal{D}\big(\dot{r}_{RB}\,\nabla\ln\omega_B - \dot{r}_{BR}\,\nabla\ln\omega_R + \nabla\dot{r}_R\big), \tag{7.10.13}$$

where

$$\boldsymbol{\Lambda}_R \equiv \nabla\otimes\nabla\omega_R - \nabla\omega_R\otimes\nabla\ln\omega_R, \tag{7.10.14}$$

which can be expressed as

$$\boldsymbol{\Lambda}_R = \omega_R\,\nabla\otimes\nabla\ln\omega_R. \tag{7.10.15}$$

Note that the divergence of $\boldsymbol{\Lambda}_R$ on the right-hand side of (7.10.13) is multiplied by the square of the diffusivity.

Substituting into (7.10.13) the expression for the interaction force given in (7.3.8), repeated below for convenience,

$$\boldsymbol{\phi}_R = \kappa\,\rho\,\mathcal{D}\,\nabla\omega_R, \tag{7.10.16}$$

we obtain in the absence of species conversion

$$\nabla\cdot\boldsymbol{\sigma}_R = -\rho\,\mathcal{D}^2\,\nabla\cdot\boldsymbol{\Lambda}_R - \rho_R\,\mathbf{g} - \iota_R\,\mathbf{b}_R - \kappa\,\rho\,\mathcal{D}\nabla\omega_R. \tag{7.10.17}$$

In the absence of significant body forces, we integrate this equation and obtain

$$\boldsymbol{\sigma}_R = -\rho\,\mathcal{D}^2\,\boldsymbol{\Lambda}_R - \tfrac{1}{2}\,\kappa\,\rho\,\mathcal{D}\,(\omega_R - \omega_B)\,\mathbf{I} + \boldsymbol{\pi}_R, \tag{7.10.18}$$

where $\mathbf{I}$ is the identity matrix and $\boldsymbol{\pi}_R$ is a solenoidal tensor field,

$$\nabla\cdot\boldsymbol{\pi}_R = \mathbf{0}. \tag{7.10.19}$$

Similar equations can be derived for the blue species without any approximation. The counterpart of Eq. (7.10.18) is

$$\boldsymbol{\sigma}_{\rm B} = -\rho \mathcal{D}_{\rm RB}^2 \,\boldsymbol{\Lambda}_{\rm B} - \tfrac{1}{2}\,\kappa\,\rho\, D_{\rm RB}\,(\omega_{\rm B} - \omega_{\rm R})\,\mathbf{I} + \boldsymbol{\pi}_{\rm B},$$

where

$$\boldsymbol{\Lambda}_{\rm B} \equiv \boldsymbol{\nabla} \otimes \boldsymbol{\nabla}\omega_{\rm B} - \boldsymbol{\nabla}\omega_{\rm B} \otimes \boldsymbol{\nabla} \ln \omega_{\rm B} = \omega_{\rm B}\, \boldsymbol{\nabla} \otimes \boldsymbol{\nabla} \ln \omega_{\rm B}, \tag{7.10.20}$$

and $\boldsymbol{\pi}_{\rm B}$ is a solenoidal tensor field,

$$\boldsymbol{\nabla} \cdot \boldsymbol{\pi}_{\rm B} = \mathbf{0}. \tag{7.10.21}$$

7.10.3 Mixture Stress Tensor

Now adding the expressions for the red and blue stresses, and using expressions (7.10.15) and (7.10.20) to write

$$\boldsymbol{\Lambda}_{\rm R} + \boldsymbol{\Lambda}_{\rm B} = (\frac{1}{\omega_{\rm R}} + \frac{1}{\omega_{\rm B}})\, \boldsymbol{\nabla}\omega_{\rm R} \otimes \boldsymbol{\nabla}\omega_{\rm R}, \tag{7.10.22}$$

we obtain

$$\boldsymbol{\sigma}_{\rm R} + \boldsymbol{\sigma}_{\rm B} = \rho\, \mathcal{D}^2\, (\frac{1}{\omega_{\rm R}} + \frac{1}{\omega_{\rm B}})\, \boldsymbol{\nabla}\omega_{\rm R} \otimes \boldsymbol{\nabla}\omega_{\rm R} + \boldsymbol{\pi}_{\rm R} + \boldsymbol{\pi}_{\rm B}, \tag{7.10.23}$$

where $\iota = \iota_{\rm R} + \iota_{\rm B}$ and $\iota \mathbf{b} = \iota_{\rm R}\mathbf{b}_{\rm R} + \iota_{\rm B} + \mathbf{b}_B$. Substituting this sum into the right-hand side of (7.9.14), repeated below for convenience,

$$\boldsymbol{\sigma} = \boldsymbol{\sigma}_{\rm R} + \boldsymbol{\sigma}_{\rm B} - (\frac{1}{\rho_{\rm R}} + \frac{1}{\rho_{\rm B}})\, \boldsymbol{j}_{\rm R} \otimes \boldsymbol{j}_{\rm R}, \tag{7.10.24}$$

we obtain

$$\boldsymbol{\pi}_{\rm R} + \boldsymbol{\pi}_{\rm B}. \tag{7.10.25}$$

Under normal circumstances, we expect that

$$\boldsymbol{\pi}_{\rm R} + \boldsymbol{\pi}_{\rm B} = -\pi_0\, \mathbf{I}, \tag{7.10.26}$$

where π_0 is a constant, yielding a uniform pressure distribution in a macroscopically stationary mixture.

7.10.4 One-Dimensional Self-diffusion

In the case of one-dimensional self-diffusion along the x axis, Eq. (7.10.18) yields

$$(\sigma_{\rm R})_{xx} = -\rho\, \mathcal{D}^2 \left(\omega_{\rm R} \frac{{\rm d}^2 \ln \omega_{\rm R}}{{\rm d}x^2} + \tfrac{1}{2} \frac{\kappa}{\mathcal{D}} (\omega_{\rm R} - \omega_{\rm B}) \right) + \pi_{\rm R}, \tag{7.10.27}$$

where $\pi_{\rm R}$ is a constant. Rearranging the term inside the tall parentheses, we obtain

$$(\sigma_R)_{xx} = -\rho\,\mathcal{D}^2\left(\frac{d^2\omega_R}{dx^2} - \frac{1}{\omega_R}\frac{d\omega_R}{dx} + \frac{\kappa}{\mathcal{D}}(\omega_R - \tfrac{1}{2})\right) + \pi_R. \tag{7.10.28}$$

This is the normal stress distribution acting on the vertical white lines shown in Fig. 4.6.3a, b.

Equation (7.10.20) provides us with a corresponding expression for the blue species,

$$(\sigma_B)_{xx} = -\rho\,\mathcal{D}^2\left(\omega_B\frac{d^2\ln\omega_B}{dx^2} + \tfrac{1}{2}\frac{\kappa}{D_{RB}}(\omega_B - \omega_R)\right) + \pi_B, \tag{7.10.29}$$

where π_B is a constant. This is the normal stresses acting on the vertical yellow lines shown in Fig. 4.6.3c, d.

Adding Eqs. (7.10.27) and (7.10.29), we obtain

$$(\sigma_R)_{xx} + (\sigma_B)_{xx} = -\rho\,\mathcal{D}^2\left(\frac{1}{\omega_R} + \frac{1}{\omega_B}\right)\left(\frac{d\,\omega_R}{dx}\right)^2 + \pi_R + \pi_B, \tag{7.10.30}$$

which illustrates that the sum of the normal stresses is not constant, even though the normal stress of the mixture is constant,

$$\sigma_{xx} = \pi_R + \pi_B. \tag{7.10.31}$$

An example of a red species stress distribution for a given red mass fraction distribution

$$\omega_R = \tfrac{1}{2} - \frac{1}{\pi}\arctan\frac{x}{a} \tag{7.10.32}$$

is shown in Fig. 7.1 for $\kappa a^2/\mathcal{D} = 0$ and 0.5, where a is a reference length and $p_R \equiv -\sigma_R$ is an effective red pressure. The asymmetry of the stress distribution can be contrasted with the symmetry of the mass fraction distribution profile.

7.10.5 Exercise

7.10.1 Generate the counterpart of Fig. 7.1 for another sensible distribution of ω_R of your choice.

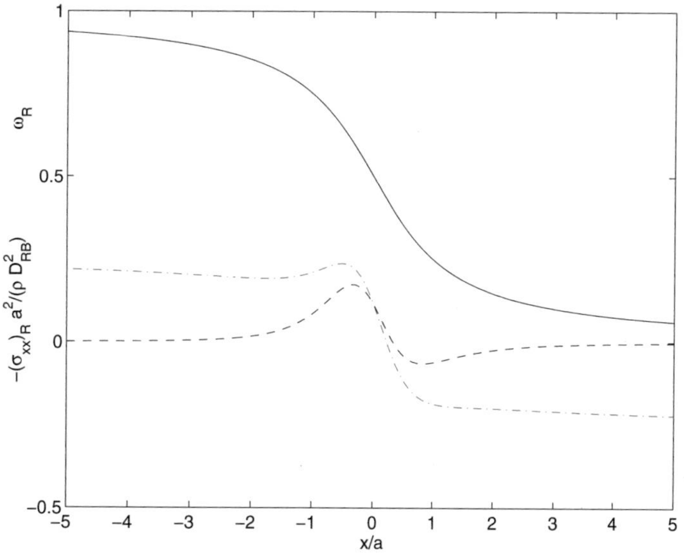

Fig. 7.1 The solid line shows the distribution of the red species mass fraction described by $\omega_R = \frac{1}{2} - \frac{1}{\pi}\arctan(x/a)$. The associated distributions of $-\sigma_R$ scaled by $\rho\mathcal{D}^2/a^2$ in one-dimensional self-diffusion is shown with the dashed line for $\kappa a^2/\mathcal{D} = 0$ and with the dot-dashed line for 0.5

Mixture Energy Balances

8

The total energy of a fluid parcel is comprised of the (a) potential energy due to an external, universal, or species-specific body force field, (b) kinetic energy due to the motion of the fluid, and (c) internal energy defined in the framework of thermodynamics, as discussed in Appendix B.

In Chap. 3, we derived expressions for rate of change of individual and combined energies for a homogeneous fluid by combining Cauchy's equation of motion with the laws of thermodynamics. In this chapter, we extend the procedures to mixtures, inevitably making certain steps that require weak or strong assumptions and an occasional leap of faith.

8.1 Kinetic Energy Balance

The red species equation of motion shown in (7.5.3) is repeated below for convenience,

$$\rho_R \left(\frac{\partial \mathbf{u}_R}{\partial t} + \mathbf{u}_R \cdot \nabla \mathbf{u}_R \right) = \nabla \cdot \boldsymbol{\sigma}_R + \rho_R \, \mathbf{g} + \iota_R \, \mathbf{b}_R + \boldsymbol{\phi}_R + \boldsymbol{\eta}_R, \tag{8.1.1}$$

where $\mathbf{g}$ is the gravitational acceleration, $\mathbf{b}_R$ is a red species body force associated with the physical constant ι_R, $\boldsymbol{\phi}_R$ is the red–blue interaction force, and $\boldsymbol{\eta}_R = \dot{r}_{BR} \, (\mathbf{u}_B - \mathbf{u}_R)$ expresses the effect of momentum transfer due to a chemical reaction.

An evolution equation for the kinetic energy of the red species can be derived by projecting equation (8.1.1) onto the red velocity, $\mathbf{u}_R$, at a point, obtaining

$$\frac{1}{2} \rho_R \left(\frac{\partial \, |\mathbf{u}_R|^2}{\partial t} + \mathbf{u}_R \cdot \nabla |\mathbf{u}_R|^2 \right) = \mathbf{u}_R \cdot (\nabla \cdot \boldsymbol{\sigma}_R) + \rho_R \, \mathbf{g} \cdot \mathbf{u}_R + \iota_R \, \mathbf{b}_R \cdot \mathbf{u}_R + \boldsymbol{\phi}_R \cdot \mathbf{u}_R + \boldsymbol{\eta}_R \cdot \mathbf{u}_R, \tag{8.1.2}$$

where

$$|\mathbf{u}_R|^2 = \mathbf{u}_R \cdot \mathbf{u}_R \tag{8.1.3}$$

is the square of the magnitude of the red-species velocity. Comparing Eq. (8.1.2) with Eq. (2.9.3) for a homogeneous fluid, we note the presence of two additional terms on the right-hand side.

C. Pozrikidis, *Transport Processes Primer*,
https://doi.org/10.1007/978-1-4939-9909-5_8

Now combining equation (8.1.2) with the red-species density evolution equation (7.5.2), repeated below for convenience,

$$\frac{\partial \rho_R}{\partial t} + \nabla \cdot (\rho_R \mathbf{u}_R) = \dot{r}_R, \tag{8.1.4}$$

we express the left-hand side in conservative form, where $\dot{r}_R$ is the rate of red mass production due to a chemical reaction. Also resolving the first term on the right-hand side of (8.1.2) into two parts, we obtain the red-species kinetic energy balance

$$\begin{aligned} \frac{\partial}{\partial t}\big(\tfrac{1}{2}\, \rho_R\, |\mathbf{u}_R|^2 \big) + \nabla \cdot \big(\tfrac{1}{2}\, \rho_R\, |\mathbf{u}_R|^2\, \mathbf{u}_R \big) &= \nabla \cdot (\boldsymbol{\sigma}_R \cdot \mathbf{u}_R) \\ &- \boldsymbol{\sigma}_R : \nabla \mathbf{u}_R + (\rho_R\, \mathbf{g} + \iota_R\, \mathbf{b}_R + \boldsymbol{\phi}_R + \mathbf{w}_R) \cdot \mathbf{u}_R, \end{aligned} \tag{8.1.5}$$

where

$$\mathbf{w}_R \equiv \frac{1}{2}\, \dot{r}_R\, \mathbf{u}_R + \boldsymbol{\eta}_R. \tag{8.1.6}$$

The first term on the right-hand side of the expression for $\mathbf{w}_R$ originates in the species continuity equation (8.1.4).

Setting $\boldsymbol{\eta}_R = \dot{r}_{BR}\, (\mathbf{u}_B - \mathbf{u}_R)$ and $\dot{r}_R = \dot{r}_{BR} - \dot{r}_{RB}$, we obtain the expression

$$\mathbf{w}_R = -\frac{1}{2}\, (\dot{r}_{BR} + \dot{r}_{RB})\, \mathbf{u}_R + \dot{r}_{BR}\, \mathbf{u}_B, \tag{8.1.7}$$

which conveys the compound effect of a chemical reaction. At chemical reaction equilibrium, $\dot{r}_{BR} = \dot{r}_{RB}$, and thus $\mathbf{w}_R = \dot{r}_{BR}\, (\mathbf{u}_B - \mathbf{u}_R)$.

No approximations or additional assumptions are made in the derivation of the preceding equations.

8.1.1 Effective Species Force

To simplify the notation, we define an effective species force,

$$\boldsymbol{\xi}_R \equiv \boldsymbol{\phi}_R + \mathbf{w}_R, \tag{8.1.8}$$

and combine the last two terms on the right-hand side of (8.1.5) to obtain

$$\begin{aligned} \frac{\partial}{\partial t}\big(\tfrac{1}{2}\, \rho_R\, |\mathbf{u}_R|^2 \big) + \nabla \cdot \big(\tfrac{1}{2}\, \rho_R\, |\mathbf{u}_R|^2\, \mathbf{u}_R \big) &= \nabla \cdot (\boldsymbol{\sigma}_R \cdot \mathbf{u}_R) \\ &- \boldsymbol{\sigma}_R : \nabla \mathbf{u}_R + (\, \rho_R\, \mathbf{g} + \iota_R\, \mathbf{b}_R + \boldsymbol{\xi}_R\,) \cdot \mathbf{u}_R. \end{aligned} \tag{8.1.9}$$

The three terms inside the last set of parentheses on the right-hand side express the effect of an effective volume force.

Similar equations can be written for the blue species or any species in a multicomponent solution. Equation (8.1.9) and its counterpart for the blue species will be invoked in Sect. 8.3 to derive an evolution equation for the species internal energy by way of the first law of thermodynamics.

8.1.2 Mixture Kinetic Energy

We may assume that the kinetic energy of the mixture is governed by Eq. (2.9.10) for an effectively homogeneous fluid, repeated below for convenience,

$$\frac{\partial}{\partial t}\left(\tfrac{1}{2}\,\rho\,|\mathbf{u}|^2\right)+\nabla\cdot\left(\tfrac{1}{2}\,\rho\,|\mathbf{u}|^2\,\mathbf{u}\right)=\nabla\cdot(\boldsymbol{\sigma}\cdot\mathbf{u})-\boldsymbol{\sigma}:\nabla\mathbf{u}+(\rho\,\mathbf{g}+\iota\,\mathbf{b})\cdot\mathbf{u}, \tag{8.1.10}$$

where $\iota\mathbf{b} \equiv \iota_R\mathbf{b}_R + \iota_B\mathbf{b}_B$ represents the combined effect of the species-specific body force. Equation (8.1.10) will be invoked in Sect. 8.3 to derive an evolution equation for the mixture internal energy.

8.1.3 Exercise

8.1.1 Express (8.1.5) in Lagrangian form involving the red material derivative of $\frac{1}{2}\rho_R\,|\mathbf{u}_R|^2$.

8.2 Kinetic–Internal Energy Balance

We may introduce a specific internal energy associated with the red species, u_R, and define the specific red kinetic–internal energy.

$$\varepsilon_R \equiv \frac{1}{2}\,|\mathbf{u}_R|^2 + u_R. \tag{8.2.1}$$

In the present context, the internal energy, u_R, includes the effect of chemical formation.

The evolution of ε for a homogeneous fluid is governed by Eq. (3.5.4), repeated below for convenience,

$$\frac{\partial(\rho\,\varepsilon)}{\partial t}+\nabla\cdot\big(\rho\,\varepsilon\,\mathbf{u}\big)=-\nabla\cdot\mathbf{q}+\nabla\cdot(\boldsymbol{\sigma}\cdot\mathbf{u})+(\rho\,\mathbf{g}+\iota\,\mathbf{b})\cdot\mathbf{u}+\dot{q}, \tag{8.2.2}$$

where $\dot{q}$ is a volumetric rate of interior energy generation. Taking a leap of faith, we write the counterpart of this equation for the red species including the effect of the species-specific body force,

$$\frac{\partial(\rho_R\,\varepsilon_R)}{\partial t}+\nabla\cdot\big(\rho_R\,\varepsilon_R\,\mathbf{u}_R\big)=-\nabla\cdot\mathbf{q}_R+\nabla\cdot(\boldsymbol{\sigma}_R\cdot\mathbf{u}_R)+(\rho_R\,\mathbf{g}+\iota_R\,\mathbf{b}_R)\cdot\mathbf{u}_R+\dot{\varepsilon}_R, \tag{8.2.3}$$

where $\mathbf{q}_R$ is a heat flux and $\dot{\varepsilon}_R$ is the volumetric rate of interior energy production associated with the red species. A similar equation can be written for the blue species or any species in a multicomponent solution.

8.2.1 Species Kinetic–Internal Energy in Terms of Diffusion Fluxes

In preparation for developing mixture from species equations, we rearrange the second term on the left-hand side of (8.2.3), and also the second and penultimate terms on the right-hand side to introduce the mixture velocity, $\mathbf{u}$, and red mass flux, $\boldsymbol{j}_R$, obtaining

$$\frac{\partial(\rho_R\,\varepsilon_R)}{\partial t} + \nabla\cdot\big(\rho_R\,\varepsilon_R\,\mathbf{u}\big) = -\nabla\cdot\big(\varepsilon_R\,\boldsymbol{j}_R\big) - \nabla\cdot\mathbf{q}_R$$
$$+\nabla\cdot(\frac{1}{\rho_R}\boldsymbol{\sigma}_R\cdot\boldsymbol{j}_R) + \nabla\cdot(\boldsymbol{\sigma}_R\cdot\mathbf{u}) + \rho_R\,\mathbf{g}\cdot\mathbf{u}_R + \frac{\iota_R}{\rho_R}\mathbf{b}_R\cdot\boldsymbol{j}_R + \iota_R\,\mathbf{b}_R\cdot\mathbf{u} + \dot{\varepsilon}_R. \tag{8.2.4}$$

Three velocities are involved in this equation: the mixture velocity, $\mathbf{u}$, the red velocity, $\mathbf{u}_R$, and the red diffusion velocity, $\boldsymbol{j}_R/\rho_R$. A similar equation can be written for the blue species or any species in a multicomponent mixture.

8.2.2 Mixture-Specific Kinetic–Internal Energy

Now we define the specific kinetic–internal energy of the mixture, ε, as a weighted average,

$$\varepsilon \equiv \omega_R\varepsilon_R + \omega_B\varepsilon_B, \tag{8.2.5}$$

so that

$$\rho\,\varepsilon \equiv \rho_R\varepsilon_R + \rho_B\varepsilon_B. \tag{8.2.6}$$

This definition is analogous to that of the mass velocity. We recall from Eq. (5.9.9) that

$$\rho_R|\mathbf{u}_R|^2 + \rho_B|\mathbf{u}_B|^2 = \rho\,|\mathbf{u}|^2 + \frac{1}{\rho_R}\,|\boldsymbol{j}_R|^2 + \frac{1}{\rho_B}\,|\boldsymbol{j}_B|^2, \tag{8.2.7}$$

and obtain

$$\rho\,\varepsilon = \tfrac{1}{2}\,\rho\,|\mathbf{u}|^2 + (\rho_R u_R + \tfrac{1}{2}\,\frac{1}{\rho_R}\,|\boldsymbol{j}_R|^2) + (\rho_B u_B + \tfrac{1}{2}\,\frac{1}{\rho_B}\,|\boldsymbol{j}_B|^2). \tag{8.2.8}$$

The sum on the right-hand side can be extended to any number of species in a multicomponent solution,

$$\varepsilon \equiv \sum_X \omega_X\varepsilon_X. \tag{8.2.9}$$

Making substitutions, we find that

$$\varepsilon = \tfrac{1}{2}|\mathbf{u}|^2 + \sum_X \omega_X\big(u_X + \tfrac{1}{2}\,\frac{1}{\rho_X^2}\,|\boldsymbol{j}_X|^2\big), \tag{8.2.10}$$

where $\omega_X \equiv \rho_X/\rho$ is the species mass fraction.

8.2.3 Interspecies Energy Transfer

Energy conservation requires that the interspecies volumetric rates of interior energy generation add to zero,

$$\dot{\varepsilon}_R + \dot{\varepsilon}_B = 0, \tag{8.2.11}$$

where the sum can be extended to an arbitrary number of species. In fact, this constraint is embedded into the definition of the species internal energies, u_R and u_B, which include the energy of chemical formation.

8.2.4 From Species to Mixture

Adding the red evolution equation (8.2.4) and its counterpart for the blue species, we obtain

$$\begin{aligned}\frac{\partial(\rho\,\varepsilon)}{\partial t} + \nabla\cdot\big(\rho\,\varepsilon\,\mathbf{u}\big) &= -\nabla\cdot\big(\varepsilon_R\,\boldsymbol{j}_R + \varepsilon_B\,\boldsymbol{j}_B\big) - \nabla\cdot(\mathbf{q}_R+\mathbf{q}_B)\\ &+ \nabla\cdot(\frac{1}{\rho_R}\boldsymbol{\sigma}_R\cdot\boldsymbol{j}_R + \frac{1}{\rho_B}\boldsymbol{\sigma}_B\cdot\boldsymbol{j}_B) + \nabla\cdot\big((\boldsymbol{\sigma}_R+\boldsymbol{\sigma}_B)\cdot\mathbf{u}\big)\\ &+ \frac{\iota_R}{\rho_R}\mathbf{b}_R\cdot\boldsymbol{j}_R + \frac{\iota_B}{\rho_B}\mathbf{b}_B\cdot\boldsymbol{j}_B + (\rho\,\mathbf{g}+\iota\mathbf{b})\cdot\mathbf{u},\end{aligned} \tag{8.2.12}$$

where $\iota\mathbf{b} \equiv \iota_R\,\mathbf{b}_R + \iota_B\,\mathbf{b}_B$. Substituting into the fourth term on the right-hand side the expression

$$\boldsymbol{\sigma}_R + \boldsymbol{\sigma}_B = \boldsymbol{\sigma} + \frac{1}{\rho_R}\boldsymbol{j}_R\otimes\boldsymbol{j}_R + \frac{1}{\rho_B}\boldsymbol{j}_B\otimes\boldsymbol{j}_B, \tag{8.2.13}$$

and rearranging, we obtain

$$\begin{aligned}\frac{\partial(\rho\,\varepsilon)}{\partial t} + \nabla\cdot\big(\rho\,\varepsilon\,\mathbf{u}\big) &= -\nabla\cdot\big(\varepsilon_R\,\boldsymbol{j}_R + \varepsilon_B\,\boldsymbol{j}_B\big) - \nabla\cdot(\mathbf{q}_R+\mathbf{q}_B)\\ &+ \nabla\cdot(\frac{1}{\rho_R}\boldsymbol{\sigma}_R\cdot\boldsymbol{j}_R + \frac{1}{\rho_B}\boldsymbol{\sigma}_B\cdot\boldsymbol{j}_B) + \nabla\cdot\big(\boldsymbol{\sigma}\cdot\mathbf{u}\big)\\ &+ \nabla\cdot\Big(\big(\frac{1}{\rho_R}\boldsymbol{j}_R\otimes\boldsymbol{j}_R + \frac{1}{\rho_B}\boldsymbol{j}_B\otimes\boldsymbol{j}_B\big)\cdot\mathbf{u}\Big)\\ &+ \frac{\iota_R}{\rho_R}\mathbf{b}_R\cdot\boldsymbol{j}_R + \frac{\iota_B}{\omega_B}\mathbf{b}_B\cdot\boldsymbol{j}_B + (\rho\,\mathbf{g}+\iota\,\mathbf{b})\cdot\mathbf{u}.\end{aligned} \tag{8.2.14}$$

The first and fifth terms on the right-hand side can be combined into the negative of a divergence,

$$\nabla\cdot\big(\varepsilon_R\,\boldsymbol{j}_R + \varepsilon_B\,\boldsymbol{j}_B\big) + \nabla\cdot\Big(\big(\frac{1}{\rho_R}\boldsymbol{j}_R\otimes\boldsymbol{j}_R + \frac{1}{\rho_B}\boldsymbol{j}_B\otimes\boldsymbol{j}_B\big)\cdot\mathbf{u}\Big) \equiv \nabla\cdot\boldsymbol{\zeta}, \tag{8.2.15}$$

where

$$\boldsymbol{\zeta} \equiv (\varepsilon_R - \frac{1}{\rho_R}\boldsymbol{j}_R\cdot\mathbf{u})\,\boldsymbol{j}_R + (\varepsilon_B - \frac{1}{\rho_B}\boldsymbol{j}_R\cdot\mathbf{u})\,\boldsymbol{j}_B. \tag{8.2.16}$$

Substituting the definitions of ε_R and ε_B, we obtain

$$\zeta = (u_R + \tfrac{1}{2}\,|\mathbf{u}_R|^2 - \frac{1}{\rho_R}\,\boldsymbol{j}_R \cdot \mathbf{u})\,\boldsymbol{j}_R + (u_B + \tfrac{1}{2}\,|\mathbf{u}_B|^2 - \frac{1}{\rho_B}\,\boldsymbol{j}_B \cdot \mathbf{u})\,\boldsymbol{j}_B. \tag{8.2.17}$$

Now we write

$$|\mathbf{u}_R|^2 = \left|\mathbf{u} + \frac{1}{\rho_R}\,\boldsymbol{j}_R\right|^2 = |\mathbf{u}|^2 + \frac{2}{\rho_R}\,\boldsymbol{j}_R \cdot \mathbf{u} + \frac{1}{\rho_R^2}\,|\boldsymbol{j}_R|, \tag{8.2.18}$$

and a corresponding equation for the blue species, recall that $\boldsymbol{j}_R + \boldsymbol{j}_B = \underline{0}$, and simplify to obtain

$$\zeta = (u_R + \tfrac{1}{2}\,\frac{1}{\rho_R^2}|\,\boldsymbol{j}_R|^2)\,\boldsymbol{j}_R + (u_B + \tfrac{1}{2}\,\frac{1}{\rho_B^2}|\,\boldsymbol{j}_B|^2)\,\boldsymbol{j}_B. \tag{8.2.19}$$

The sum on the right-hand side can be extended to an arbitrary number of species.

8.2.5 Summary

In summary, the evolution equation for a multicomponent mixture takes the form

$$\begin{aligned}\frac{\partial(\rho\,\varepsilon)}{\partial t} + \nabla\cdot(\rho\,\varepsilon\,\mathbf{u}) &= -\nabla\cdot\Big(\sum_X \mathcal{E}_X\,\boldsymbol{j}_X\Big) - \nabla\cdot\Big(\sum_X \mathbf{q}_X\Big)\\ &+ \nabla\cdot\Big(\sum_X \frac{1}{\rho_X}\,\boldsymbol{\sigma}_X\cdot\boldsymbol{j}_X\Big) + \nabla\cdot(\boldsymbol{\sigma}\cdot\mathbf{u}) + \dot{\mathcal{E}}_b + (\rho\,\mathbf{g} + \iota\,\mathbf{b})\cdot\mathbf{u}.\end{aligned} \tag{8.2.20}$$

The coefficient $\mathcal{E}_X$ in the first term on the right-hand side expresses the species internal–diffusion velocity specific energy,

$$\mathcal{E}_X \equiv u_X + \tfrac{1}{2}\,\frac{1}{\rho_X^2}|\,\boldsymbol{j}_X|^2, \tag{8.2.21}$$

the term

$$\dot{\mathcal{E}}_b \equiv \sum_X \frac{\iota_X}{\rho_X}\,\mathbf{b}_X\cdot\boldsymbol{j}_X \tag{8.2.22}$$

is the rate of change of a specific potential energy associated with the species-specific body force, and we have defined

$$\iota\mathbf{b} = \sum \iota_X\mathbf{b}_X \tag{8.2.23}$$

in the last term on the right-hand side.

8.2.6 Exercise

8.2.1 Derive the expression for ζ given in (8.2.16).

8.3 Total Energy Flux

To simplify the notation, we express the mixture kinetic–internal energy evolution equation (8.2.20) for an arbitrary number of species in the form

$$\frac{\partial(\rho\,\varepsilon)}{\partial t} + \nabla\cdot\big(\rho\,\varepsilon\,\mathbf{u}\big) = -\nabla\cdot\mathbf{q}^{\rm tot} + \nabla\cdot\big(\boldsymbol{\sigma}\cdot\mathbf{u}\big) + \dot{\mathcal{E}}_b + (\rho\,\mathbf{g} + \iota\,\mathbf{b})\cdot\mathbf{u}, \tag{8.3.1}$$

where

$$\mathbf{q}^{\rm tot} \equiv \sum_{\rm X} \mathbf{q}_{\rm X}^{\rm tot} \tag{8.3.2}$$

is the mixture total energy flux and

$$\mathbf{q}_{\rm X}^{\rm tot} \equiv \mathbf{q}_{\rm X} - \frac{1}{\rho_{\rm X}}\,\boldsymbol{\sigma}_{\rm X}\cdot\boldsymbol{j}_{\rm X} + \mathcal{E}_{\rm X}\,\boldsymbol{j}_{\rm X} \tag{8.3.3}$$

is the corresponding species flux.

Using the continuity equation for the mixture to simplify the left-hand side, we obtain the associated Lagrangian form

$$\rho\,\frac{\mathrm{D}\varepsilon}{\mathrm{D}t} = -\nabla\cdot\mathbf{q}^{\rm tot} + \nabla\cdot\big(\boldsymbol{\sigma}\cdot\mathbf{u}\big) + \dot{\mathcal{E}}_b + (\rho\,\mathbf{g} + \iota\,\mathbf{b})\cdot\mathbf{u}, \tag{8.3.4}$$

where $\mathrm{D}/\mathrm{D}t$ is the material derivative defined with respect to the mixture velocity, $\mathbf{u}$.

8.3.1 Definitions

Further notational simplification can be achieved by writing

$$\mathbf{q}^{\rm tot} = \mathbf{q} + \sum_{\rm X} \mathbf{h}_{\rm X}\cdot\boldsymbol{j}_{\rm X}, \tag{8.3.5}$$

where

$$\mathbf{q} \equiv \sum_{\rm X} \mathbf{q}_{\rm X} \tag{8.3.6}$$

is the total heat flux,

$$\mathbf{h}_{\rm X} \equiv -\frac{1}{\rho_{\rm X}}\,\boldsymbol{\sigma}_{\rm X} + \mathcal{E}_{\rm X}\,\mathbf{I}, \tag{8.3.7}$$

is a composite tensor defined with respect to the species stress tensor, and $\mathbf{I}$ is the identity matrix.

8.3.2 Inviscid Species Tensors

In the event that the species stress tensor is inviscid, we set

$$\boldsymbol{\sigma}_X = -p_X\,\mathbf{I}, \tag{8.3.8}$$

where p_X is the species pressure, we obtain

$$\mathbf{h}_X = (\frac{p_X}{\rho_X} + \mathcal{E}_X)\,\mathbf{I}. \tag{8.3.9}$$

Substituting the definition of $\mathcal{E}_X$, we obtain

$$\mathbf{h}_X = (\frac{p_X}{\rho_X} + u_X + \tfrac{1}{2}\,\frac{1}{\rho_X^2}|\,\boldsymbol{j}_X|^2)\,\mathbf{I}, \tag{8.3.10}$$

and thus

$$\mathbf{h}_X = (h_X + \tfrac{1}{2}\,\frac{1}{\rho_X^2}|\,\boldsymbol{j}_X|^2)\,\mathbf{I}, \tag{8.3.11}$$

where

$$h_X \equiv u_X + \frac{p_X}{\rho_X} \tag{8.3.12}$$

is the species-specific enthalpy.

Substituting (8.3.11) into (8.3.2), we derive an expression for the total mixture flux,

$$\mathbf{q}^{\text{tot}} = \mathbf{q} + \sum_X (h_X + \tfrac{1}{2}\,\frac{1}{\rho_X^2}|\,\boldsymbol{j}_X|^2)\,\boldsymbol{j}_X. \tag{8.3.13}$$

When the effect of diffusion is weak, the summed term involving the mass flux, $\boldsymbol{j}_X$, can be neglected as an approximation, yielding

$$\mathbf{q}^{\text{tot}} \simeq \mathbf{q} + \sum_X h_X\,\boldsymbol{j}_X. \tag{8.3.14}$$

The sum represents the contribution of an inviscid species enthalpy flux.

The assumptions that led us to expression (8.3.13) for the total mixture energy flux are that

$$\varepsilon = \sum_X \omega_X \varepsilon_X, \tag{8.3.15}$$

and the species stress tensors are inviscid.

8.3.3 Exercise

8.3.1 Discuss whether the approximation (8.3.14) is reasonable in the case of self-diffusion.

8.4 Internal-Energy Balance

The specific kinetic–internal energy of the mixture, ε, was given in Eq. (8.2.10), repeated below for convenience,

$$\varepsilon = \tfrac{1}{2}|\mathbf{u}|^2 + \sum_{\mathrm{X}} \omega_{\mathrm{X}} \left(u_{\mathrm{X}} + \tfrac{1}{2} \frac{1}{\rho_{\mathrm{X}}^2} |\boldsymbol{j}_{\mathrm{X}}|^2 \right). \tag{8.4.1}$$

In the context of the thermodynamics of irreversible processes, the sum on the right-hand side is the specific internal energy of the mixture,

$$u \equiv \sum_{\mathrm{X}} \omega_{\mathrm{X}} \left(u_{\mathrm{X}} + \tfrac{1}{2} \frac{1}{\rho_{\mathrm{X}}^2} |\boldsymbol{j}_{\mathrm{X}}|^2 \right). \tag{8.4.2}$$

Note that u is *not* a weighted average of the specific internal energies of the species, u_{X}. In fact, we will see that the weighted average is the thermal part of the specific internal energy of the mixture.

As a consequence of the preceding definitions,

$$\varepsilon = \tfrac{1}{2}\, |\mathbf{u}|^2 + u. \tag{8.4.3}$$

Note that the specific internal energy of the mixture, u, is defined with respect to the mixture velocity, $\mathbf{u}$.

8.4.1 Thermal Part of the Mixture Internal Energy

The definition (8.4.2) reveals that the specific internal energy of the mixture, u, consists of a thermal part associated with the species internal energies, u_{X}, and a kinetic part due to the species diffusion fluxes. The thermal part of the mixture internal energy is a weighted average of the species internal energies,

$$u_{\text{thermal}} = \sum_{\mathrm{X}} \omega_{\mathrm{X}} u_{\mathrm{X}}. \tag{8.4.4}$$

As a consequence of this definition,

$$u \equiv u_{\text{thermal}} + \tfrac{1}{2} \sum_{\mathrm{X}} \omega_{\mathrm{X}} \frac{1}{\rho_{\mathrm{X}}^2} |\boldsymbol{j}_{\mathrm{X}}|^2, \tag{8.4.5}$$

and thus

$$\varepsilon = \tfrac{1}{2}\, |\mathbf{u}|^2 + u_{\text{thermal}} + \tfrac{1}{2} \sum_{\mathrm{X}} \omega_{\mathrm{X}} \frac{1}{\rho_{\mathrm{X}}^2} |\boldsymbol{j}_{\mathrm{X}}|^2. \tag{8.4.6}$$

Evolution equations for u and u_{thermal} will be derived.

8.4.2 Mixture Internal Energy

From a macroscopic point of view, the kinetic energy of a mixture is governed by Eq. (8.1.10), repeated below for convenience,

$$\frac{\partial}{\partial t}\big(\tfrac{1}{2}\,\rho\,|\mathbf{u}|^2\big)+\nabla\cdot\big(\tfrac{1}{2}\,\rho\,|\mathbf{u}|^2\,\mathbf{u}\big)=\nabla\cdot(\boldsymbol{\sigma}\cdot\mathbf{u})-\boldsymbol{\sigma}:\nabla\mathbf{u}+\rho\,\mathbf{g}\cdot\mathbf{u}+\iota\mathbf{b}\cdot\mathbf{u}, \tag{8.4.7}$$

where $\iota\mathbf{b}\equiv\iota_{\mathrm{R}}\,\mathbf{b}_{\mathrm{R}}+\iota_{\mathrm{B}}\,\mathbf{b}_{\mathrm{B}}$. Subtracting this equation from Eq. (8.3.1) governing the combined kinetic–internal energy, we obtain an evolution equation for u,

$$\frac{\partial(\rho\,u)}{\partial t}+\nabla\cdot\big(\rho\,u\,\mathbf{u}\big)=-\nabla\cdot\mathbf{q}^{\mathrm{tot}}+\boldsymbol{\sigma}:\nabla\mathbf{u}+\dot{\mathcal{E}}_b, \tag{8.4.8}$$

where $\mathbf{q}^{\mathrm{tot}}$ is defined in (8.3.2) and $\dot{\mathcal{E}}_b$ is defined in (8.2.22).

The corresponding Lagrangian form involving the mixture material derivative, $\mathrm{D}/\mathrm{D}t$, follows by using the continuity equation for the mixture to simplify the left-hand side, obtaining

$$\rho\,\frac{\mathrm{D}u}{\mathrm{D}t}=-\nabla\cdot\mathbf{q}^{\mathrm{tot}}+\boldsymbol{\sigma}:\nabla\mathbf{u}+\dot{\mathcal{E}}_b \tag{8.4.9}$$

for a compressible or incompressible mixture.

Equation (8.4.8) can be compared with Eq. (3.4.2) with $\dot{q}=0$ governing the transport of the specific internal energy of a homogeneous fluid, u, repeated below for convenience,

$$\frac{\partial(\rho u)}{\partial t}+\nabla\cdot\big(\rho\,u\,\mathbf{u}\big)=-\nabla\cdot\mathbf{q}+\boldsymbol{\sigma}:\nabla\mathbf{u}. \tag{8.4.10}$$

The comparison illustrates the effects of diffusion and chemical reaction.

8.4.3 Species Internal Energy Balance

Subtracting the species kinetic energy balance shown in Eq. (8.2.3) from the species kinetic–internal energy balance shown in Eq. (8.1.9), we obtain the species internal energy balance,

$$\frac{\partial(\rho_{\mathrm{R}}u_{\mathrm{R}})}{\partial t}+\nabla\cdot\big(\rho_{\mathrm{R}}u_{\mathrm{R}}\,\mathbf{u}_{\mathrm{R}}\big)=-\nabla\cdot\mathbf{q}_{\mathrm{R}}+\boldsymbol{\sigma}_{\mathrm{R}}:\nabla\mathbf{u}_{\mathrm{R}}-\boldsymbol{\xi}_{\mathrm{R}}\cdot\mathbf{u}_{\mathrm{R}}+\dot{\varepsilon}_{\mathrm{R}}. \tag{8.4.11}$$

Introducing the mixture velocity, $\mathbf{u}$, in the second term on the left-hand side, we obtain

$$\frac{\partial(\rho_{\mathrm{R}}u_{\mathrm{R}})}{\partial t}+\nabla\cdot\big(\rho_{\mathrm{R}}u_{\mathrm{R}}\,\mathbf{u}\big)=-\nabla\cdot\mathbf{q}_{\mathrm{R}}-\nabla\cdot\big(u_{\mathrm{R}}\boldsymbol{j}_{\mathrm{R}}\big)+\boldsymbol{\sigma}_{\mathrm{R}}:\nabla\mathbf{u}_{\mathrm{R}}-\boldsymbol{\xi}_{\mathrm{R}}\cdot\mathbf{u}_{\mathrm{R}}+\dot{\varepsilon}_{\mathrm{R}}. \tag{8.4.12}$$

A similar equation can be written for the blue species or any species in a multicomponent solution.

8.4.4 Mixture Thermal Internal Energy Balance

Adding the species balances, and recalling that $\dot{\varepsilon}_R + \dot{\varepsilon}_B = 0$, we obtain a thermal internal energy balance equation. For a multicomponent mixture, we find that

$$\frac{\partial(\rho\, u_{\text{thermal}})}{\partial t} + \nabla \cdot \big(\rho\, u_{\text{thermal}}\, \mathbf{u}\big) = -\nabla \cdot \mathbf{q} - \nabla \cdot \big(\sum_X u_X\, \boldsymbol{j}_X\big) + \sum_X (\boldsymbol{\sigma}_X : \nabla \mathbf{u}_X) - \sum_X (\boldsymbol{\xi}_X \cdot \mathbf{u}_X), \tag{8.4.13}$$

where $\mathbf{q} = \sum \mathbf{q}_X$. The penultimate term on the right-hand side represents the collective effect of the species mechanical energy loss or gain due to the interaction of the species velocity and stress fields. In terms of the mixture material derivative,

$$\rho \frac{\mathrm{D}u_{\text{thermal}}}{\mathrm{D}t} = -\nabla \cdot \mathbf{q} - \nabla \cdot \big(\sum_X u_X\, \boldsymbol{j}_X\big) + \sum_X (\boldsymbol{\sigma}_X : \nabla \mathbf{u}_X) - \sum_X (\boldsymbol{\xi}_X \cdot \mathbf{u}_X), \tag{8.4.14}$$

for incompressible or compressible mixtures.

8.4.5 Temperature Evolution

By analogy with Eq. (3.7.7) for a homogeneous fluid, we may set

$$\rho \frac{\mathrm{D}u_{\text{thermal}}}{\mathrm{D}t} = \rho\, c_v \frac{\mathrm{D}T}{\mathrm{D}t} + \Big(T \Big(\frac{\partial p}{\partial T}\Big)_v - p\Big) \nabla \cdot \mathbf{u}, \tag{8.4.15}$$

where c_v is the mixture heat capacity under constant volume. Combining this expression with (8.4.14), we obtain an evolution equation for the temperature,

$$\rho\, c_v \frac{\mathrm{D}T}{\mathrm{D}t} = -\nabla \cdot \mathbf{q} - \nabla \cdot \big(\sum_X u_X\, \boldsymbol{j}_X\big) + \sum_X (\boldsymbol{\sigma}_X : \nabla \mathbf{u}_X) + \Big(p - T \Big(\frac{\partial p}{\partial T}\Big)_v\Big) \nabla \cdot \mathbf{u} - \sum_X (\boldsymbol{\xi}_X \cdot \mathbf{u}_X). \tag{8.4.16}$$

An alternative evolution equation based on the mixture heat capacity under constant volume, c_p, will be derived in Sect. 8.5 with regard to the specific enthalpy.

8.4.6 Exercise

8.4.1 Discuss possible reasons for not learning from one's mistakes.

8.5 Enthalpy Balance

The specific enthalpy of the mixture, h, is defined with respect to the mixture density, ρ, the specific internal energy of the mixture, u, and the mixture pressure, p, by the equations

$$h = u + \frac{p}{\rho}, \qquad u = h - \frac{p}{\rho}. \tag{8.5.1}$$

Substituting the expression for u in terms of h into the evolution equation (8.4.9) for the mixture internal energy, we obtain

$$\rho \frac{\mathrm{D}h}{\mathrm{D}t} - \rho \frac{\mathrm{D}}{\mathrm{D}t}\Big(\frac{p}{\rho}\Big) = -\nabla \cdot \mathbf{q}^{\mathrm{tot}} + \boldsymbol{\sigma} : \nabla \mathbf{u} + \mathcal{E}_b, \tag{8.5.2}$$

where $\mathbf{q}^{\mathrm{tot}}$ is defined in (8.3.2) and $\mathcal{E}_b$ is defined in (8.2.22).

Working as in Sect. 3.6 for a homogeneous fluid, we obtain the evolution equation

$$\rho \frac{\mathrm{D}h}{\mathrm{D}t} = \frac{\mathrm{D}p}{\mathrm{D}t} - \nabla \cdot \mathbf{q}^{\mathrm{tot}} + \boldsymbol{\sigma}^{\mathrm{dev}} : \nabla \mathbf{u} + \mathcal{E}_b, \tag{8.5.3}$$

where $\boldsymbol{\sigma}^{\mathrm{dev}} \equiv p\,\mathbf{I} + \boldsymbol{\sigma}$ is the deviatoric part of the mixture stress tensor and

$$\mathbf{q}^{\mathrm{tot}} \simeq \mathbf{q} + \sum_{\mathrm{X}} h_{\mathrm{X}}\, \boldsymbol{j}_{\mathrm{X}}, \tag{8.5.4}$$

as shown in (8.3.14).

8.5.1 First Leap of Faith

Now taking a leap of faith, we assume that the specific enthalpy of a binary mixture is a weighted average of the specific enthalpies of the species,

$$h = \omega_{\mathrm{R}} h_{\mathrm{R}} + \omega_{\mathrm{B}} h_{\mathrm{B}}. \tag{8.5.5}$$

Taking the material derivative defined with respect to the mixture velocity, $\mathbf{u}$, we obtain

$$\frac{\mathrm{D}h}{\mathrm{D}t} = \omega_{\mathrm{R}} \frac{\mathrm{D}h_{\mathrm{R}}}{\mathrm{D}t} + \omega_{\mathrm{B}} \frac{\mathrm{D}h_{\mathrm{B}}}{\mathrm{D}t} + \frac{\mathrm{D}\omega_{\mathrm{R}}}{\mathrm{D}t} h_{\mathrm{R}} + \frac{\mathrm{D}\omega_{\mathrm{B}}}{\mathrm{D}t} h_{\mathrm{B}}. \tag{8.5.6}$$

The sum of the first two terms on the right-hand side define a weighted average of the species rates of change, denoted by

$$\dot{h} \equiv \omega_{\mathrm{R}} \frac{\mathrm{D}h_{\mathrm{R}}}{\mathrm{D}t} + \omega_{\mathrm{B}} \frac{\mathrm{D}h_{\mathrm{B}}}{\mathrm{D}t}. \tag{8.5.7}$$

Substituting into (8.5.5) the definition of the specific enthalpy in terms of the specific internal energy, we obtain

$$u = \omega_{\mathrm{R}} u_{\mathrm{R}} + \omega_{\mathrm{B}} u_{\mathrm{B}} + \frac{p_{\mathrm{R}} + p_{\mathrm{B}} - p}{\rho}. \tag{8.5.8}$$

The numerator of the fraction on the right-hand side is typically small.

To evaluate the last two terms on the right-hand side of (8.5.6), we invoke the mass fraction evolution equations (5.11.9) and (5.11.10), repeated below for convenience,

$$\rho \frac{D\omega_R}{Dt} = -\nabla \cdot \boldsymbol{j}_R + \dot{r}_R, \qquad \rho \frac{D\omega_B}{Dt} = -\nabla \cdot \boldsymbol{j}_B + \dot{r}_B, \tag{8.5.9}$$

and obtain

$$\rho \Big(\frac{D\omega_R}{Dt} h_R + \frac{D\omega_B}{Dt} h_B \Big) = -h_R \nabla \cdot \boldsymbol{j}_R - h_B \nabla \cdot \boldsymbol{j}_B + \dot{r}_R h_R + \dot{r}_B h_B. \tag{8.5.10}$$

The negative of the sum of the last two terms on the right-hand side is the volumetric rate of enthalpy generation due to a chemical reaction,

$$h_r \equiv -\dot{r}_R h_R - \dot{r}_B h_B. \tag{8.5.11}$$

In the case of an exothermic reaction, $h_r > 0$. Substituting (8.5.10) into (8.5.6) and recalling the definition (8.5.7), we find that

$$\rho \frac{Dh}{Dt} = \rho \dot{h} - h_r - h_R \nabla \cdot \boldsymbol{j}_R - h_B \nabla \cdot \boldsymbol{j}_B. \tag{8.5.12}$$

Considering the third term on the right-hand side, we write

$$h_R \nabla \cdot \boldsymbol{j}_R = \nabla \cdot (h_R \boldsymbol{j}_R) - \boldsymbol{j}_R \cdot \nabla h_R. \tag{8.5.13}$$

When the red flux is perpendicular to the gradient of the specific enthalpy of the red species, the term $\boldsymbol{j}_R \cdot \nabla h_R$ on the left-hand side is identically zero. A similar equation can be written for the blue species.

8.5.2 Second Leap of Faith

Now taking a second leap of faith, we set by analogy with Eq. (3.7.15) for a homogeneous fluid

$$\dot{h} = \rho\, c_p \frac{DT}{Dt} + (1 - \widehat{\alpha}\, T) \frac{Dp}{Dt}, \tag{8.5.14}$$

where c_p is the mixture heat capacity under constant pressure,

$$\widehat{\alpha} \equiv \frac{1}{v} \Big(\frac{\partial v}{\partial T} \Big)_p = -\frac{1}{\rho} \Big(\frac{\partial \rho}{\partial T} \Big)_p \tag{8.5.15}$$

is the mixture coefficient of thermal expansion, and $v = 1/\rho$ is the mixture specific volume.

Substituting the right-hand side of (8.5.14) for the first term on the right-hand side of (8.5.12), we obtain

$$\rho \frac{Dh}{Dt} = \rho\, c_p \frac{DT}{Dt} + (1 - \widehat{\alpha}\, T) \frac{Dp}{Dt} - h_r - h_R \nabla \cdot \boldsymbol{j}_R - h_B \nabla \cdot \boldsymbol{j}_B. \tag{8.5.16}$$

The sum of the last two terms on the right-hand side can be extended to an arbitrary number of species in a multicomponent solution.

8.5.3 Temperature Evolution

To obtain an evolution equation for the temperature, we set the right-hand side of (8.5.3) equal to the right-hand side of (8.5.16), and rearrange to obtain

$$\rho\, c_p\, \frac{\mathrm{D}T}{\mathrm{D}t} = -\nabla \cdot \mathbf{q}^{\mathrm{tot}} + \boldsymbol{\sigma}^{\mathrm{dev}} : \nabla \mathbf{u} + \widehat{\alpha}\, T\, \frac{\mathrm{D}p}{\mathrm{D}t} + \boldsymbol{\lambda} + h_{\mathrm{R}} \nabla \cdot \boldsymbol{j}_{\mathrm{R}} + h_{\mathrm{B}} \nabla \cdot \boldsymbol{j}_{\mathrm{B}} + h_{\mathrm{r}}. \tag{8.5.17}$$

Species specific enthalpies and diffusive fluxes are involved in this equation.

The last three terms on the right-hand side of (8.5.17) can be combined into a sum. For an arbitrary number of species X in a multicomponent solution, we recall that

$$h_{\mathrm{r}} \equiv -\sum_{\mathrm{X}} \dot{r}_{\mathrm{X}} h_{\mathrm{X}} \tag{8.5.18}$$

and find that the sum is

$$\sum_{\mathrm{X}} h_{\mathrm{X}} \left(\nabla \cdot \boldsymbol{j}_{\mathrm{X}} - \dot{r}_{\mathrm{X}} \right) = \sum_{\mathrm{X}} h_{\mathrm{X}} \left(\nabla \cdot \boldsymbol{j}_{\mathrm{X}} - \frac{\dot{r}_{\mathrm{X}}^{*}}{M_{\mathrm{X}}} \right), \tag{8.5.19}$$

where $\dot{r}_{\mathrm{X}}^{*}$ is the rate of mole generation due to a chemical reaction and M_{X} is the species molar mass.

In Sect. 8.6, we will see that the species enthalpies, h_{X}, are, in fact, the partial species enthalpies defined in terms of appropriate mixture derivatives.

8.5.4 Exercise

8.5.1 Use (8.5.13) to modify the right-hand side of (8.5.17).

8.5.2 Demonstrate that the temperature evolution equation derived in the text is consistent with Eq. (3.6.12) for a homogeneous fluid.

8.6 Partial Species Enthalpies

The specific enthalpy of a red–blue binary mixture, h, can be regarded a function of the mixture pressure, p, temperature, T, and red-species mass fraction, ω_{R}. To signify this dependence, we write

$$h(T, p, \omega_{\mathrm{R}}). \tag{8.6.1}$$

Skipping the dependence on ω_{B} is justified by the constraint $\omega_{\mathrm{B}} = 1 - \omega_{\mathrm{R}}$. Consequently, we may write

$$\frac{\mathrm{D}h}{\mathrm{D}t} = \left(\frac{\partial h}{\partial T}\right)_{p,\omega_{\mathrm{R}}} \frac{\mathrm{D}T}{\mathrm{D}t} + \left(\frac{\partial h}{\partial p}\right)_{T,\omega_{\mathrm{R}}} \frac{\mathrm{D}p}{\mathrm{D}t} + \left(\frac{\partial h}{\partial \omega_{\mathrm{R}}}\right)_{p,T} \frac{\mathrm{D}\omega_{\mathrm{R}}}{\mathrm{D}t}, \tag{8.6.2}$$

where $\mathrm{D}/\mathrm{D}t$ is the material derivative defined with respect to the mixture velocity, $\mathbf{u}$.

The first two partial derivatives on the right-hand side of (8.6.1) can be evaluated using formulas (B.6.6) and (B.6.25), Appendix B,

$$\left(\frac{\partial h}{\partial T}\right)_{p,\omega_R} = c_p, \qquad \left(\frac{\partial h}{\partial p}\right)_{T,\omega_R} = v\,(1-\widehat{\alpha}\,T), \tag{8.6.3}$$

where $v = 1/\rho$ is the specific volume,

$$\widehat{\alpha} = -\left(\frac{\partial \ln \rho/\rho_0}{\partial \ln T/T_0}\right)_p \tag{8.6.4}$$

is the coefficient of thermal expansion, ρ_0 is a reference density, and T_0 is a reference temperature.

We will show that the third term on the left-hand side of (8.6.2) can be evaluated from the expression

$$\left(\frac{\partial h}{\partial \omega_R}\right)_{p,T} \frac{\mathrm{D}\omega_R}{\mathrm{D}t} = h_R\,\frac{\mathrm{D}\omega_R}{\mathrm{D}t} + h_B\,\frac{\mathrm{D}\omega_B}{\mathrm{D}t}, \tag{8.6.5}$$

where

$$h_R \equiv \frac{\partial}{\partial \rho_R}\Big(\rho\,h(T,p,\omega_R)\Big), \qquad h_B \equiv \frac{\partial}{\partial \rho_B}\Big(\rho\,h(T,p,\omega_R)\Big), \tag{8.6.6}$$

and the partial specific enthalpies of the red and blue species, the partial derivatives are taken at constant T, p, and ρ_B or ρ_R, and

$$\rho = \rho_R + \rho_B, \qquad \omega_R = \frac{\rho_R}{\rho_R + \rho_B}. \tag{8.6.7}$$

Expanding the derivatives, on the right-hand sides of (8.6.6), we find that

$$h_R = \frac{\partial \rho}{\partial \rho_R}\,h + \rho\,\frac{\partial h}{\partial \omega_R}\,\frac{\partial \omega_R}{\partial \rho_R}, \qquad h_B = \frac{\partial \rho}{\partial \rho_B}\,h + \rho\,\frac{\partial h}{\partial \omega_R}\,\frac{\partial \omega_R}{\partial \rho_B}. \tag{8.6.8}$$

Setting

$$\frac{\partial \rho}{\partial \rho_R} = 1, \qquad \frac{\partial \rho}{\partial \rho_B} = 1, \qquad \frac{\partial \omega_R}{\partial \rho_R} = \frac{1}{\rho} - \frac{\rho_R}{\rho^2}, \qquad \frac{\partial \omega_R}{\partial \rho_B} = -\frac{\rho_R}{\rho^2}, \tag{8.6.9}$$

we find that

$$h_R = h + \frac{\partial h}{\partial \omega_R}\left(1 - \frac{\rho_R}{\rho}\right), \qquad h_B = h - \frac{\partial h}{\partial \omega_R}\,\frac{\rho_R}{\rho}. \tag{8.6.10}$$

Subtracting these equations, we find that

$$\frac{\partial h}{\partial \omega_R} = h_R - h_B, \tag{8.6.11}$$

which can be restated as (8.6.5) by setting $\omega_B = 1 - \omega_R$.

Substituting expression (8.6.5) into the third term on the right-hand side of (8.6.2), and multiplying both sides of the resulting equation by ρ, we obtain

$$\rho \frac{\mathrm{D}h}{\mathrm{D}t} = \rho\, c_p \frac{\mathrm{D}T}{\mathrm{D}t} + (1 - \widehat{\alpha}\, T) \frac{\mathrm{D}p}{\mathrm{D}t} + h_\mathrm{R}\, \rho \frac{\mathrm{D}\omega_\mathrm{R}}{\mathrm{D}t} + h_\mathrm{B}\, \rho \frac{\mathrm{D}\omega_\mathrm{B}}{\mathrm{D}t}. \tag{8.6.12}$$

In the absence of a chemical reaction,

$$\rho \frac{\mathrm{D}\omega_\mathrm{R}}{\mathrm{D}t} = -\nabla \cdot \boldsymbol{j}_\mathrm{R}, \qquad \rho \frac{\mathrm{D}\omega_\mathrm{B}}{\mathrm{D}t} = -\nabla \cdot \boldsymbol{j}_\mathrm{B}, \tag{8.6.13}$$

as shown in (5.11.9) and (5.11.10), yielding

$$\rho \frac{\mathrm{D}h}{\mathrm{D}t} = \rho\, c_p \frac{\mathrm{D}T}{\mathrm{D}t} + (1 - \widehat{\alpha}\, T) \frac{\mathrm{D}p}{\mathrm{D}t} - h_\mathrm{R}\, \nabla \cdot \boldsymbol{j}_\mathrm{R} - h_\mathrm{B}\, \nabla \cdot \boldsymbol{j}_\mathrm{B}. \tag{8.6.14}$$

which is precisely Eq. (8.5.16) in the absence of chemical reaction.

8.6.1 Multicomponent Mixtures

The specific enthalpy of a multicomponent solution with N species is

$$h(T, p, \omega_1, \omega_2, \ldots, \omega_{N-1}). \tag{8.6.15}$$

Skipping the dependence on ω_N is justified by the interdependence

$$\omega_\mathrm{N} = 1 - \sum_{i=1}^{N-1} \omega_i. \tag{8.6.16}$$

Consequently, we may write

$$\frac{\mathrm{D}h}{\mathrm{D}t} = \left(\frac{\partial h}{\partial T}\right)_{p,\omega_\mathrm{R}} \frac{\mathrm{D}T}{\mathrm{D}t} + \left(\frac{\partial h}{\partial p}\right)_{T,\omega_\mathrm{R}} \frac{\mathrm{D}p}{\mathrm{D}t} + \sum_{i=1}^{N-1} \left(\frac{\partial h}{\partial \omega_i}\right)_{p,T} \frac{\mathrm{D}\omega_i}{\mathrm{D}t}, \tag{8.6.17}$$

where $\mathrm{D}/\mathrm{D}t$ is the material derivative defined with respect to the mixture mass velocity, $\mathbf{u}$.

We will show that the third term on the right-hand side of (8.6.17) is given by

$$\sum_{i=1}^{N-1} \left(\frac{\partial h}{\partial \omega_i}\right)_{p,T} \frac{\mathrm{D}\omega_i}{\mathrm{D}t} = \sum_{i=1}^{N} h_i \frac{\mathrm{D}\omega_i}{\mathrm{D}t}, \tag{8.6.18}$$

where

$$h_i \equiv \frac{\partial}{\partial \rho_i} \Big(\rho\, h(T, p, \omega_1, \omega_2, \ldots, \omega_{N-1}) \Big) \tag{8.6.19}$$

for $i = 1, \ldots, N$ is the partial specific enthalpy of the ith species, the partial derivative is taken at constant T, p, and ω_j for any j except $j = i$. We recall that

$$\rho = \sum_{k=1}^{N} \rho_k \qquad \omega_j = \frac{\rho_j}{\rho}. \tag{8.6.20}$$

Expanding the derivative in (8.6.19), we find that

$$h_i = h + \rho \sum_{j=1}^{N-1} \frac{\partial h}{\partial \omega_j} \frac{\partial \omega_j}{\partial \rho_i}. \tag{8.6.21}$$

Differentiating the second equation in (8.6.20) defining that mass fraction with respect to ρ_i, we find that

$$\rho \frac{\partial \omega_j}{\partial \rho_i} = \delta_{ij} - \delta_{iN} - \frac{\rho_j}{\rho} \tag{8.6.22}$$

for $j = 1, \ldots, N-1$ and $i = 1, \ldots, N$. Substituting this expression into (8.6.21), we find that

$$h_i = h + \frac{\partial h}{\partial \omega_i} - \delta_{iN} \frac{\partial h}{\partial \omega_N} - \sum_{j=1}^{N-1} \frac{\partial h}{\partial \omega_j} \omega_j \tag{8.6.23}$$

for $i = 1, \ldots, N-1$ and

$$h_{\mathrm{N}} = h - \sum_{j=1}^{N-1} \frac{\partial h}{\partial \omega_j} \omega_j. \tag{8.6.24}$$

Combining these equations, we find that

$$\frac{\partial h}{\partial \omega_i} = h_i - h_N \tag{8.6.25}$$

for $i = 1, \ldots, N-1$. Multiplying this equation by $\mathrm{D}\omega_i/\mathrm{D}t$ and summing over the first N_1 species, we obtain

$$\sum_{i=1}^{N-1} \left(\frac{\partial h}{\partial \omega_i}\right)_{p,T} \frac{\mathrm{D}\omega_i}{\mathrm{D}t} = \sum_{i=1}^{N-1} h_i \frac{\mathrm{D}\omega_i}{\mathrm{D}t} - h_{\mathrm{N}} \sum_{i=1}^{N-1} \frac{\mathrm{D}\omega_i}{\mathrm{D}t}, \tag{8.6.26}$$

which reproduces equation (8.6.18).

Substituting Eq. (8.6.18) into (8.6.17), we obtain

$$\rho \frac{\mathrm{D}h}{\mathrm{D}t} = \rho\, c_p \frac{\mathrm{D}T}{\mathrm{D}t} + (1 - \widehat{\alpha}\, T) \frac{\mathrm{D}p}{\mathrm{D}t} + \rho \sum_{\mathrm{X}} \frac{\mathrm{D}\omega_{\mathrm{X}}}{\mathrm{D}t} h_{\mathrm{X}}. \tag{8.6.27}$$

A temperature evaluation equation can be derived working as in the case of a binary mixture.

8.6.2 First Leap of Faith

For a multicomponent mixture of N species, the leap of faith expressed by (8.5.5) takes the form

$$h = \sum_{i=1}^{N} \omega_i h_i. \tag{8.6.28}$$

Substituting $\omega_i = \rho_i/\rho$, we obtain

$$\rho h = \sum_{i=1}^{N} \rho_i h_i. \tag{8.6.29}$$

By definition,

$$\mathrm{d}(\rho h) = \frac{\partial(\rho h)}{\partial T}\,\mathrm{d}T + \frac{\partial(\rho h)}{\partial p}\,\mathrm{d}p + \sum_{i=1}^{N} h_i\,\mathrm{d}\rho_i, \tag{8.6.30}$$

where the first two partial derivatives are taken at constant species densities.

Taking the differential of (8.6.29), we obtain

$$\mathrm{d}(\rho h) = \sum_{i=1}^{N} \rho_i\,\mathrm{d}h_i + \sum_{i=1}^{N} h_i\,\mathrm{d}\rho_i. \tag{8.6.31}$$

Comparing this equation with (8.6.30), we find that

$$\sum_{i=1}^{N} \rho_i\,\mathrm{d}h_i = \frac{\partial(\rho h)}{\partial T}\,\mathrm{d}T + \frac{\partial(\rho h)}{\partial p}\,\mathrm{d}p. \tag{8.6.32}$$

The right-hand side is zero under constant T and p. For example, when $N = 1$, the specific enthalpy cannot change under constant T and p, that is, $\mathrm{d}h_1 = 0$.

We may regard h_i as a function of temperature, pressure, and all species densities, and write

$$\mathrm{d}h_i = \frac{\partial h_i}{\partial T}\,\mathrm{d}T + \frac{\partial h_i}{\partial p}\,\mathrm{d}p + \sum_{j=1}^{N}\left(\frac{\partial h_i}{\partial \rho_j}\right)\mathrm{d}\rho_j. \tag{8.6.33}$$

Substituting this expansion into (8.6.32), we find that

$$\sum_{j=1}^{N}\left(\sum_{i=1}^{N}\left(\rho_i\,\frac{\partial h_i}{\partial \rho_j}\right)\right)\mathrm{d}\rho_j = 0 \tag{8.6.34}$$

for constant T and p. For this equation to be true, the inner sum must be zero, yielding

$$\sum_{i=1}^{N} \omega_i \left(\frac{\partial h_i}{\partial \rho_j}\right)_{T,p} = 0. \tag{8.6.35}$$

Now taking the differential of (8.6.28), we obtain

$$\mathrm{d}h = \sum_{i=1}^{N} \omega_i \mathrm{d}h_i + \sum_{i=1}^{N} h_i \mathrm{d}\,\omega_i. \tag{8.6.36}$$

Substituting (8.6.33) into the first sum, we obtain

$$dh = c_p \, dT + \frac{\partial h}{\partial p} \, dp + \sum_{i=1}^{N} \omega_i \Big(\sum_{j=1}^{N} \Big(\frac{\partial h_i}{\partial \rho_j} \Big) \, d\rho_j \Big) + \sum_{i=1}^{N} h_i \, d\,\omega_i, \tag{8.6.37}$$

where

$$c_p \equiv \sum_{i=1}^{N} \omega_i \frac{\partial h_i}{\partial T}, \qquad \frac{\partial h}{\partial p} \equiv \sum_{i=1}^{N} \omega_i \frac{\partial h_i}{\partial p}. \tag{8.6.38}$$

The third term on the right-hand side of (8.6.37) is zero by virtue of (8.6.35), yielding

$$dh = c_p \, dT + \frac{\partial h}{\partial p} \, dp + \sum_{i=1}^{N} h_i \, d\,\omega_i. \tag{8.6.39}$$

Under constant pressure or for an ideal gas, the second term on the right-hand side is zero, yielding

$$dh \simeq c_p \, dT + \sum_{i=1}^{N} h_i \, d\,\omega_i, \tag{8.6.40}$$

and then

$$\nabla h \simeq c_p \, \nabla T + \sum_{i=1}^{N} h_i \nabla \, \omega_i. \tag{8.6.41}$$

This equation can be used to express ∇T in terms of ∇h.

8.6.3 Exercise

8.6.1 Discuss the implications of (8.6.35) for a homogeneous fluid, $N = 1$.

8.7 Partial Molar Enthalpies

Assume that a binary mixture consists of n_R red molecules and n_B blue molecules. The specific molar enthalpy of the mixture can be regarded a function of temperature T, pressure p, and red molar fraction, $f_R = n_R/n$,

$$h^*(T, p, f_R), \tag{8.7.1}$$

where $n = n_R + n_B$. The absence of f_B is explained by the constraint $f_B = 1 - f_R$.

The red and blue specific molar enthalpies, indicated by an asterisk, are defined by the equations

$$h_R^* \equiv \frac{\partial}{\partial n_R} \Big(n \, h^*(T, p, n_R \Big) = M \, \frac{\partial}{\partial n_R} \Big(n \, h(T, p, \omega_R \Big) \tag{8.7.2}$$

and

$$h_B^* \equiv \frac{\partial}{\partial n_B}\Big(n\, h^*(T, p, n_R\Big) = M\, \frac{\partial}{\partial n_B}\Big(n\, h(T, p, \omega_R\Big), \tag{8.7.3}$$

where $M = f_R M_R + f_B M_B$ is the mixture molar mass, the partial derivatives are taken at constant T, p, and n_B or n_R,

$$\omega_R = \frac{M_R n_R}{M_R n_R + M_B n_B}, \tag{8.7.4}$$

M_R and M_B are the species molar masses, and h is the specific enthalpy discussed in Sect. 8.6.

Expanding the derivatives, in (8.7.2) and (8.7.3), we find that

$$\frac{h_R^*}{M} = \frac{\partial n}{\partial n_R}\, h + n\, \frac{\partial h}{\partial \omega_R}\, \frac{\partial \omega_R}{\partial n_R} \tag{8.7.5}$$

and

$$\frac{h_B^*}{M} = \frac{\partial n}{\partial n_B}\, h + n\, \frac{\partial h}{\partial \omega_R}\, \frac{\partial \omega_R}{\partial n_B}. \tag{8.7.6}$$

The required partial derivatives on the right-hand sides are given by

$$\frac{\partial n}{\partial n_R} = 1, \qquad \frac{\partial n}{\partial n_B} = 1, \tag{8.7.7}$$

and also by

$$\frac{\partial \omega_R}{\partial n_R} = \frac{M_R}{M_R n_R + M_B n_B} - \frac{M_R^2}{(M_R n_R + M_B n_B)^2}\, n_R = \frac{1}{n}\, \frac{M_R}{M}\, (1 - \omega_R) \tag{8.7.8}$$

and

$$\frac{\partial \omega_R}{\partial n_B} = -\frac{M_R M_B}{(M_R n_R + M_B n_B)^2}\, n_R = -\frac{1}{n}\, \frac{M_B}{M}\, \omega_R, \tag{8.7.9}$$

where $n M = n_R M_R + n_B M_B$. Substituting these expressions into (8.7.5) and (8.7.6), and comparing the result with the expressions given in (8.6.10), we find that

$$h_R = \frac{h_R^*}{M_R}, \qquad h_B = \frac{h_B^*}{M_B}. \tag{8.7.10}$$

A similar derivation can be carried out for an arbitrary number of species in a multicomponent solution.

8.7.1 Exercise

8.7.1 Show that Eq. (8.7.10) applies for an arbitrary number of species.

8.8 Entropy Balance

Gibbs' fundamental equation of thermodynamics for a homogeneous fluid is shown in (3.8.3) in terms of the material derivative, repeated below for convenience,

$$\frac{\mathrm{D}u}{\mathrm{D}t} = T\,\frac{\mathrm{D}s}{\mathrm{D}t} + \frac{p}{\rho^2}\frac{\mathrm{D}\rho}{\mathrm{D}t}. \tag{8.8.1}$$

An extra term is added to the right-hand side in the case of a red and blue binary mixture, yielding

$$\frac{\mathrm{D}u}{\mathrm{D}t} = T\,\frac{\mathrm{D}s}{\mathrm{D}t} + \frac{p}{\rho^2}\frac{\mathrm{D}\rho}{\mathrm{D}t} + \dot{\mu}, \tag{8.8.2}$$

where

$$\dot{\mu} = \mu_{\mathrm{R}}\,\frac{\mathrm{D}\phi_{\mathrm{R}}}{\mathrm{D}t} + \mu_{\mathrm{B}}\,\frac{\mathrm{D}\phi_{\mathrm{B}}}{\mathrm{D}t} \tag{8.8.3}$$

is the rate of change of a combined chemical potential, ϕ_{R} and ϕ_{B} are the red and blue molar-mass concentrations defined in Table 4.1, and μ_{R} and μ_{B} are red and blue chemical potentials associated with the red and blue species (e.g., Landau, L. D. & Lifsitz, E. M. (1980) *Statistical Physics, Part 1*, Third Edition, Pergamon Press, § 85.)

Setting $\phi_{\mathrm{R}} = \omega_{\mathrm{R}}/M_{\mathrm{R}}$ and $\phi_{\mathrm{B}} = \omega_{\mathrm{B}}/M_{\mathrm{B}}$, and simplifying the right-hand side of (8.8.3), we obtain

$$\dot{\mu} = \mu\,\frac{\mathrm{D}\omega_{\mathrm{R}}}{\mathrm{D}t}, \tag{8.8.4}$$

where ω_{R} and ω_{B} are the mass fractions, M_{R} and M_{B} are the red and blue molar masses, and

$$\mu \equiv \frac{\mu_{\mathrm{R}}}{M_{\mathrm{R}}} - \frac{\mu_{\mathrm{B}}}{M_{\mathrm{B}}} \tag{8.8.5}$$

is the exchange potential (e.g., Landau, L. D. & Lifsitz, E. M. (1987) *Fluid Mechanics*, Second Edition, Pergamon Press, p. 228.)

According to the species evolution equation (5.11.9), the rate of change of the red species mass fraction is given by

$$\frac{\mathrm{D}\,\omega_{\mathrm{R}}}{\mathrm{D}t} = \frac{1}{\rho}\big(-\nabla\cdot\boldsymbol{j}_{\mathrm{R}} + \dot{r}_{\mathrm{R}}\big). \tag{8.8.6}$$

The right-hand side of this equation can be substituted for the material derivative on the right-hand side of (8.8.4), yielding

$$\dot{\mu} = \frac{\mu}{\rho}\big(-\nabla\cdot\boldsymbol{j}_{\mathrm{R}} + \dot{r}_{\mathrm{R}}\big). \tag{8.8.7}$$

Substituting this expression into Eq. (8.8.2), and using the continuity equation to write

$$\frac{\mathrm{D}\rho}{\mathrm{D}t} = -\rho\,\nabla\cdot\mathbf{u}, \tag{8.8.8}$$

we obtain

$$\frac{\mathrm{D}u}{\mathrm{D}t} = T\,\frac{\mathrm{D}s}{\mathrm{D}t} - \frac{p}{\rho}\,\nabla\cdot\mathbf{u} + \frac{\mu}{\rho}\big(-\nabla\cdot\boldsymbol{j}_{\mathrm{R}} + \dot{r}_{\mathrm{R}}\big). \tag{8.8.9}$$

In the absence of diffusion and chemical reaction, we recover the transport equation for the specific internal energy of a homogeneous fluid.

Other terms associated with different driving forces can be added to the right-hand sides of the fundamental law of thermodynamics expressed by (8.8.2).

8.8.1 Entropy Transport

An evolution equation for the specific entropy of the mixture can be derived by substituting into (8.8.9) the expression for the rate of change of the specific internal energy given in (8.4.9),

$$\rho\,\frac{\mathrm{D}u}{\mathrm{D}t} = -\nabla\cdot\mathbf{q}^{\mathrm{tot}} + \boldsymbol{\sigma} : \nabla\mathbf{u} + \dot{\mathcal{E}}_b. \tag{8.8.10}$$

We find that

$$\rho\,T\,\frac{\mathrm{D}s}{\mathrm{D}t} = -\nabla\cdot\mathbf{q}^{\mathrm{tot}} + \boldsymbol{\sigma}^{\mathrm{dev}} : \nabla\mathbf{u} + \mu\,\big(\nabla\cdot\boldsymbol{j}_{\mathrm{R}} - \dot{r}_{\mathrm{R}}\big) + \dot{\mathcal{E}}_b, \tag{8.8.11}$$

which shows that transport of entropy in a binary mixture is affected by the total heat flux, $\mathbf{q}^{\mathrm{tot}}$, and the diffusion flux, $\boldsymbol{j}_{\mathrm{R}}$. In the absence of diffusion and chemical reaction, we obtain the regular entropy transport equation for a homogeneous fluid.

A slight rearrangement of the third term on the right-hand side of (8.8.11) yields

$$\rho\,T\,\frac{\mathrm{D}s}{\mathrm{D}t} = -\nabla\cdot(\mathbf{q}^{\mathrm{tot}} - \mu\,\boldsymbol{j}_{\mathrm{R}}) + \boldsymbol{\sigma}^{\mathrm{dev}} : \nabla\mathbf{u} - \boldsymbol{j}_{\mathrm{R}}\cdot\nabla\mu - \mu\,\dot{r}_{\mathrm{R}} + \dot{\mathcal{E}}_b. \tag{8.8.12}$$

Dividing this equation by the temperature, T, and rearranging the first term on the right-hand side, we obtain

$$\begin{aligned}\rho\,\frac{\mathrm{D}s}{\mathrm{D}t} = &-\nabla\cdot\Big(\frac{1}{T}\,(\mathbf{q}^{\mathrm{tot}} - \mu\,\boldsymbol{j}_{\mathrm{R}})\Big) - \frac{1}{T^2}(\mathbf{q}^{\mathrm{tot}} - \mu\,\boldsymbol{j}_{\mathrm{R}})\cdot\nabla T\\ &+ \frac{1}{T}\,\boldsymbol{\sigma}^{\mathrm{dev}} : \nabla\mathbf{u} - \frac{1}{T}\,\boldsymbol{j}_{\mathrm{R}}\cdot\nabla\mu - \frac{1}{T}\,\mu\,\dot{r}_{\mathrm{R}} + \frac{1}{T}\,\dot{\mathcal{E}}_b.\end{aligned} \tag{8.8.13}$$

Integrating this evolution equation over a nominal parcel volume and applying the divergence theorem, we obtain a generalized parcel entropy evolution equation,

$$\begin{aligned}\frac{\mathrm{d}}{\mathrm{d}t}\iiint_{\mathrm{parcel}}\rho s\,\mathrm{d}V = &-\iint_{\mathrm{parcel}}\frac{1}{T}\,\mathbf{n}\cdot(\mathbf{q}^{\mathrm{tot}} - \mu\,\boldsymbol{j}_{\mathrm{R}})\,\mathrm{d}S\\ &-\iiint_{\mathrm{parcel}}\frac{1}{T^2}\,(\mathbf{q} - \mu\,\boldsymbol{j}_{\mathrm{R}})\cdot\nabla T\,\mathrm{d}V + \iiint_{\mathrm{parcel}}\frac{1}{T}\,\boldsymbol{\sigma}^{\mathrm{dev}} : \nabla\mathbf{u}\,\mathrm{d}V\\ &-\iiint_{\mathrm{parcel}}\frac{1}{T}\,\boldsymbol{j}_{\mathrm{R}}\cdot\nabla\mu\,\mathrm{d}V + \iiint_{\mathrm{parcel}}\frac{1}{T}\,(-\mu\,\dot{r}_{\mathrm{R}} + \dot{\mathcal{E}}_b)\,\mathrm{d}V.\end{aligned} \tag{8.8.14}$$

A corresponding transport equation over a stationary or frozen control volume can be derived by using the Reynolds transport equation to resolve the left-hand side into two parts expressing accumulation and convection.

8.8.2 Enthalpy, Helmholtz Free Energy, and Gibbs Energy

The evolution of the specific enthalpy of the mixture, h, is governed by the equation

$$\frac{\mathrm{D}h}{\mathrm{D}t} = T\,\frac{\mathrm{D}s}{\mathrm{D}t} + \frac{1}{\rho}\frac{\mathrm{D}p}{\mathrm{D}t} + \dot{\mu}. \tag{8.8.15}$$

The evolution of the specific Helmholtz free energy, a, is governed by the equation

$$\frac{\mathrm{D}a}{\mathrm{D}t} = \frac{p}{\rho^2}\frac{\mathrm{D}\rho}{\mathrm{D}t} - s\,\frac{\mathrm{D}T}{\mathrm{D}t} + \dot{\mu}. \tag{8.8.16}$$

The evolution of the specific Gibbs energy, g, is governed by the equation

$$\frac{\mathrm{D}g}{\mathrm{D}t} = \frac{1}{\rho}\frac{\mathrm{D}p}{\mathrm{D}t} - s\,\frac{\mathrm{D}T}{\mathrm{D}t} + \dot{\mu}. \tag{8.8.17}$$

Note that the specific entropy of the mixture, s, is involved in all of these equations.

When the pressure and temperature are held constant, $\dot{\mu}$ is the rate of change of the specific Gibbs energy following a mixture point particle. At dynamic equilibrium, this rate of change is zero.

Other terms associated with different driving forces can be added to the right-hand sides of Eqs. (8.8.2) and (8.8.15)–(8.8.17).

8.8.3 Advanced Formulations

The evolution and transport equations derived in this section rely on a generalization of an expression for the rate of change of the specific internal energy, $\mathrm{D}u/\mathrm{D}t$, for an effectively homogeneous fluid given in (3.4.3). Advanced entropy equations and discussions of possible formulations and constitutive equations are available (e.g., Bothe, D. & Dreyer, W. (2014) Continuum thermodynamics of chemically reacting fluid mixtures. *Acta Mechanica* **226**, 1757–1805.).

8.8.4 Exercises

8.8.1 Discuss the exchange potential for a binary mixture of two isotopes.

8.8.2 Derive an integral balance equation for the mixture entropy.

Summary of Transport Equations

A

In the main part of this book, we introduced fundamental concepts of continuum mechanics and thermodynamics and discussed the properties and behavior of single-component, binary, and multicomponent mixtures with regard to volume, mass, momentum, angular momentum, kinetic energy, potential energy, internal energy, enthalpy, and entropy transport.

By invoking mass conservation, Newton's second law of motion, and the first law of thermodynamics, we formulated evolution equations for material parcel properties and derived associated differential equations governing the evolution of corresponding scalar and vectors fields in terms of material derivatives. Species material derivatives were employed in the case of fluid mixtures.

A.1 Eulerian and Lagrangian Evolution Equations

Eulerian evolution equations involve the time derivative $\partial/\partial t$ at a fixed position, whereas Lagrangian evolution equations involve the material derivative $\mathrm{D}/\mathrm{D}t$ pertaining to point particle motion. The conservative form of Eulerian evolution equations involves the divergence of a convective flux instead of a directional derivative inherent in the material derivative.

In the remainder of this appendix, we summarize and unify the conservative form of evolution equations derived. All equations presented in this appendix apply for incompressible as well as compressible fluids. The stress tensor, $\boldsymbol{\sigma}$, is not necessarily symmetric.

A.2 Density

The density of a fluid, ρ, is governed by the continuity equation written in the conservative form

$$\frac{\partial \rho}{\partial t} + \nabla \cdot (\rho\, \mathbf{u}) = 0, \qquad \text{(A.2.1)}$$

where $\mathbf{u}$ is the velocity.

C. Pozrikidis, *Transport Processes Primer*,
https://doi.org/10.1007/978-1-4939-9909-5

A.3 Momentum

The momentum is governed by Cauchy's equation of motion written in the conservative form

$$\frac{\partial(\rho\,\mathbf{u})}{\partial t} + \nabla\cdot(\rho\,\mathbf{u}\otimes\mathbf{u}) = \nabla\cdot\boldsymbol{\sigma} + \rho\,\mathbf{g} + \iota\,\mathbf{b}, \tag{A.3.1}$$

where $\boldsymbol{\sigma}$ is the stress tensor, $\otimes$ denotes the tensor product, $\mathbf{g}$ is the gravitational acceleration, and $\mathbf{b}$ is an arbitrary body force field associated with the physical constant ι.

A.4 Angular Momentum

The angular momentum is governed by the equation

$$\frac{\partial(\rho\,(\mathbf{x}-\mathbf{x}_0)\times\mathbf{u})}{\partial t} + \nabla\cdot(\rho\,(\mathbf{x}-\mathbf{x}_0)\times\mathbf{u}\otimes\mathbf{u})$$
$$= (\mathbf{x}-\mathbf{x}_0)\times\nabla\cdot\boldsymbol{\sigma} + (\mathbf{x}-\mathbf{x}_0)\times(\rho\,\mathbf{g}+\iota\,\mathbf{b}) + \lambda\,\mathbf{c}, \tag{A.4.1}$$

where $\mathbf{c}$ is a torque-inducing field associated with a physical constant, λ.

A.5 Kinetic Energy

The kinetic energy is governed by the evolution equation

$$\frac{\partial}{\partial t}\left(\tfrac{1}{2}\,\rho\,|\mathbf{u}|^2\right) + \nabla\cdot\left(\tfrac{1}{2}\,\rho\,|\mathbf{u}|^2\,\mathbf{u}\right) = \nabla\cdot(\boldsymbol{\sigma}\cdot\mathbf{u}) - \boldsymbol{\sigma} : \nabla\mathbf{u} + (\rho\,\mathbf{g}+\iota\,\mathbf{b})\cdot\mathbf{u}, \tag{A.5.1}$$

where $\nabla\mathbf{u}$ is the velocity gradient tensor.

A.6 Total Energy

The specific total energy,

$$\vartheta \equiv \tfrac{1}{2}\,|\mathbf{u}|^2 - \mathbf{g}\cdot\mathbf{x} + u, \tag{A.6.1}$$

is governed by the evolution equation

$$\frac{\partial(\rho\,\vartheta)}{\partial t} + \nabla\cdot\left(\rho\,\vartheta\,\mathbf{u}\right) = -\nabla\cdot\mathbf{q} + \nabla\cdot(\boldsymbol{\sigma}\cdot\mathbf{u}) + \iota\,\mathbf{b}\cdot\mathbf{u} + \dot{q}, \tag{A.6.2}$$

where u is the specific internal energy, $\mathbf{q}$ is the heat flux, and $\dot{q}$ is the volumetric rate of interior heat generation.

A.7 Internal Energy

The specific internal energy is governed by the evolution equation

$$\frac{\partial(\rho u)}{\partial t}+\nabla\cdot\big(\rho\,u\,\mathbf{u}\big)=-\nabla\cdot\mathbf{q}+\boldsymbol{\sigma}:\nabla\mathbf{u}+\dot{q}, \tag{A.7.1}$$

where $\nabla\mathbf{u}$ is the velocity gradient tensor.

A.8 Kinetic–Internal Energy

The sum of the specific kinetic and internal energies,

$$\varepsilon \equiv \tfrac{1}{2}\,|\mathbf{u}|^2 + u, \tag{A.8.1}$$

is governed by the evolution equation

$$\frac{\partial(\rho\,\varepsilon)}{\partial t}+\nabla\cdot\big(\rho\,\varepsilon\,\mathbf{u}\big)=-\nabla\cdot\mathbf{q}+\nabla\cdot(\boldsymbol{\sigma}\cdot\mathbf{u})+\mathbf{u}\cdot(\rho\,\mathbf{g}+\iota\,\mathbf{b})+\dot{q}. \tag{A.8.2}$$

Note that the body force appears on the right-hand side.

A.9 Enthalpy

The specific enthalpy, h, is governed by the evolution equation

$$\frac{\partial(\rho\,h)}{\partial t}+\nabla\cdot\big(\rho\,h\,\mathbf{u}\big)=\frac{\mathrm{D}p}{\mathrm{D}t}-\nabla\cdot\mathbf{q}+\boldsymbol{\sigma}^{\mathrm{dev}}:\nabla\mathbf{u}+\dot{q}, \tag{A.9.1}$$

where $\boldsymbol{\sigma}^{\mathrm{dev}} \equiv \boldsymbol{\sigma} + p\,\mathbf{I}$ is the deviatoric stress tensor.

A.10 Entropy

The specific entropy, s, is governed by the evolution equation

$$\frac{\partial(\rho\,s)}{\partial t}+\nabla\cdot(\rho\,s\,\mathbf{u})=\frac{1}{T}\big(-\nabla\cdot\mathbf{q}+\boldsymbol{\sigma}^{\mathrm{dev}}:\nabla\mathbf{u}+\dot{q}\,\big), \tag{A.10.1}$$

where s is the specific entropy and $\boldsymbol{\sigma}^{\mathrm{dev}}$ is the deviatoric stress tensor.

A.11 Unification

The notion of transport phenomena relies on the observation that the conservative form of a generic evolution equation takes the form

$$\frac{\partial(\rho\,\boldsymbol{\psi})}{\partial t} + \nabla \cdot \boldsymbol{j} = \dot{\boldsymbol{\pi}}, \qquad \text{(A.11.1)}$$

where ρ is the density, $\boldsymbol{\psi}$ is a transported scalar, vectorial, or tensorial field, $\boldsymbol{j}$ is an associated compound flux, and $\dot{\boldsymbol{\pi}}$ is the associated rate of generation in the interior of the fluid.

For example, to recover the evolution equation (A.8.2) for the combined specific kinetic–internal energy, $\varepsilon \equiv \frac{1}{2}\,|\mathbf{u}|^2 + u$, we make the substitutions

$$\boldsymbol{\psi} \rightarrow \varepsilon, \qquad \boldsymbol{j} \rightarrow \rho\,\varepsilon\mathbf{u} + \mathbf{q} - \boldsymbol{\sigma}\cdot\mathbf{u}, \qquad \dot{\boldsymbol{\pi}} \rightarrow \mathbf{u}\cdot(\rho\,\mathbf{g} + \iota\,\mathbf{b}) + \dot{q}. \qquad \text{(A.11.2)}$$

In this case, the flux, $\boldsymbol{j}$, involves three constituents.

A.12 Integral Balances

An integral balance over an arbitrary control volume, $\mathcal{V}_c$, that is bounded by a surface or a collection of surfaces, $\mathcal{S}_c$, arises either directly from parcel evolution equations or indirectly by integrating the differential equations in their Eulerian conservative form over the control volume.

In the direct approach, the Reynolds transport equation is used to bring the time derivative inside the rate-of-change integrals over evolving parcel volumes. In the indirect approach, the Gauss divergence theorem is applied to convert volume integrals into surface integrals expressing integrated fluxes.

The typical balance for a transported quantity over a stationary control volume shown in Table 1.1 is repeated below for convenience:

(rate of accumulation in a stationary control or frozen volume)

=

(rate of inward convection with velocity **u**)

\+ (rate of short-range surface action or molecular transport)

\+ (rate of interior production or loss)

\+ (rate of long-range surface production or loss)

A.13 Summary

A summary of parcel property evolution laws along with their physical origin, the associated evolution differential equations, and the corresponding integral balances are presented in Tables A.1, A.2, A.3, A.4, A.5, A.6, A.7, A.8, A.9, A.10, A.11, A.12, and A.13. The normal unit vector, $\mathbf{n}$, points outward from a parcel. The normal unit vector $\mathbf{n}^{\text{in}}$ points into a control volume, $\mathcal{V}_c$, which is bounded by a contiguous or discontiguous surface, $\mathcal{S}_c$.

A.14 Exercise

A.1.1 Define the quantities $\boldsymbol{\psi}$, $\boldsymbol{j}$, and $\dot{\boldsymbol{\pi}}$ of the generalized form (A.11.1) for the entropy.

Table A.1 Parcel volume evolution equation, associated differential evolution equations, and corresponding integral balance

Volume	
Parcel volume	$V \equiv \iiint \mathrm{d}V \equiv \iiint \mathcal{J} \mathrm{d}V(\boldsymbol{\alpha})$
Kinematics	$\dfrac{\mathrm{d}V}{\mathrm{d}t} = \iint \mathbf{n} \cdot \mathbf{u} \, \mathrm{d}S$
Evolution equation	$\dfrac{\mathrm{D}\mathcal{J}}{\mathrm{D}t} = (\nabla \cdot \mathbf{u}) \, \mathcal{J}$
Evolution equation	$\dfrac{\partial \mathcal{J}}{\partial t} + \nabla \cdot (\mathcal{J}\mathbf{u}) = 2 \mathcal{J} \, \nabla \cdot \mathbf{u}$
Integral balance	$\iiint_{\mathcal{V}_c} \dfrac{\partial \mathcal{J}}{\partial t} \, \mathrm{d}V = \iint_{\mathcal{S}_c} \mathcal{J} \, \mathbf{n}^{\text{in}} \cdot \mathbf{u} \, \mathrm{d}S + 2 \iiint_{\mathcal{V}_c} \mathcal{J} \, \nabla \cdot \mathbf{u} \, \mathrm{d}V$

The volumetric coefficient, $\mathcal{J}$, is the determinant of the Lagrangian mapping of labeling space, $\boldsymbol{\alpha}$, to physical space

Table A.2 Parcel mass evolution equation, associated differential evolution equation, and corresponding integral balance, where ρ is the fluid density

Mass	
Parcel mass	$m \equiv \iiint \rho \, \mathrm{d}V$
Mass conservation	$\dfrac{\mathrm{d}m}{\mathrm{d}t} = 0$
Continuity equation	$\dfrac{\mathrm{D}\rho}{\mathrm{D}t} + \rho \, \nabla \cdot \mathbf{u} = 0$
Continuity equation	$\rho \, \dfrac{\mathrm{D}}{\mathrm{D}t}\left(\dfrac{1}{\rho}\right) = \nabla \cdot \mathbf{u}$
Continuity equation	$\dfrac{\partial \rho}{\partial t} + \nabla \cdot (\rho \mathbf{u}) = 0$
Integral balance	$\iiint_{\mathcal{V}_c} \dfrac{\partial \rho}{\partial t} \, \mathrm{d}V = \iint_{\mathcal{S}_c} \rho \, \mathbf{n}^{\text{in}} \cdot \mathbf{u} \, \mathrm{d}S$

The mathematical definition of an incompressible fluid is $\mathrm{D}\rho/\mathrm{D}t = 0$; a direct consequence is that $\nabla \cdot \mathbf{u} = 0$

Table A.3 Parcel momentum evolution equation, associated differential evolution equation, and corresponding integral balance

Momentum	
Parcel momentum	$\mathbf{M} \equiv \iiint \rho\, \mathbf{u}\, \mathrm{d}V$
Newton's law	$\dfrac{\mathrm{d}\mathbf{M}}{\mathrm{d}t} = \iint \mathbf{n} \cdot \boldsymbol{\sigma}\, \mathrm{d}S + \iiint (\rho\, \mathbf{g} + \iota\, \mathbf{b})\, \mathrm{d}V$
Cauchy equation	$\rho \dfrac{\mathrm{D}\mathbf{u}}{\mathrm{D}t} = \dfrac{\partial(\rho \mathbf{u})}{\partial t} + \boldsymbol{\nabla} \cdot (\rho\, \mathbf{u} \otimes \mathbf{u})$ $= \boldsymbol{\nabla} \cdot \boldsymbol{\sigma} + \rho\, \mathbf{g} = -\boldsymbol{\nabla} p + \boldsymbol{\nabla} \cdot \boldsymbol{\sigma}^{\mathrm{dev}} + \rho\, \mathbf{g} + \iota\, \mathbf{b}$
Integral balance	$\iiint_{\mathcal{V}_c} \dfrac{\partial(\rho \mathbf{u})}{\partial t}\, \mathrm{d}V = \iint_{\mathcal{S}_c} \rho\, \mathbf{u}\, (\mathbf{n}^{\mathrm{in}} \cdot \mathbf{u})\, \mathrm{d}S$ $- \iint_{\mathcal{S}_c} \mathbf{n}^{\mathrm{in}} \cdot \boldsymbol{\sigma}\, \mathrm{d}S + \iiint_{\mathcal{V}_c} (\rho\, \mathbf{g} + \iota\, \mathbf{b})\, \mathrm{d}V$
Integral balance	$\iiint_{\mathcal{V}_c} \dfrac{\partial(\rho \mathbf{u})}{\partial t}\, \mathrm{d}V = \iint_{\mathcal{S}_c} \rho\, \mathbf{u}\, (\mathbf{n}^{\mathrm{in}} \cdot \mathbf{u})\, \mathrm{d}S + \iint_{\mathcal{S}_c} p\, \mathbf{n}^{\mathrm{in}}\, \mathrm{d}S$ $- \iint_{\mathcal{S}_c} \mathbf{n}^{\mathrm{in}} \cdot \boldsymbol{\sigma}^{\mathrm{dev}}\, \mathrm{d}S + \iiint_{\mathcal{V}_c} (\rho\, \mathbf{g} + \iota\, \mathbf{b})\, \mathrm{d}V$

The stress tensor, $\boldsymbol{\sigma}$, can be resolved into an isotropic component involving the pressure, p, and a deviatoric component $\boldsymbol{\sigma}^{\mathrm{dev}}$, so that $\boldsymbol{\sigma} = -p\, \mathbf{I} + \boldsymbol{\sigma}^{\mathrm{dev}}$, where $\mathbf{I}$ is the identity matrix and, by definition, the trace of $\boldsymbol{\sigma}^{\mathrm{dev}}$ is zero. In the equations shown, the stress tensor is not necessarily symmetric

Table A.4 Parcel angular momentum evolution equation, associated differential evolution equation, and corresponding integral balance, where $\mathbf{c}$ is a torque-inducing field and λ is an associated physical constant

Angular momentum	
Parcel angular momentum	$\mathbf{A} \equiv \iiint \rho\,(\mathbf{x}-\mathbf{x}_0)\times\mathbf{u}\,\mathrm{d}V$
Newton's law	$\frac{\mathrm{d}\mathbf{A}}{\mathrm{d}t} = \iint (\mathbf{x}-\mathbf{x}_0)\times(\mathbf{n}\cdot\boldsymbol{\sigma})\,\mathrm{d}S + \iiint (\mathbf{x}-\mathbf{x}_0)\times(\rho\,\mathbf{g}+\iota\,\mathbf{b})\,\mathrm{d}V + \iiint \lambda\,\mathbf{c}\,\mathrm{d}V$
Cauchy equation	$\rho\,(\mathbf{x}-\mathbf{x}_0)\times\frac{\mathrm{D}\mathbf{u}}{\mathrm{D}t} = \frac{\partial\big(\rho\,(\mathbf{x}-\mathbf{x}_0)\times\mathbf{u}\big)}{\partial t} + \nabla\cdot\big(\rho\,(\mathbf{x}-\mathbf{x}_0)\times\mathbf{u}\otimes\mathbf{u}\big) = (\mathbf{x}-\mathbf{x}_0)\times\nabla\cdot\boldsymbol{\sigma} + (\mathbf{x}-\mathbf{x}_0)\times(\rho\,\mathbf{g}+\iota\,\mathbf{b})$
Integral balance	$\iiint_{\mathcal{V}_c} \frac{\partial\big(\rho(\mathbf{x}-\mathbf{x}_0)\times\mathbf{u}\big)}{\partial t}\,\mathrm{d}V = \iint_{\mathcal{S}_c} \rho\,(\mathbf{x}-\mathbf{x}_0)\times\mathbf{u}\,(\mathbf{n}^{\mathrm{in}}\cdot\mathbf{u})\,\mathrm{d}S - \iint_{\mathcal{S}_c} (\mathbf{x}-\mathbf{x}_0)\times(\mathbf{n}^{\mathrm{in}}\cdot\boldsymbol{\sigma})\,\mathrm{d}S + \iiint_{\mathcal{V}_c} (\mathbf{x}-\mathbf{x}_0)\times(\rho\,\mathbf{g}+\iota\mathbf{b})\,\mathrm{d}V + \iiint_{\mathcal{V}_c} \lambda\,\mathbf{c}\,\mathrm{d}V$

A torque field can be applied in the case of a magnetized ferrofluid

Table A.5 Parcel kinetic energy evolution equation, associated differential evolution equation, and corresponding integral balance

Kinetic energy	
Parcel kinetic energy	$\mathcal{E}_{\mathrm{kin}} = \frac{1}{2}\iiint \rho\,\lvert\mathbf{u}\rvert^2\,\mathrm{d}V$
Evolution equation	$\frac{\mathrm{d}\mathcal{E}_{\mathrm{kin}}}{\mathrm{d}t} = \iiint \nabla\cdot(\boldsymbol{\sigma}\cdot\mathbf{u})\,\mathrm{d}V - \iiint \boldsymbol{\sigma} : \nabla\mathbf{u}\,\mathrm{d}V + \iiint (\rho\,\mathbf{g}+\iota\,\mathbf{b})\cdot\mathbf{u}\,\mathrm{d}V$
Cauchy equation	$\rho\,\frac{\mathrm{D}(\frac{1}{2}\lvert\mathbf{u}\rvert^2)}{\mathrm{D}t} = \frac{\partial\rho\,(\frac{1}{2}\lvert\mathbf{u}\rvert^2)}{\partial t} + \nabla\cdot\big(\rho\,\frac{1}{2}\lvert\mathbf{u}\rvert^2\,\mathbf{u}\big) = \nabla\cdot(\boldsymbol{\sigma}\cdot\mathbf{u}) - \boldsymbol{\sigma} : \nabla\mathbf{u} + (\rho\,\mathbf{g}+\iota\,\mathbf{b})\cdot\mathbf{u} = -\nabla\cdot(p\mathbf{u}) + \nabla\cdot(\boldsymbol{\sigma}^{\mathrm{dev}}\cdot\mathbf{u}) + p\nabla\cdot\mathbf{u} - \boldsymbol{\sigma}^{\mathrm{dev}} : \nabla\mathbf{u} + (\rho\,\mathbf{g}+\iota\,\mathbf{b})\cdot\mathbf{u}$
Integral balance	$\iiint_{\mathcal{V}_c} \frac{\partial(\rho\,(\frac{1}{2}\lvert\mathbf{u}\rvert^2)}{\partial t}\,\mathrm{d}V = \iint_{\mathcal{S}_c} \rho\,(\frac{1}{2}\lvert\mathbf{u}\rvert^2)\,(\mathbf{n}^{\mathrm{in}}\cdot\mathbf{u})\,\mathrm{d}S - \iint_{\mathcal{S}_c} (\mathbf{n}^{\mathrm{in}}\cdot\boldsymbol{\sigma})\cdot\mathbf{u}\,\mathrm{d}S - \iiint_{\mathcal{V}_c} \boldsymbol{\sigma} : \nabla\mathbf{u}\,\mathrm{d}V + \iiint_{\mathcal{V}_c} (\rho\,\mathbf{g}+\iota\,\mathbf{b})\cdot\mathbf{u}\,\mathrm{d}V$

Table A.6 Parcel potential energy evolution equation, associated differential evolution equation, and corresponding integral balance, where $\mathbf{g}$ is the acceleration of gravity and $\mathbf{X}$ is the position of a point particle; by definition, $\mathrm{D}\mathbf{X}/\mathrm{D}t = \mathbf{u}$

Potential energy	
Parcel gravitational potential energy	$\mathcal{E}^g_{\text{pot}} \equiv - \iiint \rho\, \mathbf{g} \cdot \mathbf{X}\, \mathrm{d}V$
Evolution equation	$\dfrac{\mathrm{d}\mathcal{E}^g_{\text{pot}}}{\mathrm{d}t} = - \iiint \rho\, \mathbf{g} \cdot \mathbf{u}\, \mathrm{d}V$
Parcel body-force potential energy	$\mathcal{E}^b_{\text{pot}} \equiv - \iiint \iota\, f^b_{\text{pot}}(\mathbf{X}, t)\, \mathrm{d}V$ $\dfrac{\mathrm{D}}{\mathrm{D}t}\Big(\dfrac{\iota}{\rho}\, f^b_{\text{pot}}(\mathbf{x}, t)\Big) = \dfrac{\iota}{\rho}\, \mathbf{b} \cdot \mathbf{u}$
Evolution equation	$\dfrac{\mathrm{d}\mathcal{E}^b_{\text{pot}}}{\mathrm{d}t} = - \iiint_{\text{parcel}} \iota\, \mathbf{b} \cdot \mathbf{u}\, \mathrm{d}V$

A potential energy can be attributed to a field other than gravity, such as an electrical field applied to an ionic fluid

Table A.7 Parcel mechanical energy evolution equation, associated differential evolution equation, and corresponding integral balance for a body force field exclusively due to gravity

Mechanical energy	
Parcel mechanical energy	$\mathcal{E}_{\text{mech}} = \mathcal{E}_{\text{kin}} + \mathcal{E}_{\text{pot}}$ $= \iiint \rho\, (\frac{1}{2}\, \lvert\mathbf{u}\rvert^2 - \mathbf{g} \cdot \mathbf{x} - \dfrac{\iota}{\rho}\, f^b_p)\, \mathrm{d}V$
Parcel evolution equation	$\dfrac{\mathrm{d}(\mathcal{E}_{\text{kin}} + \mathcal{E}_{\text{pot}})}{\mathrm{d}t}$ $= \iint (\mathbf{n} \cdot \boldsymbol{\sigma}) \cdot \mathbf{u}\, \mathrm{d}S - \iiint \boldsymbol{\sigma} : \nabla\mathbf{u}\, \mathrm{d}V$
Evolution equation	$\rho\, \dfrac{\mathrm{D}(\frac{1}{2}\, \lvert\mathbf{u}\rvert^2 - \mathbf{g} \cdot \mathbf{X} - \frac{\iota}{\rho}\, f^b_p)}{\mathrm{D}t}$ $= \dfrac{\partial\big(\rho\, (\frac{1}{2}\, \lvert\mathbf{u}\rvert^2 - \mathbf{g} \cdot \mathbf{X} - \frac{\iota}{\rho}\, f^b_p)\big)}{\partial t}$ $+\nabla \cdot \big(\rho\, (\dfrac{1}{2}\, \lvert\mathbf{u}\rvert^2 - \mathbf{g} \cdot \mathbf{X} - \dfrac{\iota}{\rho}\, f^b_p)\, \mathbf{u}\big)$ $= \nabla \cdot (\boldsymbol{\sigma} \cdot \mathbf{u}) - \boldsymbol{\sigma} : \nabla\mathbf{u}$
Integral balance	$\iiint_{\mathcal{V}_c} \dfrac{\partial(\rho\, (\frac{1}{2}\, \lvert\mathbf{u}\rvert^2 - \mathbf{g} \cdot \mathbf{x} - \frac{\iota}{\rho}\, f^b_p)}{\partial t}\, \mathrm{d}V$ $= \iint_{\mathcal{S}_c} \rho\, (\dfrac{1}{2}\, \lvert\mathbf{u}\rvert^2 - \mathbf{g} \cdot \mathbf{x} - \dfrac{\iota}{\rho}\, f^b_p)\, (\mathbf{n}^{\text{in}} \cdot \mathbf{u})\, \mathrm{d}S$ $- \iint_{\mathcal{S}_c} (\mathbf{n}^{\text{in}} \cdot \boldsymbol{\sigma}) \cdot \mathbf{u}\, \mathrm{d}S - \iiint_{\mathcal{V}_c} \boldsymbol{\sigma} : \nabla\mathbf{u}\, \mathrm{d}V$

The mechanical energy is the sum of the kinetic energy and the potential energy due to the gravitational or any other field, as shown in Table A.6

Table A.8 Parcel total energy evolution equation, associated differential evolution equation, and corresponding integral balance

Total energy	
Parcel total energy	$\mathcal{E}_{\text{tot}} \equiv \mathcal{E}_{\text{kin}} + \mathcal{E}^{g}_{\text{pot}} + \mathcal{E}_{\text{int}} = \iiint \rho\,(\frac{1}{2}\,\lvert\mathbf{u}\rvert^2 - \mathbf{g}\cdot\mathbf{x} + u)\,\mathrm{d}V$
Cauchy equation and second law of thermodynamics	$\frac{\mathrm{d}\mathcal{E}_{\text{tot}}}{\mathrm{d}t} = -\iint \mathbf{n}\cdot\mathbf{q}\,\mathrm{d}S + \iint \mathbf{u}\cdot(\mathbf{n}\cdot\boldsymbol{\sigma})\,\mathrm{d}S + \iiint \dot{q}\,\mathrm{d}V + \iiint \iota\,\mathbf{b}\cdot\mathbf{u}\,\mathrm{d}V$
Evolution equation	$\rho\,\frac{\mathrm{D}\,(\frac{1}{2}\,\lvert\mathbf{u}\rvert^2 - \mathbf{g}\cdot\mathbf{x} + u)}{\mathrm{D}t} = \frac{\partial\big(\rho\,(\frac{1}{2}\,\lvert\mathbf{u}\rvert^2 - \mathbf{g}\cdot\mathbf{x} + u)\big)}{\partial t} + \boldsymbol{\nabla}\cdot\big(\rho\,(\frac{1}{2}\,\lvert\mathbf{u}\rvert^2 - \mathbf{g}\cdot\mathbf{x} + u)\,\mathbf{u}\big) = -\boldsymbol{\nabla}\cdot\mathbf{q} + \boldsymbol{\nabla}\cdot(\boldsymbol{\sigma}\cdot\mathbf{u}) + \dot{q} + \iota\,\mathbf{b}\cdot\mathbf{u}$
Integral balance	$\iiint_{\mathcal{V}_c} \frac{\partial\big(\rho\,(\frac{1}{2}\,\lvert\mathbf{u}\rvert^2 - \mathbf{g}\cdot\mathbf{x} + u)\big)}{\partial t}\,\mathrm{d}V = \iint_{\mathcal{S}_c} \rho\,(\frac{1}{2}\,\lvert\mathbf{u}\rvert^2 - \mathbf{g}\cdot\mathbf{x} + u)\,(\mathbf{n}^{\text{in}}\cdot\mathbf{u})\,\mathrm{d}S + \iint_{\mathcal{S}_c} \mathbf{n}^{\text{in}}\cdot\mathbf{q}\,\mathrm{d}S - \iint_{\mathcal{S}_c} (\mathbf{n}^{\text{in}}\cdot\boldsymbol{\sigma})\cdot\mathbf{u}\,\mathrm{d}S + \iiint_{\mathcal{V}_c} \dot{q}\,\mathrm{d}V + \iiint_{\mathcal{V}_c} \iota\,\mathbf{b}\cdot\mathbf{u}\,\mathrm{d}V$

Table A.9 Parcel internal energy evolution equation, associated differential evolution equation, and corresponding integral balance, where u is the specific internal energy and $\dot{q}$ is the local rate of volumetric interior heat production due, for example, to a chemical reaction

Internal energy	
Parcel internal energy	$\mathcal{E}_{\text{int}} = \iiint \rho\,u\,\mathrm{d}V$
Cauchy equation and second law of thermodynamics	$\frac{\mathrm{d}\mathcal{E}_{\text{int}}}{\mathrm{d}t} = -\iint \mathbf{n}\cdot\mathbf{q}\,\mathrm{d}S + \iiint \boldsymbol{\sigma} : \boldsymbol{\nabla}\mathbf{u}\,\mathrm{d}V + \iiint \dot{q}\,\mathrm{d}V$
Evolution equation	$\rho\,\frac{\mathrm{D}u}{\mathrm{D}t} = \frac{\partial(\rho u)}{\partial t} + \nabla\cdot(\rho u\,\mathbf{u}) = -\boldsymbol{\nabla}\cdot\mathbf{q} + \boldsymbol{\sigma} : \boldsymbol{\nabla}\mathbf{u} + \dot{q} = -\boldsymbol{\nabla}\cdot\mathbf{q} - p\boldsymbol{\nabla}\cdot\mathbf{u} + \boldsymbol{\sigma}^{\text{dev}} : \boldsymbol{\nabla}\mathbf{u} + \dot{q}$
Integral balance	$\iiint_{\mathcal{V}_c} \frac{\partial(\rho\,u)}{\partial t}\,\mathrm{d}V = \iint_{\mathcal{S}_c} \rho\,u\,\mathbf{n}^{\text{in}}\cdot\mathbf{u}\,\mathrm{d}S + \iint_{\mathcal{S}_c} \mathbf{n}^{\text{in}}\cdot\mathbf{q}\,\mathrm{d}S + \iiint_{\mathcal{V}_c} \boldsymbol{\sigma} : \boldsymbol{\nabla}\mathbf{u}\,\mathrm{d}V + \iiint_{\mathcal{V}_c} \dot{q}\,\mathrm{d}V$

Table A.10 Parcel enthalpy evolution equation, associated differential evolution equation, and corresponding integral balance, where p is the pressure and η is the specific enthalpy defined as enthalpy divided by an infinitesimal mass

Enthalpy	
Parcel enthalpy	$H \equiv \iiint \rho\,\eta\,\mathrm{d}V$
Cauchy equation and second law of thermodynamics	$\dfrac{\mathrm{d}H}{\mathrm{d}t} = -\iint \mathbf{q}\cdot\mathbf{n}\,\mathrm{d}S + \iiint \boldsymbol{\sigma}^{\mathrm{dev}} : \boldsymbol{\nabla}\mathbf{u}\,\mathrm{d}V + \iiint \dfrac{\mathrm{D}p}{\mathrm{D}t}\,\mathrm{d}V + \iiint \dot{q}\,\mathrm{d}V$
Evolution equation	$\rho\,\dfrac{\mathrm{D}\eta}{\mathrm{D}t} = \dfrac{\partial(\rho\eta)}{\partial t} + \boldsymbol{\nabla}\cdot(\rho\,\eta\,\mathbf{u}) = -\boldsymbol{\nabla}\cdot\mathbf{q} + \boldsymbol{\sigma}^{\mathrm{dev}} : \boldsymbol{\nabla}\mathbf{u} + \dfrac{\mathrm{D}p}{\mathrm{D}t} + \dot{q}$
Integral balance	$\iiint_{\mathcal{V}_c} \dfrac{\partial(\rho\,\eta)}{\partial t}\,\mathrm{d}V = \iint_{\mathcal{S}_c} \rho\,\eta\,(\mathbf{n}^{\mathrm{in}}\cdot\mathbf{u})\,\mathrm{d}S + \iint_{\mathcal{S}_c} \mathbf{n}^{\mathrm{in}}\cdot\mathbf{q}\,\mathrm{d}S + \iiint_{\mathcal{V}_c} \boldsymbol{\sigma}^{\mathrm{dev}} : \boldsymbol{\nabla}\mathbf{u}\,\mathrm{d}V + \iiint_{\mathcal{V}_c} \dfrac{\mathrm{D}p}{\mathrm{D}t}\,\mathrm{d}V + \iiint_{\mathcal{V}_c} \dot{q}\,\mathrm{d}V$

Table A.11 Parcel enthalpy evolution equation, associated differential evolution equation, and corresponding integral balance, where s is the specific entropy defined as entropy divided by an infinitesimal mass

Entropy	
Parcel entropy	$S \equiv \iiint \rho\,s\,\mathrm{d}V$
Cauchy equation and second law of thermodynamics	$\dfrac{\mathrm{d}S}{\mathrm{d}t} = -\iint \dfrac{1}{T}\,\mathbf{q}\cdot\mathbf{n}\,\mathrm{d}S - \iiint \dfrac{1}{T^2}\,\mathbf{q}\cdot\boldsymbol{\nabla}T\,\mathrm{d}V + \iiint \dfrac{1}{T}\,\boldsymbol{\sigma}^{\mathrm{dev}} : \boldsymbol{\nabla}\mathbf{u}\,\mathrm{d}V + \iiint \dfrac{1}{T}\,\dot{q}\,\mathrm{d}V$
Evolution equation	$\rho\,\dfrac{\mathrm{D}s}{\mathrm{D}t} = \dfrac{\partial(\rho s)}{\partial t} + \boldsymbol{\nabla}\cdot(\rho s\,\mathbf{u}) = -\boldsymbol{\nabla}\cdot\left(\dfrac{1}{T}\,\mathbf{q}\right) - \dfrac{1}{T^2}\,\mathbf{q}\cdot\boldsymbol{\nabla}T + \dfrac{1}{T}\,\boldsymbol{\sigma}^{\mathrm{dev}} : \boldsymbol{\nabla}\mathbf{u} + \dfrac{1}{T}\,\dot{q}$
Integral balance	$\iiint_{\mathcal{V}_c} \dfrac{\partial(\rho\,s)}{\partial t}\,\mathrm{d}V = \iint_{\mathcal{S}_c} \rho\,s\,(\mathbf{n}^{\mathrm{in}}\cdot\mathbf{u})\,\mathrm{d}S + \iint_{\mathcal{S}_c} \dfrac{1}{T}\,\mathbf{n}^{\mathrm{in}}\cdot\mathbf{q}\,\mathrm{d}S - \iiint_{\mathcal{V}_c} \dfrac{1}{T^2}\,\mathbf{q}\cdot\boldsymbol{\nabla}T\,\mathrm{d}S + \iiint_{\mathcal{V}_c} \dfrac{1}{T}\,\boldsymbol{\sigma}^{\mathrm{dev}} : \boldsymbol{\nabla}\mathbf{u}\,\mathrm{d}V + \iiint_{\mathcal{V}_c} \dfrac{1}{T}\,\dot{q}\,\mathrm{d}V$

Table A.12 Parcel of species X mass evolution equation, associated differential evolution equation, and corresponding integral balance, where $\dot{r}_X$ is the volumetric rate of mass generation

Mass of species X	
Mass of parcel X	$m_X \equiv \iiint \rho_X \, dV$
Mass conservation	$\frac{dm_X}{dt} = \iiint_{X\text{ parcel}} \dot{r}_X \, dV$
Evolution equation	$\frac{\partial \rho_X}{\partial t} + \nabla \cdot (\rho_X \mathbf{u}_X) = \dot{r}_X$ $\frac{\partial \rho_X}{\partial t} + \nabla \cdot (\rho_X \mathbf{u}) = -\nabla \cdot \boldsymbol{j}_X + \dot{r}_X$
Integral balance	$\iiint_{\mathcal{V}_c} \frac{\partial \rho_X}{\partial t} \, dV = \iint_{\mathcal{S}_c} \rho_X \, \mathbf{n}^{\text{in}} \cdot \mathbf{u}_X \, dS + \iiint_{\mathcal{V}_c} \dot{r}_X \, dV$
Integral balance	$\iiint_{\mathcal{V}_c} \frac{\partial \rho_X}{\partial t} \, dV = \iint_{\mathcal{S}_c} \rho_X \, \mathbf{n}^{\text{in}} \cdot \mathbf{u} \, dS$ $+ \iiint_{\mathcal{V}_c} \mathbf{n}^{\text{in}} \cdot \boldsymbol{j}_X \, dS + \iiint_{\mathcal{V}_c} \dot{r}_X \, dV$

The first integral balance arises from an X parcel, while the second integral balance arises from a mixture parcel

Table A.13 Parcel of species X mole evolution equation, associated differential evolution equation, and corresponding integral balance, where $\dot{r}_X$ is the volumetric rate of mass generation and M_X is the molar mass

Moles of species X	
Moles of parcel X	$n_X \equiv \iiint c_X \, dV$
Mass conservation	$\dfrac{dn_X}{dt} = \dfrac{1}{M_X} \iiint_{X \text{ parcel}} \dot{r}_X \, dV$
Evolution equation	$\dfrac{\partial c_X}{\partial t} + \nabla \cdot (c_X \mathbf{u}_X) = \dfrac{1}{M_X} \dot{r}_X$ $\dfrac{\partial c_X}{\partial t} + \nabla \cdot (c_X \mathbf{u}^*) = -\nabla \cdot j_X^* + \dfrac{1}{M_X} \dot{r}_X$
Integral balance	$\iiint_{\mathcal{V}_c} \dfrac{\partial c_X}{\partial t} dV = \iint_{\mathcal{S}_c} c_X \mathbf{n}^{\text{in}} \cdot \mathbf{u}_X \, dS + \dfrac{1}{M_X} \iiint_{\mathcal{V}_c} \dot{r}_X \, dV$
Integral balance	$\iiint_{\mathcal{V}_c} \dfrac{\partial c_X}{\partial t} dV = \iint_{\mathcal{S}_c} c_X \mathbf{n}^{\text{in}} \cdot \mathbf{u}^* \, dS$ $+ \iint_{\mathcal{S}_c} \mathbf{n}^{\text{in}} \cdot j_X^* \, dS + \dfrac{1}{M_X} \iiint_{\mathcal{V}_c} \dot{r}_X \, dV$

The first integral balance arises from an X parcel, while the second integral balance arises from a mixture parcel

Elements of Thermodynamics

B

The motion of a gas is governed by (a) the continuity equation expressing mass conservation, and (b) Cauchy's equation of motion with a constitutive equation that relates stress to velocity for a compressible fluid. In the framework of fluid mechanics and transport processes, a point particle of a gas is regarded as a fluid parcel with infinitesimal dimensions embedded in a gaseous environment acting as a virtual container.

For closure, an equation relating the specific internal energy of a material point particle to the thermodynamic state of the particle is required. Fundamental pertinent concepts from classical thermodynamics are reviewed in this appendix. An important underlying assumption is that the motion of the fluid does not affect the thermodynamic behavior of the material.

B.1 Equations of State

The pressure at a point in a fluid in motion, p, is defined in terms of the trace of the Cauchy stress tensor, $\boldsymbol{\sigma}$, as

$$p \equiv -\tfrac{1}{3}\,\mathrm{trace}(\boldsymbol{\sigma}), \tag{B.1.1}$$

where the trace is the sum of the diagonal elements hosting the three normal stresses. Equation (B.1.1) identifies the pressure with the negative of the arithmetic average of the three normal stresses acting on three mutually perpendicular planes.

In thermodynamics, the pressure is related to the specific volume of a compressible gas, $v \equiv 1/\rho$, and to the absolute temperature, T, by an equation

$$\mathcal{S}(p, v, T) = 0, \tag{B.1.2}$$

where ρ is the density. The specific volume, v, has units of volume divided by mass.

B.1.1 Ideal Gas

In the case of an ideal gas, the equation of state takes the form

$$\mathcal{S}^{\text{ideal gas}}(p, v, T) = p\,v - \frac{R}{M}\,T, \tag{B.1.3}$$

C. Pozrikidis, *Transport Processes Primer*,
https://doi.org/10.1007/978-1-4939-9909-5

where

$$R = 8.314 \times 10^3 \frac{\text{kg m}^2}{\text{sec}^2 \cdot \text{kmole} \cdot \text{K}} \tag{B.1.4}$$

is the ideal gas constant and M is the molar mass with units of mass over mole. One mole of a homogeneous substance, denoted by the standard unit symbol *mol*, is defined as a collection of N_A molecules, where

$$N_A = 6.022140857 \times 10^{23} \tag{B.1.5}$$

is the Avogadro number.

Consider an amount of gas with volume V comprised of n moles. Setting in (B.1.3) $v = V/(nM)$, we obtain

$$\mathcal{S}^{\text{ideal gas}}(p, v, T) = \frac{1}{nM}(pV - nRT). \tag{B.1.6}$$

Setting the expression inside the parentheses on the right-hand side to zero provides us with the familiar ideal-gas law,

$$pV = nRT \qquad \text{or} \qquad p = cRT, \tag{B.1.7}$$

where $c = n/V$ is the molar concentration.

B.1.2 PvT Surface

An equation of state defines a surface in the pvT space in that, if a doublet of p and T is specified, the equation of state provides us with a unique corresponding value for v by way of the implicit-function theorem. Similar statements can be made for other pairs of the triplet (p, v, T).

B.1.3 pv Diagram

For fixed temperature, the equation of state provides us with a relation between p and v, represented by a curve parametrized by T in a pv diagram, as shown in Fig. B.1, where $T_1 > T_2$. The higher the temperature, the higher the specific volume for a fixed pressure, p. The higher the pressure, the lower the specific volume for a fixed temperature, T.

It is important to remember that the equation of state applies at *equilibrium*, that is, when the rate of change of a system is sufficiently slow.

B.1.4 Euler's Cyclic Relation

Suppose that three variables, x, y, and z, are related by an equation, $f(x, y, z) = 0$. The implicit-function theorem ensures that this equation can be solved for x, y, or z, yielding three somewhat related functions of two variables each,

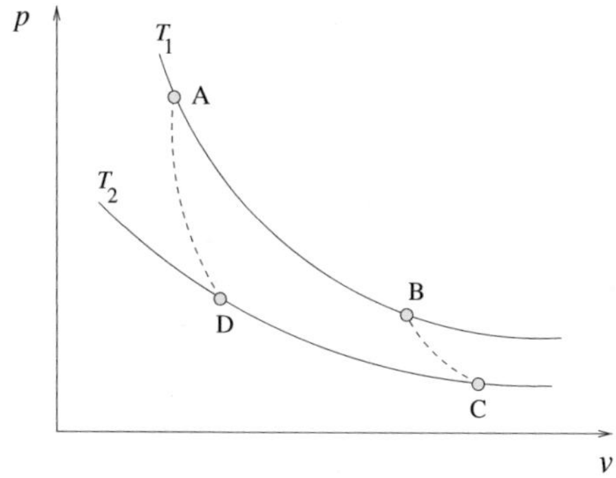

Fig. B.1 Illustration of a pv diagram corresponding to a certain equation of state, parametrized by T, where p is the pressure, v is the specific volume, and $T_1 > T_2$. The Carnot cycle runs from A to B to C to D and then back to A. The segments AB and CD are reversible and isothermal, while the segments BC and DA are reversible and adiabatic (isentropic). All four segment processes are reversible

$$x = X(y, z), \qquad y = Y(z, x), \qquad z = Z(x, y). \tag{B.1.8}$$

For example, x, y, and z can be p, v, and T, and $f(x, y, z) = 0$ can be an equation of state.

Using the definition of the partial derivative and that of the inverse function, we derive three relations,

$$X_y Y_x = 1, \qquad Y_z Z_y = 1, \qquad Z_x X_z = 1, \tag{B.1.9}$$

where a subscript denotes a partial derivative. For example, X_y is the partial derivative of $X(y, z)$ with respect to y keeping z constant. Miscellaneous relations between first and second derivatives can be derived by differentiating these equations.

The partial derivatives of the functions X, Y, and Z with respect to their respective arguments satisfy compatibility conditions. To see this, we observe that a triplet of differential changes, $\mathrm{d}\mathbf{x} = (\mathrm{d}x, \mathrm{d}y, \mathrm{d}z)$, is related by

$$\begin{gathered} \mathrm{d}x = X_y\,\mathrm{d}y + X_z\,\mathrm{d}z, \qquad \mathrm{d}y = Y_z\,\mathrm{d}z + Y_x\,\mathrm{d}x, \\ \mathrm{d}z = Z_x\,\mathrm{d}x + Z_y\,\mathrm{d}y. \end{gathered} \tag{B.1.10}$$

These three equations can be assembled into a matrix–vector form of a homogeneous linear system,

$$\begin{bmatrix} -1 & X_y & X_z \\ Y_x & -1 & Y_z \\ Z_x & Z_y & -1 \end{bmatrix} \cdot \begin{bmatrix} \mathrm{d}x \\ \mathrm{d}y \\ \mathrm{d}z \end{bmatrix} = \mathbf{0}. \tag{B.1.11}$$

For this linear system to have a nonzero (nontrivial) solution for $\mathrm{d}x$, $\mathrm{d}y$, and $\mathrm{d}z$, the determinant of the matrix on the left-hand side must be zero. Expanding the determinant with respect to the first row, we obtain

$$-1 \times (1 - Y_z Z_y) - X_y \times (-Y_x - Y_z Z_x) + X_z \times (Y_x Z_y + Z_x) = 0. \tag{B.1.12}$$

Using Eq. (B.1.9) to simplify this expression, we obtain

$$X_y Y_z Z_x + X_z Y_x Z_y = -2. \tag{B.1.13}$$

Note that the first term on the left-hand side involves a cyclic permutation of x, y, z with respect to the indices, whereas the second term on the left-hand side involves an anti-cyclic permutation. Using once again Eq. (B.1.9), we obtain

$$X_y Y_z Z_x + \frac{1}{Z_x X_y Y_z} = -2, \tag{B.1.14}$$

which is a quadratic equation for the product $X_y Y_z Z_x$ with a double solution expressed by Euler's cyclic relation

$$X_y Y_z Z_x = -1. \tag{B.1.15}$$

The derivative X_y can be replaced by $1/Y_x$, the derivative Y_z can be replaced by $1/Z_y$, and the derivative Z_x can be replaced by $1/X_z$ on the left-hand side.

As an example, we consider the ideal-gas equation of state, $f(x, y, z) = xy - cz = 0$, and deduce the functions

$$X = c\,\frac{z}{y}, \qquad Y = c\,\frac{z}{x}, \qquad Z = \frac{1}{c}\,xy, \tag{B.1.16}$$

where c is a constant. We compute the partial derivatives

$$X_y = -c\,\frac{z}{y^2}, \qquad Y_z = c\,\frac{1}{x}, \qquad Z_x = \frac{1}{c}\,y, \tag{B.1.17}$$

and find that

$$X_y Y_z Z_x = -c\,\frac{z}{xy} = -1 \tag{B.1.18}$$

according to the equation of state.

Miscellaneous relations between first and second derivatives can be derived by differentiating equation (B.1.15).

B.1.5 Moduli of Pressure and Volume

Given an equation of state, we may compute: (a) a pressure expansion modulus with units of pressure defined as a scaled partial derivative,

$$K_T \equiv -v\left(\frac{\partial p}{\partial v}\right)_T, \tag{B.1.19}$$

and (b) a coefficient of thermal expansion with units of inverse temperature,

$$\widehat{\alpha} \equiv \frac{1}{v}\left(\frac{\partial v}{\partial T}\right)_p = -\frac{1}{\rho}\left(\frac{\partial \rho}{\partial T}\right)_p = -\left(\frac{\partial \ln \rho/\rho_0}{\partial T}\right)_p, \tag{B.1.20}$$

where ρ_0 is a reference density. Using Euler's cyclic relation (B.1.15), we find that

$$\widehat{\alpha}\, K_T = -\left(\frac{\partial v}{\partial T}\right)_p \left(\frac{\partial p}{\partial v}\right)_T = \frac{1}{\left(\frac{\partial T}{\partial p}\right)_v}, \tag{B.1.21}$$

which shows that

$$\widehat{\alpha}\, K_T = \left(\frac{\partial p}{\partial T}\right)_v. \tag{B.1.22}$$

Equations (B.1.19), (B.1.20), and (B.1.21) provide us with expressions for the partial derivatives of one of p, v, and T, with respect to any other, keeping the third constant.

In the case of an ideal gas, we find that $K_T = p$ and $\widehat{\alpha} = 1/T$. The partial derivative in (B.1.21) is simply p/T.

B.1.6 Redlich–Kwong Equation of State

The Redlich–Kwong equation of state is an improvement of the ideal-gas law. The equation prescribes that

$$p = \frac{RT}{v-b} - \frac{a}{v(v+b)\sqrt{T}}, \tag{B.1.23}$$

where a and b are two physical constants related to the critical temperature, T_c, and critical pressure, p_c, by the equations

$$a = 0.4278\,\frac{R^2\, T_c^{5/2}}{p_c}, \qquad b = 0.0867\,\frac{R\, T_c}{p_c}. \tag{B.1.24}$$

The critical temperature is the minimum temperature above which a substance or mixture cannot be in the liquid phase, no matter how high the pressure. The minimum pressure necessary to liquefy a fluid at the critical temperature is the critical pressure. When $a = 0$ and $b = 0$, we recover the ideal gas law.

B.1.7 Exercises

B.1.1 Confirm Euler's cyclic relation for $f(x, y, z) = xy - cz^n = 0$, where c and n are two constants.

B.1.2 Confirm that $K_T = p$ and $\widehat{\alpha} = 1/T$ for an ideal gas.

B.2 State Functions

A thermodynamic state function attributed to a point particle depends on the current thermodynamic state of the point particle, and is independent of the point-particle history or evolution. For example, the state function can be the instantaneous specific internal energy, u.

A thermodynamic state function can be regarded as a function of any two chosen independent variables of the triplet (p, v, T). The third variable is related to the chosen doublet by an equation of state.

B.2.1 Functions $\psi(v, T)$ and $\psi(p, T)$

A state function, ψ, can be regarded a function of (T, v), so that

$$d\psi = \left(\frac{\partial \psi}{\partial T}\right)_v dT + \left(\frac{\partial \psi}{\partial v}\right)_T dv. \tag{B.2.1}$$

Alternatively, ψ can be regarded a function of (T, p), so that

$$d\psi = \left(\frac{\partial \psi}{\partial T}\right)_p dT + \left(\frac{\partial \psi}{\partial p}\right)_T dp. \tag{B.2.2}$$

The four partial derivatives in these equations are related by two equations. This means that only two of these partial derivatives are independent.

To derive the two equations relating the four partial derivatives, we write

$$\left(\frac{\partial \psi}{\partial v}\right)_T = \left(\frac{\partial \psi}{\partial p}\right)_T \left(\frac{\partial p}{\partial v}\right)_T, \tag{B.2.3}$$

which can be rearranged into

$$\left(\frac{\partial \psi}{\partial v}\right)_T = -\frac{K_T}{v} \left(\frac{\partial \psi}{\partial p}\right)_T, \tag{B.2.4}$$

where K_T is the pressure expansion modulus defined in (B.1.19). Note that all partial derivatives in these equations are taken under constant T.

The second relation arises by dividing figuratively (B.2.2) by dT to obtain

$$\left(\frac{\partial \psi}{\partial T}\right)_v = \left(\frac{\partial \psi}{\partial T}\right)_p + \left(\frac{\partial \psi}{\partial p}\right)_T \left(\frac{\partial p}{\partial T}\right)_v, \tag{B.2.5}$$

which can be written as

$$\left(\frac{\partial \psi}{\partial T}\right)_v = \left(\frac{\partial \psi}{\partial T}\right)_p + \widehat{\alpha}\, K_T \left(\frac{\partial \psi}{\partial p}\right)_T, \tag{B.2.6}$$

where $\widehat{\alpha}$ is the coefficient of thermal expansion defined in (B.1.20).

Equations (B.2.4) and (B.2.6) relate the four partial derivatives involved in (B.2.1) and (B.2.2). Substituting (B.2.4) into (B.2.6), we obtain

$$\left(\frac{\partial \psi}{\partial T}\right)_v = \left(\frac{\partial \psi}{\partial T}\right)_p - \widehat{\alpha}\, v \left(\frac{\partial \psi}{\partial v}\right)_T, \tag{B.2.7}$$

which can be used instead of (B.2.4) or (B.2.6). This equation also arises by dividing figuratively (B.2.1) by $d\,T$.

In fact, Eqs. (B.2.4) and (B.2.6) can be derived formally by rewriting (B.2.1) as

$$d\psi = \left(\frac{\partial \psi}{\partial T}\right)_v dT + \left(\frac{\partial \psi}{\partial v}\right)_T \left(\left(\frac{\partial v}{\partial T}\right)_p dT + \left(\frac{\partial v}{\partial p}\right)_T dp\right). \tag{B.2.8}$$

Comparing this equation with Eq. (B.2.2), immediately suggests relations (B.2.4) and (B.2.6).

B.2.2 Internal Energy

The specific internal energy, u, or any other state function, can be regarded a function of another state function, ψ, and v, so that

$$du = \left(\frac{\partial u}{\partial \psi}\right)_v d\psi + \left(\frac{\partial u}{\partial v}\right)_\psi dv. \tag{B.2.9}$$

Specifying the partial derivatives on the right-hand side defines the function ψ up to an arbitrary constant.

B.2.3 Heat Capacity Under Constant Volume

The heat capacity under constant volume is the rate of the change of the specific internal energy, u, with respect to temperature, T,

$$c_v \equiv \left(\frac{\partial u}{\partial T}\right)_v. \tag{B.2.10}$$

The units of c_v are length squared over time squared over absolute temperature $(m^2/(s^2T)$. Using the chain rule, we find that

$$c_v = \left(\frac{\partial u}{\partial \psi}\right)_v \left(\frac{\partial \psi}{\partial T}\right)_v, \tag{B.2.11}$$

for any state function, ψ.

B.2.4 Ideal Gas

In the case of an ideal gas, the specific internal energy is given by

$$u^{\text{ideal gas}} = \widehat{c}_v \frac{R}{M} T, \tag{B.2.12}$$

where R is the ideal-gas constant, M is the molar mass, and $\widehat{c}_v$ is a dimensionless constant that takes the value $3/2$ for a mono-atomic gas, $5/2$ for a diatomic gas, and $7/2$ for a triatomic gas. The associated heat capacity under constant volume is

$$c_v^{\text{ideal gas}} = \widehat{c}_v \frac{R}{M}, \tag{B.2.13}$$

independent of the specific volume, v.

B.2.5 Exercise

B.2.1 Explain in physical terms why the specific internal energy of an ideal gas depends on the temperature alone and increases with the number of atoms in a molecule.

B.3 Entropy

The specific entropy, s, is a state function defined judiciously by Gibbs' fundamental equation of thermodynamics. The fundamental equation relates corresponding small changes in the specific internal energy of a small material parcel, $\mathrm{d}u$, and specific volume, $\mathrm{d}v$, by the equation

$$\mathrm{d}u = T\,\mathrm{d}s - p\,\mathrm{d}v, \tag{B.3.1}$$

which shows that

$$\left(\frac{\partial u}{\partial s}\right)_v = T, \qquad \left(\frac{\partial u}{\partial v}\right)_s = -p. \tag{B.3.2}$$

Compatibility requires the Maxwell relation

$$\left(\frac{\partial T}{\partial v}\right)_s = -\left(\frac{\partial p}{\partial s}\right)_v = \frac{\partial^2 u}{\partial v \partial s}, \tag{B.3.3}$$

where the specific internal energy, temperature, and pressure are regarded as functions of s and v.

Dividing figuratively equation (B.3.1) by $\mathrm{d}v$, we obtain

$$\left(\frac{\partial u}{\partial v}\right)_T = T\left(\frac{\partial s}{\partial v}\right)_T - p, \tag{B.3.4}$$

which can be rearranged into

$$\left(\frac{\partial s}{\partial v}\right)_T = \frac{1}{T}\left(\frac{\partial u}{\partial v}\right)_T + \frac{p}{T}. \tag{B.3.5}$$

This equation can be derived formally, as discussed in Sect. B.2.

Applying Eq. (B.2.11) for $\psi = s$, we find that

$$\left(\frac{\partial s}{\partial T}\right)_v = \frac{c_v}{T}, \tag{B.3.6}$$

where c_v is the heat capacity under constant volume.

B.3.1 Ideal Gas

In the case of an ideal gas, we use expression (B.2.12) for the specific internal energy and find that

$$\left(\frac{\partial s}{\partial T}\right)_v = \widehat{c}_v \frac{R}{MT}, \qquad \left(\frac{\partial s}{\partial v}\right)_T = 0, \tag{B.3.7}$$

where $\widehat{c}_v$ is defined in Sect. B.2.

B.3.2 Power Laws

As an example, we assume that the specific internal energy is given by a pv power law,

$$u = c\, p^{\alpha} v^{\beta}, \tag{B.3.8}$$

where c, α, and β are constants. Differentiating, we obtain

$$\mathrm{d}u = c\,\alpha p^{\alpha-1}\, v^{\beta}\, \mathrm{d}p + c\,\beta p^{\alpha} v^{\beta-1}\, \mathrm{d}v. \tag{B.3.9}$$

Comparing this equation with (B.3.1), we deduce that

$$\mathrm{d}s = c\,\alpha\, \frac{p^{\alpha-1}\, v^{\beta}}{T}\, \mathrm{d}p + \frac{c\,\beta p^{\alpha} v^{\beta-1} + p}{T}\, \mathrm{d}v, \tag{B.3.10}$$

which requires that

$$\left(\frac{\partial s}{\partial p}\right)_v = c\,\alpha\, \frac{p^{\alpha-1}\, v^{\beta}}{T}, \qquad \left(\frac{\partial s}{\partial v}\right)_p = \frac{c\,\beta p^{\alpha} v^{\beta-1} + p}{T}. \tag{B.3.11}$$

These equations are compatible only if

$$c\,\alpha \left(\frac{\partial}{\partial v} \frac{p^{\alpha-1}\, v^{\beta}}{T}\right)_p = \left(\frac{\partial}{\partial p} \frac{c\beta p^{\alpha} v^{\beta-1} + p}{T}\right)_v. \tag{B.3.12}$$

We may assume the additional power-law dependence

$$T = c'\, p^{\alpha'} v^{\beta'}, \tag{B.3.13}$$

where c', α', and β' are constants. Substituting this expression into (B.3.12), we obtain

$$c\alpha(\beta - \beta')\, p^{\alpha-\alpha'-1} v^{\beta-\beta'-1} = c\beta(\alpha - \alpha')\, p^{\alpha-\alpha'-1} v^{\beta-\beta'-1} + (1 - \alpha')\, p^{-\alpha'} v^{-\beta'}. \tag{B.3.14}$$

Simplifying, we obtain

$$c\alpha(\beta - \beta')\, p^{\alpha-1} v^{\beta-1} = c\beta(\alpha - \alpha')\, p^{\alpha-1} v^{\beta-1} + 1 - \alpha'. \tag{B.3.15}$$

This equation is satisfied only when $\alpha' = 1$ and

$$\alpha(\beta - \beta') = \beta(\alpha - 1). \tag{B.3.16}$$

In the case of an ideal gas, $\beta' = 1$, we find that $\alpha = \beta = 1$.

B.3.3 Entropy of an Ideal Gas

The specific internal energy of an ideal gas is $u = c_v T$, where c_v is the heat capacity at constant volume. Substituting this expression along with the ideal gas law into Gibbs' fundamental equation, repeated below for convenience,

$$\mathrm{d}u = T\, \mathrm{d}s - p\, \mathrm{d}v, \tag{B.3.17}$$

we obtain

$$c_v \, \mathrm{d} \ln T = \mathrm{d}s - \frac{R}{M} \, \mathrm{d} \ln v. \tag{B.3.18}$$

Integrating between two states labeled A and B, and rearranging, we obtain

$$s_\mathrm{B} - s_\mathrm{A} = c_v \ln \frac{T_\mathrm{B}}{T_\mathrm{A}} + \frac{R}{M} \ln \frac{v_\mathrm{B}}{v_\mathrm{A}}. \tag{B.3.19}$$

The two terms on the right-hand side can be accommodated into a common logarithm, yielding

$$s_\mathrm{B} - s_\mathrm{A} = \ln \Big(\Big(\frac{T_\mathrm{B}}{T_\mathrm{A}}\Big)^{c_v} \Big(\frac{v_\mathrm{B}}{v_\mathrm{A}}\Big)^{R/M} \Big). \tag{B.3.20}$$

The temperature, T, or the specific volume, v, in this equation can be replaced by the pressure using the ideal-gas equation of state.

B.3.4 Exercise

B.3.1 Derive (B.3.18).

B.4 First Law of Thermodynamics

The first law of thermodynamics requires that any change in the specific internal energy of a material point particle is given by

$$\mathrm{d}u = \mathrm{d}q - \mathrm{d}w^{\mathrm{performed}}, \tag{B.4.1}$$

where q is the specific heat *supplied* to the point particle, defined as heat added divided by the density, and $\mathrm{d}w^{\mathrm{performed}}$ is the specific work *performed* by the point particle, defined as work performed divided by the density.

An alternative expression in terms of the specific work *received* by the material is

$$\mathrm{d}u = \mathrm{d}q + \mathrm{d}w^{\mathrm{received}}. \tag{B.4.2}$$

Work received is the negative of the work performed.

An infinity of combinations of $\mathrm{d}q$ and $\mathrm{d}w$ corresponding to different processes yields the same change in internal specific energy, $\mathrm{d}u$. We may reach the top floor of a tower by taking the stairs, by riding the elevator, or by a combination.

B.4.1 Expansion of a Gas

If a gas inside a cylinder performs work by expansion by pushing a piston, $\mathrm{d}w^{\mathrm{performed}} > 0$, heat will enter the gas from the ambience, $\mathrm{d}q > 0$, so that $-\mathrm{d}w^{\mathrm{performed}} < \mathrm{d}u < 0$.

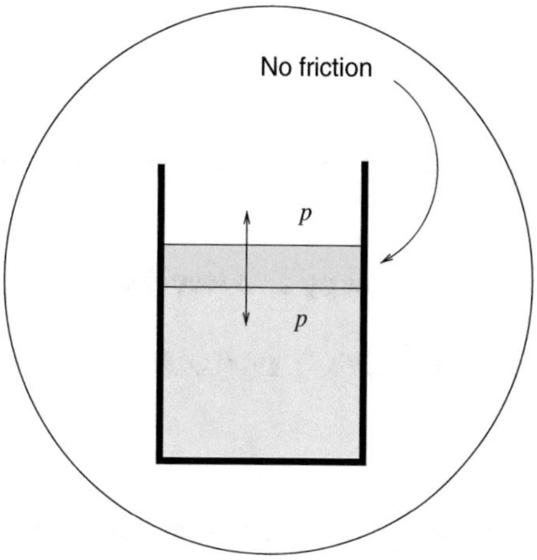

Fig. B.2 Illustration of reversible expansion or contraction of a gas in a cylinder due to the up or down motion of a piston

The minimum possible heat addition, $\mathrm{d}q$, denoted by $\mathrm{d}q^{\text{reversible}}$, corresponds to a *reversible* process where

$$\mathrm{d}w^{\text{performed,reversible}} = p\,\mathrm{d}v, \tag{B.4.3}$$

yielding

$$\mathrm{d}u = \mathrm{d}q^{\text{reversible}} - p\,\mathrm{d}v. \tag{B.4.4}$$

The physical conditions underlying a reversible process are depicted in Fig. B.2.

B.4.2 Reversibility

Reversibility requires a frictionless piston and equal pressures on either side of the piston, so that the force acting on the piston is zero at any time and the piston moves with constant infinitesimal velocity. If these conditions are not met, the process is irreversible and p in (B.4.3) must be replaced by the pressure of the surroundings.

B.4.3 Compression

Conversely, if a gas inside a cylinder receives work by compression by way of pushing a piston, $\mathrm{d}w^{\text{received}} > 0$, heat will escape from the material, $\mathrm{d}q < 0$, so that $0 < \mathrm{d}u < \mathrm{d}w^{\text{received}}$.

The minimum possible heat loss, $\mathrm{d}q$, denoted by $\mathrm{d}q^{\text{reversible}}$, corresponds to a *reversible* process with

$$\mathrm{d}w^{\text{received,reversible}} = -p\,\mathrm{d}v, \tag{B.4.5}$$

and thus

$$du = dq^{\text{reversible}} - p\, dv. \tag{B.4.6}$$

We recall that the physical conditions underlying a reversible process are depicted in Fig. B.2.

B.4.4 Entropy and Reversibility

Comparing (B.4.4) and (B.4.6) with (B.3.1), we find that

$$ds = \frac{1}{T}\, d\, q^{\text{reversible}} \tag{B.4.7}$$

in expansion or compression. This equation provides us with a venue for computing the change in entropy from one state to another in terms of a reversible process, disregarding the actual process that might have occurred.

B.4.5 Exercise

B.4.1 Discuss an example of a reversible macroscopic process.

B.5 Carnot Cycle

The Carnot cycle ABCD is described in the caption of Fig. B.1. Steps AB and CD are isothermal, occurring under constant temperature, T_1 or T_2, where $T_1 > T_2$. Heat does *not* enter or exit the gas during the adiabatic steps BC and DA.

By construction,

$$(\Delta s)_{\text{AB}} = \frac{q_{\text{AB}}}{T_1} > 0, \qquad (\Delta w^{\text{performed}})_{\text{AB}} > 0 \tag{B.5.1}$$

for the isothermal expansion step AB,

$$q_{\text{BC}} = 0, \qquad (\Delta s)_{\text{BC}} = 0, \qquad (\Delta w^{\text{performed}})_{\text{BC}} > 0 \tag{B.5.2}$$

for the adiabatic step expansion BC,

$$(\Delta s)_{\text{CD}} = \frac{q_{\text{CD}}}{T_2} < 0, \qquad (\Delta w^{\text{performed}})_{\text{CD}} < 0 \tag{B.5.3}$$

for the isothermal compression step CD, and

$$q_{\text{DA}} = 0, \qquad (\Delta s)_{\text{DA}} = 0. \qquad (\Delta w^{\text{performed}})_{\text{DA}} < 0 \tag{B.5.4}$$

for the adiabatic compression step, DA.

Since the system returns to the initial state after a cycle, $(\Delta u)_{\text{ABCD}} = 0$, it must be that

$$(\Delta w^{\text{performed}})_{\text{ABCD}} = q_{\text{ABCD}}, \tag{B.5.5}$$

where

$$q_{\text{ABCD}} \equiv q_{\text{AB}} + q_{\text{CD}} \tag{B.5.6}$$

and

$$(\Delta w^{\text{performed}})_{\text{ABCD}} \equiv (\Delta w^{\text{performed}})_{\text{AB}} + (\Delta w^{\text{performed}})_{\text{BC}} + (\Delta w^{\text{performed}})_{\text{CD}} + (\Delta w^{\text{performed}})_{\text{DA}}, \tag{B.5.7}$$

which is represented by the area of the Carnot circuit ABCD shown in Fig. B.1.

The efficiency of the Carnot cycle is defined as

$$e \equiv \frac{(\Delta w^{\text{performed}})_{\text{ABCD}}}{q_{\text{ABCD}}} = \frac{q_{\text{AB}} + q_{\text{CD}}}{q_{\text{AB}}}. \tag{B.5.8}$$

The total change in entropy after a complete cycle is null,

$$(\Delta s)_{\text{AB}} + (\Delta s)_{\text{CD}} = \frac{q_{\text{AB}}}{T_1} + \frac{q_{\text{CD}}}{T_2} = 0. \tag{B.5.9}$$

B.5.1 Exercise

B.5.1 Provide an estimate for the Carnot efficiency.

B.6 Enthalpy

Combining the definition of the specific enthalpy,

$$h \equiv u + pv, \tag{B.6.1}$$

with Gibbs' fundamental equation of thermodynamics, repeated below for convenience,v

$$\mathrm{d}u = T\,\mathrm{d}s - p\,\mathrm{d}v, v \tag{B.6.2}$$

we obtain

$$\mathrm{d}h = T\,\mathrm{d}s + v\,\mathrm{d}p, \tag{B.6.3}$$

which shows that

$$\left(\frac{\partial h}{\partial s}\right)_p = T, \qquad \left(\frac{\partial h}{\partial p}\right)_s = v. \tag{B.6.4}$$

Compatibility requires the Maxwell relation

$$\left(\frac{\partial T}{\partial p}\right)_s = \left(\frac{\partial v}{\partial s}\right)_p = \frac{\partial^2 h}{\partial p \partial s}, \tag{B.6.5}$$

where the specific enthalpy is regarded a function of s and p.

B.6.1 Heat Capacity Under Constant Pressure

The heat capacity under constant pressure is defined as

$$c_p \equiv \left(\frac{\partial h}{\partial T}\right)_p. \tag{B.6.6}$$

Writing

$$c_p = \left(\frac{\partial h}{\partial s}\right)_p \left(\frac{\partial s}{\partial T}\right)_p, \tag{B.6.7}$$

and using the first relation in (B.6.4), we find that

$$\left(\frac{\partial s}{\partial T}\right)_p = \frac{c_p}{T}, \tag{B.6.8}$$

which is the counterpart of Eq. (B.3.6) involving c_v, repeated below for comparison,

$$\left(\frac{\partial s}{\partial T}\right)_v = \frac{c_v}{T}. \tag{B.6.9}$$

B.6.2 Relations Between c_v and c_p

Combining equation (B.6.8) with Eq. (B.6.9), we obtain

$$\frac{c_p}{c_v} = \frac{(\partial s/\partial T)_p}{(\partial s/\partial T)_v}. \tag{B.6.10}$$

Using the Euler cyclic rule, we obtain

$$\frac{c_p}{c_v} = \frac{(\partial T/\partial v)_s(\partial v/\partial s)_T}{(\partial T/\partial p)_s(\partial p/\partial s)_T} = \left(\frac{\partial v}{\partial p}\right)_T \left(\frac{\partial p}{\partial v}\right)_s. \tag{B.6.11}$$

The simplification is implemented by use of the chain rule.

Taking the difference between Eqs. (B.6.8) and (B.6.9), we obtain

$$\frac{c_p - c_v}{T} = \left(\frac{\partial s}{\partial T}\right)_p - \left(\frac{\partial s}{\partial T}\right)_v. \tag{B.6.12}$$

Applying (B.2.7) for $\psi = s$, we obtain

$$\frac{c_p - c_v}{T} = \widehat{\alpha}\, v \left(\frac{\partial s}{\partial v}\right)_T, \tag{B.6.13}$$

where $\widehat{\alpha}$ is the coefficient of thermal expansion defined in (B.1.20).

Next, we introduce the Maxwell relation (B.7.5), repeated below for convenience,

$$\left(\frac{\partial s}{\partial v}\right)_T = \left(\frac{\partial p}{\partial T}\right)_v. \tag{B.6.14}$$

Substituting this relation into (B.6.12), we obtain

$$\frac{c_p - c_v}{T} = \widehat{\alpha}\, v \left(\frac{\partial p}{\partial T}\right)_v = \left(\frac{\partial v}{\partial T}\right)_p \left(\frac{\partial p}{\partial T}\right)_v. \tag{B.6.15}$$

Applying Euler's cyclic rule discussed in Sect. B.1 for the last partial derivative on the right-hand side, we obtain

$$\frac{c_p - c_v}{T} = -\left(\frac{\partial v}{\partial T}\right)_p \frac{1}{\left(\frac{\partial T}{\partial v}\right)_p \left(\frac{\partial v}{\partial p}\right)_T}, \tag{B.6.16}$$

yielding

$$c_p - c_v = vT\, \frac{1}{\beta_T}\, \widehat{\alpha}^2, \tag{B.6.17}$$

where

$$\beta_T \equiv -\frac{1}{v} \left(\frac{\partial v}{\partial p}\right)_T \tag{B.6.18}$$

is the isothermal compressibility.

B.6.3 Ideal Gas

In the case of an ideal gas, the specific enthalpy is a function of temperature alone,

$$h^{\text{ideal gas}} = (\widehat{c}_v + 1)\, \frac{R}{M}\, T \tag{B.6.19}$$

independent of the specific volume, v, where $\widehat{c}_v$ is a dimensionless constant that takes the value $3/2$ for a mono-atomic gas, $5/2$ for a diatomic gas, and $7/2$ for a triatomic gas. Consequently,

$$c_p^{\text{ideal gas}} = (\widehat{c}_v + 1)\, \frac{R}{M}. \tag{B.6.20}$$

B.6.4 Dependence of Enthalpy on Pressure

We may regard h a function of s and p, and write

$$\mathrm{d}h = \left(\frac{\partial h}{\partial s}\right)_p \mathrm{d}s + \left(\frac{\partial h}{\partial p}\right)_s \mathrm{d}p. \tag{B.6.21}$$

Dividing figuratively both sides of this equation by $\mathrm{d}p$, we obtain

$$\left(\frac{\partial h}{\partial p}\right)_T = \left(\frac{\partial h}{\partial s}\right)_p \left(\frac{\partial s}{\partial p}\right)_T + \left(\frac{\partial h}{\partial p}\right)_s. \tag{B.6.22}$$

Using expressions (B.6.4), we obtain

$$\left(\frac{\partial h}{\partial p}\right)_T = T\left(\frac{\partial s}{\partial p}\right)_T + v. \tag{B.6.23}$$

Next, we introduce the Maxwell relation (B.7.5), repeated below for convenience,

$$\left(\frac{\partial s}{\partial v}\right)_T = \left(\frac{\partial p}{\partial T}\right)_v. \tag{B.6.24}$$

Substituting this relation into (B.6.23), we obtain

$$\left(\frac{\partial h}{\partial p}\right)_T = v\,(1 - \widehat{\alpha}\,T), \tag{B.6.25}$$

where

$$\widehat{\alpha}\,T = -\left(\frac{\partial \ln \rho/\rho_0}{\partial \ln T/T_0}\right)_p \tag{B.6.26}$$

where ρ_0 is a reference density and T_0 is a reference temperature. In the case of an ideal gas, because $\widehat{\alpha}\,T = 1$, the right-hand side of (B.6.25) is zero.

B.6.5 Exercise

B.6.1 Provide a rigorous derivation for (B.6.22).

B.7 Free Energies

Combining the definition of the specific Helmholtz free energy

$$a \equiv u - Ts, \tag{B.7.1}$$

with Gibbs' fundamental equation of thermodynamics, repeated below for convenience,

$$\mathrm{d}u = T\,\mathrm{d}s - p\,\mathrm{d}v, \tag{B.7.2}$$

we obtain

$$\mathrm{d}a = -s\,\mathrm{d}T - p\,\mathrm{d}v, \tag{B.7.3}$$

which shows that

$$\left(\frac{\partial a}{\partial T}\right)_v = -s, \qquad \left(\frac{\partial a}{\partial v}\right)_T = -p. \tag{B.7.4}$$

Compatibility requires the Maxwell relation

$$\left(\frac{\partial s}{\partial v}\right)_T = \left(\frac{\partial p}{\partial T}\right)_v = -\frac{\partial^2 a}{\partial p \partial v}, \tag{B.7.5}$$

where the specific Helhmoltz free energy is regarded as a function of T and v. Substituting this relation into (B.3.4), we obtain

$$\left(\frac{\partial u}{\partial v}\right)_T = T\left(\frac{\partial p}{\partial T}\right)_v - p. \tag{B.7.6}$$

B.7.1 Gibbs Free Energy

Combining the definition of the specific Gibbs free energy,

$$g \equiv u + pv - Ts, \tag{B.7.7}$$

with Gibbs' fundamental equation of thermodynamics, repeated below for convenience,

$$\mathrm{d}u = T\,\mathrm{d}s - p\,\mathrm{d}v, \tag{B.7.8}$$

we obtain

$$\mathrm{d}g = -s\,\mathrm{d}T + v\,\mathrm{d}p, \tag{B.7.9}$$

which shows that

$$\left(\frac{\partial g}{\partial T}\right)_p = -s, \qquad \left(\frac{\partial g}{\partial p}\right)_T = v. \tag{B.7.10}$$

Compatibility requires the Maxwell relation

$$\left(\frac{\partial s}{\partial p}\right)_T = -\left(\frac{\partial v}{\partial T}\right)_p = -\widehat{\alpha}\ v = -\frac{\partial^2 g}{\partial T \partial p}, \tag{B.7.11}$$

where $\widehat{\alpha}$ is the coefficient of thermal expansion defined in (B.1.20). The specific Gibbs free energy is regarded as a function of T and p.

B.7.2 Exercise

B.7.1 Discuss the differences between the two free energies.

B.8 Maxwell Relations

The four Maxwell relations derived previously in this appendix as compatibility conditions are summarized below:

$$\left(\frac{\partial T}{\partial v}\right)_s = -\left(\frac{\partial p}{\partial s}\right)_v = \frac{\partial^2 u}{\partial v \partial s}, \tag{B.8.1}$$

$$\left(\frac{\partial T}{\partial p}\right)_s = \left(\frac{\partial v}{\partial s}\right)_p = \frac{\partial^2 \eta}{\partial p \partial s}, \tag{B.8.2}$$

$$\left(\frac{\partial s}{\partial v}\right)_T = \left(\frac{\partial p}{\partial T}\right)_v = -\frac{\partial^2 a}{\partial p \partial v}, \tag{B.8.3}$$

$$\left(\frac{\partial s}{\partial p}\right)_T = -\left(\frac{\partial v}{\partial T}\right)_p = -\frac{\partial^2 g}{\partial T \partial p} \equiv -\widehat{\alpha}\, v, \tag{B.8.4}$$

where $\widehat{\alpha}$ is the coefficient of thermal expansion defined in (B.1.20).

A bewildering array of other relations between partial derivatives of state functions can be derived using compatibility conditions, the chain rule, and Euler's cyclic rule.

B.8.1 Exercise

B.8.1 Derive a thermodynamic relation of your choice not discussed in the text.

Index

A
Acceleration, 42, 110
Accumulation, 22, 34
Advection, 20, 108
Alternating tensor, 6
Angular
  momentum, 49, 51, 246
    species, 146
    transport, 54
  velocity, 7
Avogadro number, 90, 258

B
Balance
  of energy, 59, 63
  integral, 20
    over an evolving volume, 34
  of mass, 23, 29
  of mechanical energy, 55, 57
  of momentum, 48
  of total energy, 69
Barodiffusion, 176
Barycentric velocity, 100
Bayesian analysis, 181
Bernoulli probability, 118, 124, 128
Binary fluid, 89
Binomial expansion, 125
Blood, 103, 104
Blue species, 89
Brinkmann law, 199

C
Carnot cycle, 268
Cauchy
  angular momentum equation, 53
    species, 208
  equation of motion, 41
    species, 203
  stress tensor, 37
    partial, 193
    species, 193
Chemical
  potential, 80, 241
  reaction, 133
  solution, 103
Clausius–Duhem inequality, 83, 85
Coffee, 103
Compressibility, 271
Concentration
  mass, 92
  mixture, 143
  molar, 92, 159
    evolution equation, 142
  molar-mass, 91
Conduction, 84
Conductivity, 84
Conservative form, 20, 43
Continuity equation, 24
Continuum approximation, 1, 2, 90
Contraction, 8
Control volume, 21
  evolving, 22
Convection, 19
  –diffusion equation, 119
Convective transport, 19, 22
Coyote, 170
Curl, 6, 7

D
Darcy law, 199
Deformation, 8
  gradient, 10
  rate of, 6
De Moivre–Laplace theorem, 126
Density, 3
  evolution, 25
    species, 136, 157
  species, 92
Derivative
  directional, 12
  Eulerian, 9
  material, 11
    species, 109

C. Pozrikidis, *Transport Processes Primer*,
https://doi.org/10.1007/978-1-4939-9909-5

Deviatoric stress tensor, 39, 57, 69, 71, 73, 195
Diffusion
  essence of, 115
  flux
    mass, 149
    molar, 160
  fractional, 177, 183
  in a gas, 117
  law, 163
    validation, 152, 162
  in a liquid, 117
  in a solid, 117
  velocity, 103, 153, 165
Diffusivity, 130, 163
  Einstein, 118
  fractional, 177
Dilatation, 8
Distribution
  Gaussian, 120
  heavy-tailed, 184
Divergence of the velocity, 5
Double-dot product, 56
Dpd, 125, 186

E
Einstein diffusivity, 118
Energy
  conversion, 58, 62
  specific, 261
  total, 69
Enthalpy, 74, 232, 269
Entropy, 80, 264
  of a mixture, 242
Equation
  of motion, 41
    for a mixture, 212
    for a species, 203
  of state, 257
Euler
  cyclic relation, 258
  equation, 42
Eulerian
  derivative, 9, 12
  description, 3
Evolution
  for the density, 25
  equation, 14
  for the temperature, 76
Exchange
  of momentum, 201
  potential, 241
Expansion, 8, 156
Extensive property, 4

F
Fick's law, 163, 174, 216
  for the molar flux, 166
Fick–Stefan framework, 105
First law of thermodynamics, 63, 66, 70
Fluid
  homogeneous, 1
  ideal, 39
  incompressible, 25
  inviscid, 39
  Newtonian, 39
  parcel, 1
Flux, 33
  diffusive, 149, 160
  energy, 226
  heat, 64
  laws, 176
Fourier law of conduction, 84
Fractional
  diffusion, 177, 183
  Laplacian, 177
Free energy, 272
Fundamental
  decomposition of kinematics, 6
  equation of thermodynamics, 80
  law of kinematics, 13

G
Gaussian distribution, 120
Gibbs
  energy, 86
  fundamental equation, 80, 264
Gradient, 5
  in a binary mixture, 94
  in a multicomponent mixture, 97
  of the velocity, 5
Gravel, 103
Green's function, 119, 181
  of the unsteady diffusion equation
    in one dimension, 120

H
Heat
  capacity
    under constant pressure, 78
    under constant volume, 77, 263
  conduction, 64, 84
  flux, 64
Helmholtz energy, 85
Homogeneous fluid, 1
Hydrostatics, 39

I
Ideal
  fluid, 39
  gas, 257, 261, 263–265, 271
    constant, 258
    law, 258
Implicit-function theorem, 258

Improper pdf, 181
Incompressible fluid, 25
Inertia, 43
Integral
  balance, 20, 248
    over an evolving volume, 34
Intensive property, 4
Interaction force, 198
Internal energy, 70
  specific, 65
Inviscid fluid, 39
Isotope, 115
Isotropic material, 85

J
Jacobian matrix, 10
Jumpers, 183, 188

K
Kinematics
  fundamental decomposition, 6, 112, 155
  fundamental law, 13
  of a species, 108
Kinetic energy, 61, 104
  of a mixture, 154
Kinetics, 201
Kronecker delta, 6

L
Label, 9
Lagrangian
  mapping, 9
  volumetric coefficient, 11
Lag velocity, 35, 102
Laplacian, 168, 169
  fractional, 177
Leibnitz integration rule, 19
Levi–Civita symbol, 6
Lotka–Volterra model, 144

M
Mapping, 9
Mass
  accumulation, 29
  balance, 29
  concentration, 92
  conservation, 23
    for a closed system, 23
    for an open system, 30
  density, 92
  diffusive flux, 149
  flux, 101
  fraction, 91, 92
    evolution, 137, 158
  molar, 90
  transport, 29
  velocity, 100
  velocity of a species, 100
Material
  derivative, 11
    species, 109
  parcel, 1, 3, 14
Maxwell relation, 264, 269, 272–274
Mechanical energy
  balance, 55
  of a fluid parcel, 59
  integral balance, 57
Medical imaging technology (MRI), 84
Mixture, 35, 89
  concentration, 143
  equation of motion, 212
  molecular, 103
  multiphase, 103
  stationary, 152, 216
  stress, 212
  velocity
    molar, 143, 159
Molar
  concentration, 92
  fraction, 89
  mass, 90, 258
    mixture, 91
  -mass concentration, 91
Mole, 90, 258
Molecular fraction, 89
Momentum, 40, 144
  angular, 49
    species, 146
  exchange, 201
  of a parcel, 40
  of a species, 100
  tensor, 40, 145, 148
  transport, 48
  velocity, 101

N
Natural law, 16
Navier–Stokes equation, 42
Newton's second law, 41, 51
  for an open system, 48, 49, 54
Newtonian fluid, 39
Noninertial frame, 46
Normal
  stress, 37
  vector, 16, 19, 34

O
One-dimensional flow, 28, 45
Onsager principle, 85

P

Packet, 30
Parcel, 3, 14
  advection, 108
  expansion, 156
  momentum, 40
  motion, 105
  of a species, 105
  superposition, 135, 209
  volume, 19
Partial
  pressure, 194
  stress tensor, 193
  volume, 93
Pdf, 129
  mapped, 131
Point particle, 2, 9, 14
  advection, 155
  position, 9
  velocity, 9, 13
Population balance, 118
Potential energy, 59
Predator–prey equations, 144
Pressure, 214, 257
  critical, 261
  in hydrodynamics, 39
  in hydrostatics, 39
  partial, 194
  species, 194
Probability, 90

R

Random
  crossing, 116
  jumpers, 183
  numbers, 186
  walker, 117, 129
  walks, 123
Rankin–Hugoniot
  density condition, 27
  momentum condition, 44
  species density condition, 139
  species momentum condition, 206
Rate
  of deformation, 6, 40
    for a mixture, 115
  of expansion, 13, 113
Reaction
  chemical, 133
  rate, 201
Redlich–Kwong equation of state, 261
Red species, 89
Reversible process, 267
Reynolds transport equation, 17, 31
Riemann zeta function, 184
Rotation, 7

S

Second law of thermodynamics, 82
Self-diffusion, 116, 176
Shear
  flow, 8
  rate, 8
  stress, 37
Soret effect, 176
Species, 89
  balance, 133
  equation of motion, 203
  interaction force, 198
  stationary, 152
Specific
  internal energy, 65
  total energy, 67, 68
  volume, 25
Standard deviation, 120
State function, 261
Stationary
  mixture, 152
  species, 152
Stirling's formula, 125
Stress
  normal, 37
  shear, 37
Stress–momentum tensor, 43, 205, 213
Stress tensor, 37
  deviatoric, 39, 57, 69, 71, 73, 195
  for a mixture, 212
  partial, 193
  species, 193
  spin, 52, 207
  symmetry, 53, 209
  viscous, 39, 57

T

Temperature, 4
  critical, 261
  evolution, 76
Tensor, 37
  product, 5
Thermal
  expansion coefficient, 78, 261
  flux, 64, 84
  part of energy, 229
Thermodynamics, 72, 257
  first law, 63, 66, 70, 266
  of irreversible processes, 229
  second law, 82
Total energy
  of a fluid parcel, 63
  specific, 67, 68
  transport, 67
Tracer, 30
  diffusion, 176

Traction, 37
Translation, 7
Transport
  of angular momentum, 54
  convective, 22
  of enthalpy, 74, 232
  of entropy, 80
  equation, 20
    summary, 245
  of Gibbs energy, 86
  of Helmholtz energy, 85
  of internal energy, 70
  of kinetic–internal energy, 72
  of mass, 29
  of momentum, 48
  of total energy, 67
  Reynolds equation, 17
Traveler's time derivative, 26, 44

V
Vapor pressure, 172
Velocity, 1
  barycentric, 100
  diffusion, 103, 153, 165
  field, 2
  gradient, 156
  gradient tensor, 5
  lag, 102
  mass, 100
  mixture
    molar, 143, 159
  momentum, 101
  of a species, 100
Viscosity, 40
Viscous stress tensor, 39, 57
Volume
  control, 21
  of a parcel, 249
  specific, 25, 74
Vorticity, 6, 168
  in a mixture, 113
  tensor, 6

W
Walkers1, 120

Z
Zeta function, 184

Printed in the United States
By Bookmasters